FOOD SCIENCE AND TECHNOLOGY

FROM MEDICINAL CHEMISTRY TO FOOD SCIENCE

A TRANSFER OF *IN SILICO* METHODS APPLICATIONS

FOOD SCIENCE AND TECHNOLOGY

Additional books in this series can be found on Nova's website
under the Series tab.

Additional e-books in this series can be found on Nova's website
under the e-books tab.

FOOD SCIENCE AND TECHNOLOGY

FROM MEDICINAL CHEMISTRY TO FOOD SCIENCE

A TRANSFER OF *IN SILICO* METHODS APPLICATIONS

PIETRO COZZINI
EDITOR

nova publishers
New York

Library of Congress Cataloging-in-Publication Data

From medicinal chemistry to food science: a transfer of in silico methods applications / Pietro Cozzini (Department of Food Science, University of Parma, Parco Area delle Scienze, Parma, Italy), editor.
pages cm. -- (Food science and technology)
Includes index.
ISBN 978-1-63483-587-9 (hardcover)
1. Bioinformatics. 2. Drugs--Testing. 3. Food--Testing. I. Cozzini, Pietro, editor.
QH324.2.F74 2015
570.285--dc23
2015029237

Published by Nova Science Publishers, Inc. † New York

CONTENTS

PREFACE

The idea to describe a molecule using an equation started a long time ago. We can state that from the Hartree-Fock's work in the Thirties we have learned how to simulate the behavior of molecules, considering this as a set of balls and springs or a set of orbitals.

From a vision of molecular modeling to ball and springs and quantomechanics (orbitals and electrons), we have a wide set of possibilities to describe a molecular system depending on the complexity of the systems and the power of our computing systems.

The lack of powerful computing before the Sixties did not prevent the improving of computational methods, but the large availability of these methods started after the PC system was born in 1980.

The other driving force for computational chemistry besides CPU power has been graphical user interfaces that allow the user to "create" a molecule by hands in front of a PC.

Obviously, the first users have been computational chemists and medicinal chemists.

Two main goals were to design new molecules and to understand a mechanism of binding.

These techniques are very well established in Medicinal Chemistry where molecular modelling, virtual screening, docking/scoring, QSAR, etc. are well consolidated techniques to discovery new lead compounds chosen to be developed as promising drugs.

More recently, these techniques have been shyly introduced in the food safety field, trying to follow the directives of major food organizations about the reduction of *in vivo* and *in vitro* tests for economic and ethical reasons.

From a pure chemical point of view, the application of computational techniques to food safety is essentially the same as for medicinal chemistry compounds: the main problem the study behavior of two or more bodies (molecules) and their physico-chemical interactions, e.g., a protein, the target, a small molecule (the bullet), and other actors, like "different" waters or metals participating in the binding event.

The aim of this book is to supply a non-exhaustive overview of computational techniques to be applied to food problems. In particular, we focus on several typical approaches to discover new lead compounds in medicinal chemistry: virtual screening, docking/scoring, QSAR and molecular dynamics.

In details the first chapter describes a panorama of "computational methods applied in food safety". The authors, from Nestlé Research Centre in Lausanne, give to the reader a specific point of view of a company that daily and routinely must deal with important food problems.

The second chapter presents the classical approach using an *in vivo* test for food contaminants.

Chapter Three reports a review of *in vitro* and *in silico* digestion models.

Chapter Four shows a typical *in silico/in vitro* approach using docking/scoring and wet tests on a case study.

Chapter Five illustrates the docking paradigm from a "reverse" point of view, when the target is a small molecule instead of a protein.

Chapter Six is a general introduction to molecular dynamics to explore molecular events in food research. Chapter Seven discusses how to include protein flexibility in structure based drug design. Chapter Eight is a pretty complete review on QSAR techniques, while Chapter Nine introduces a new software technique to define protein flexibility and conformational space of a protein in the Apo form or complexed with a small molecule.

The last chapter is a quick guide for readers on how to help run molecular dynamics on high performance computers, because molecular dynamics require HPC resources to be carried out for a long time to reproduce a "real event".

We asked the authors to be "controversial": to show advantages and disadvantages, pros and cons of any technique or method they want to show.

Moreover, the chapters are intended for readers who want to have a "simple" overview of computational techniques to be applied to food molecules simulations.

In: From Medicinal Chemistry to Food Science
Editor: Pietro Cozzini
ISBN: 978-1-63483-587-9

Chapter 1

COMPUTATIONAL METHODS APPLIED TO FOOD SAFETY

Elena Lo Piparo** and Benoît Schilter
Nestlé Research Centre, Vers-Chez-Les-Blanc, Lausanne, Switzerland

Abstract

In silico approaches have been successfully applied for the screening of potential drugs in pharmaceutical discovery pipelines, and their application in the food safety sector appears very promising. However, a direct transfer of tools and technologies may not be straightforward since the nature of food safety issues differ significantly from questions raised by pharmaceutical development and drug design. In the present chapter, an overview on how and which computational methods can be used in the field of food safety is provided. The main challenges when switching from medicinal chemistry to food safety are highlighted. In addition, a structured approach integrating the use of diverse *in silico* tools in order to establish the level of safety concern of chemicals in food without experimental toxicological data is described. Finally, the application of specific *in silico* tools to identify molecules of safety concern in packaging material is presented.

Introduction

The food sector is increasingly facing cases of issues associated with chemicals for which no or little toxicological data are available. Indeed the application of continuously improving analytical methods has revealed that many man-made chemicals can enter the food chain leading to human exposure, albeit often at low levels. It is also important to note that traditional food processes result in the formation of large numbers of substances and that packaging materials constitute a source of many chemicals which can potentially migrate into food. In addition, there are thousands of naturally occurring food compounds of known structure. To identify which, amongst these chemicals, should be managed requires the understanding of their safety significance. However, for most of them, little or no

* Corresponding author: elena.lopiparo@rdls.nestle.com.

toxicological data is available. In this context, *in silico* methods have obvious advantages in terms of time, cost and animal protection. Indeed it is neither practical nor desirable to experimentally test all of them in animal studies.

To gain an overview of how computational methods are used internationally in the regulatory assessment of food chemicals, a survey was carried out by EFSA (European Food Safety Authority) and JRC (Joint Research Centre) [1]. The results showed that the majority of key players in the food safety field either do not use *in silico* methodology at all or use it in a very limited way, mainly because of a lack of expertise. When they are used, they are typically applied to support priority setting exercises or to fill information gaps on possible health concerns. At present, they are not used routinely to fill data gaps in the pre-marketing assessment of food additives, food contact substances, or pesticide and pesticide metabolite residues, but they are currently being explored, developed and rarely utilised by regulatory authorities for risk assessment purposes. Despite this situation, *in silico* tools were widely perceived as useful and important to support regulatory assessments in the field of food safety. In both industry and regulatory contexts, the *in silico* methods appear of particular interest since experimental toxicology cannot deal with emergency situations requiring fast decision making and because political authorities are more and more pushing for the application of alternative methods that **r**efine, **r**educe and **r**eplace animal tests (the **3R**s guiding principles).

In the following paper we will summarize which are the main challenges when applying these methods from medicinal chemistry to food safety. In this context the main questions to face are: i) which computational approach(s) can be applied in food safety? and ii) which endpoints are relevant in the food safety field? The paper will also report how to integrate *in silico* tools to establish a level of safety concern of chemicals in food without the need for toxicological data. In addition, it will present the application of *in silico* methods to identify chemicals of potential concern in packaging materials.

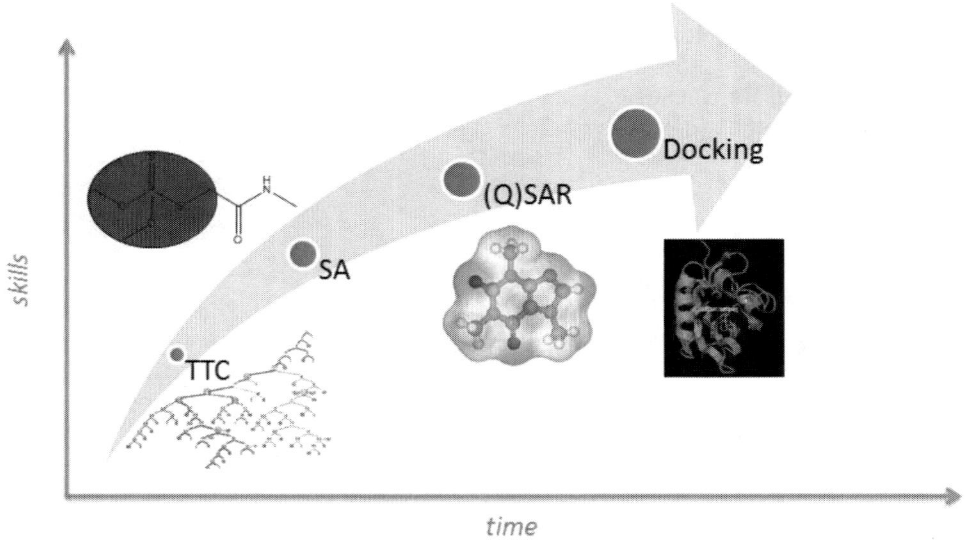

Figure 1. Graphical representation of the different computational approaches whose can be applied in food safety, ranked according to complexity and knowledge required for application.

Computational Approaches Applied in Food Safety

When moving from medicinal chemistry to food safety the first question to answer is which *in silico* approach can also be applied in food safety? Indeed *in silico* is a generic word that means "*performed on computer*" or "*via computer simulation*" and it can include many different methods. The main ones that can be applied in food safety are summarized in figure 1, where they are ranked according to complexity and knowledge required for application. Read across is not included in the picture because it potentially uses all the information collected from the other methods to fill data gaps. All these approaches are based on scientific knowledge gained from diverse research fields but all of them are built on the premise that a property (in our case a toxicological endpoint) of a chemical can be directly predicted from its molecular structure, and inferred from the properties of similar compounds whose activities are known.

Threshold of Toxicological Concern (TTC)

The Threshold of Toxicological Concern (TTC) has been designed to specifically handle situations where no toxicological data is available [2]. The TTC is a concept widely used in chemical food safety and is based on the decision tree developed by Cramer [3] for the estimation of chronic toxic hazard. Depending on its chemical features, a chemical can be assigned into one of the three classes reflecting a presumption of low, moderate or serious toxicity, corresponding to human exposure threshold values below which there are no significant risks to human health. Guidelines on the application of the TTC concept have been developed [4]. Although the TTC concept is not per se a computational predictive model, it has been the first attempt to formally combine structural chemical information with statistical processing of available oral toxicity data to establish levels of no safety concern for chemicals for which no experimental toxicological data are available.

Structural Alerts

A substructure associated with the presence of a biological activity is called a Structural Alert (SA). It represents a very simple and easily interpretable way to flag compounds with potential toxicity. Well known commercial available expert systems based on SA are MultiCASE and DEREK and more recently a freely available one, called ToxAlerts, was developed [5]. SA can also be used to define structural similar compounds for read-across and to investigate the applicability domain of a given QSAR model.

Structure–Activity Relationships (SAR) and Quantitative Structure–Activity Relationships (QSAR)

QSAR are quantitative relationship models between the chemical structure of compounds and a given property, such as a toxicological endpoint, while a Structure–Activity

Relationship (SAR) is a qualitative relationship between a molecular (sub)structure and the presence or absence of a given activity. The term substructure refers to an atom, or group of adjacently connected atoms, in a molecule. Qualitative relationships are derived from discontinuous data (classification, or yes and no answer), while quantitative relationships are derived for continuous data (e.g., toxic potency). These types of models can be used to predict the property values of chemicals and in pharma they support the design of new chemicals with given property values.

Modelling Receptor-Ligand Interactions: Docking

It is increasingly acknowledged that toxicity can be formally described through the Adverse Outcome Pathway (AOP) framework. It links knowledge on molecular initiating event (MIE) in which a chemical interacts with a biological target (e.g., a nuclear receptor) to a sequential series of cellular, anatomical and functional changes resulting in an adverse effect [6]. Importantly, since MIEs in general reflect direct interactions between chemicals and biological/biochemical targets, they appear suitable to be modelled. For this type of purpose docking softwares, predicting molecular interactions that occur between a target protein and a small molecule, can be very valuable. Their application require more time and computational knowledge compared to the methods mentioned above. Such models are extensively used in drug design as powerful approach for structure-based drug discovery. Its application to computational toxicology is more and more investigated especially in the field of endocrine active substance where Nuclear Receptors binding plays a key role in triggering toxicity (sometimes as MIE), and can be predicted [7]. It can also be used in defining similarity driven by MIE to support analogs selection for read-across in hazard characterization [8].

Read Across

It is also possible to estimate toxicological properties, either qualitative or quantitative, of chemicals by using the so-called read across approach. It is based on the grouping and comparison of chemicals. It uses existing information for one or a group of chemicals, called “source chemical(s) or analogs”, to make a prediction of the same property or toxicological endpoint for another, uncharacterized, chemical, called a “target chemical”. The target chemical and analogs must exhibit relevant similarity. The selection of suitable analogs for the prediction of a toxicological endpoint is a complex exercise which play a key role to ensure reliable read across. Ideally similarity should be based on various criteria, including structure, physicochemistry, metabolism as well as mechanistic chemistry and mode-of-action, but this is in general not the case due to lack of fundamental knowledge [9-10]. Care should be taken in grouping chemicals together and in determining chemical similarity since for a given toxicity endpoint some structural features are more important than others, and these structural feature’s behaviour differ from one toxicity endpoint to another. In addition, the analogs need to be associated with reliable and relevant experimental data. Thus, the application of grouping read across requires significant expert knowledge and judgment, in both structural chemistry and toxicology. Grouping and read across has been used within the OECD High Production Volume Chemicals Program as an alternative for experimental testing [11] and is currently being applied under REACH. One of the advantages of read across is that it can generally be applied when suitable (scientifically valid and applicable) (Q)SARs are not available. It can also be used to supplement the results of (Q)SAR analysis as part of a weight-of-evidence approach. The OECD ToolBox is the platform mostly used to perform read -across studies.

Which Endpoints Are Relevant for Risk Assessment and Food Safety?

The safety/risk assessment of a food compound involves the identification and characterization of its hazard. Classically, this involves the generation of experimental data which, according to estimated level of exposure, must address long-term (e.g., 2 years for rodents), reproductive and developmental toxicity. The most relevant and conservative value (called the pivotal value), in general a No Observed Effect Level (NOAEL), is used to establish a health based guidance value for human such as the Acceptable Daily Intake (ADI). This safe level of exposure is obtained through the application of uncertainty factors accounting for the potential differences between the test species and human as well as to cover inter-individual variability. Additional factors may be necessary, for example if the toxicological database is incomplete (e.g., no developmental data) or if a NOAEL is not available and a Lowest Observed Adverse Effect Level (LOAEL) is used instead. This suggest that in principle, to reflect the current hazard characterization process, long-term, reproductive and developmental effects should be quantitatively modelled and the most conservative value used to establish the level of concern. This may be achieved through read across, if toxicologically fully characterized analogs can be identified [12]. Prediction of mechanistic information such as MIE may also provide indirect evidence of potential developmental or reproductive effects (e.g., nuclear receptor activation for endocrine active substances). Because in most cases chronic toxicity and/or carcinogenicity have been documented to constitute the pivotal studies, the development of QSAR models has up to now focused on these specific endpoints. As mentioned earlier, toxicology is changing, putting more emphasis on mode and mechanisms of action, therefore methods predicting such biological pathways are also of importance.

As compared to the pharmaceutical area, the situation of the food sector requires the development of alternative models with some specific characteristics: quantitative rather than qualitative predictions and transparency of the models used. Indeed in the food context the most likely application of computational toxicology models would be in the establishment of the level of safety concern associated with the inadvertent/accidental presence of chemical in products. This requires not only qualitative information on the potential hazard properties of the chemicals (e.g., mutagenicity) but also quantitative information (e. g. carcinogenic potency), allowing the derivation of a margin of exposure (MOE) with the estimated intake.

Chronic Toxicity

Such studies are often the most sensitive in the food toxicology database. Indeed they provide usually the pivotal data to establish safe levels of exposure. The effect of long-term exposure to chemicals is generally addressed by feeding rodents with various doses of test materials over long periods of time up to lifetime. These chronic studies are designed to obtain a dose-response including a dose producing severe toxicity, the lowest dose exhibiting overt toxic effect (Lowest Adverse Effect Level, LOAEL) and a dose without any toxic effects (No Observed Adverse Effect Level, NOAEL). These tests are fundamental for risk assessment but the correlation between the chemical structure and the toxicological outputs has received only little attention. This is partly due to the complexity of such experimental tests that embrace a plethora of different biological effects and mechanism of actions, making QSAR studies extremely challenging. Indeed it is often considered that chronic toxicity is a far too heterogeneous end point to be encoded in a single predictive model. We have developed a model aimed to predict chronic toxicity, for which validation studies have indicated sufficient performance to justify applications in well selected and controlled situations [13]. The analysis of this model revealed that the bioavailability of the compounds drives chronic toxicity effects, constituting a baseline effect. The root mean squared error of the model was found to be 0.73 (in logarithmic scale on a leave-one-out cross validation), close to 0.64 variability observed for the experimental values.

Carcinogenic Potency

When a compound is considered to be genotoxic through DNA-reactive mechanism, the prediction of the carcinogenic potency is necessary to establish level of concern. Unfortunately most of the models that attempt to predict carcinogenicity are qualitative (classification, or yes and no answer) or give a probability that the compound will be carcinogenic. Again in the context of safety assessment such prediction cannot be used because a quantitative value is needed for the comparison with the exposure through the calculation of a Margin of Exposure (MoE, see next section). Carcinogenicity has often been considered as a too complex end point (many mechanisms of action involved and little structural commonality) to be adequately modelled and quantitatively predicted. In this contest, guidance for genotoxic impurities (GTI) was developed by US-FDA. The guidance suggests calculating cancer risk based on carcinogenic potency from a structural similar known carcinogen [14], where the carcinogenic potency is given by TD_{50} and it is defined as the daily dose in mg/kg/day that causes a tumor type in 50% of the exposed animals that otherwise would not develop the tumor in a standard lifetime [15]. Recent efforts in the (Q)SAR field have resulted in the development of local quantitative models for the prediction of carcinogenic potency, opening some hopes in this area. Indeed these models provide reasonable predictions with errors within the same order of magnitude than the estimated variability of experimental data. This promising approach is currently limited by the fact that the proposed programs are propriety [16] or commercially [17] available. Automated and reproducible read-across like models for predicting carcinogenic potency was developed *in house* [18] and will be freely available in the near future through a user friendly internet web site.

Mechanistic Endpoint Prediction

Mechanistic endpoints valuable in food safety are genotoxicity and receptor-ligand interactions. Regarding genotoxicity many models both commercial and free of charge are available [19]. Most of them rely on the propensity of certain structural motives to give rise to DNA-active electrophiles (SA). We developed an *in house* model based on a hybrid system that integrate Structure-Activity Relationship and Artificial Intelligence Systems to improve the prediction of Ames test mutagenicity [20]. The overall error of this system on an external test set of compounds was found to be quantitatively similar to the experimental error of Ames test data (based on the average interlaboratory reproducibility determined by the National Toxicology Program). Moreover, each single prediction is provided with a specific confidence level. These characteristics make the system suitable for early and rapid evaluation of the level of mutagenicity concern.

Receptor mediated endpoints such as nuclear receptor (NR) interaction for endocrine active substances are increasingly attracting attention in the field of toxicology. Moreover the Adverse Outcome Pathway (AOP) approach is more and more acknowledged as a tool for expanding the use of mechanistic data in risk assessment, giving a framework to relate information linking a Molecular Initiating Event (MIE) to adverse effects. In this context *in silico* tools such as docking can be a useful approach to identify compounds that act through the same MIE. Developing and validating a three-dimensional docking model is very laborious and require significant expertise. VirtualToxLab is a software, developed at the Biographics Laboratory 3R , for prediction of ligand-receptor interactions. It is based on an automated protocol that simulates and quantifies the binding of small molecule towards a series of Nuclear Receptors (NR) known or suspected to trigger adverse effects. It identifies the binding mode(s) of a compound by simulating its 3D interaction with a battery of different NRs: Androgen Receptor (AR), Aryl Hydrocarbon Receptor (AhR), Estrogen Receptor (ER) α and β, human Ether-à-go-go-Related Gene Receptor (hERG), Glucocorticoid Receptor (GR), Mineralocorticoid Receptor (MR), Progesterone Receptor (PR), Thyroid Receptor (TR) α and β, Liver X Receptor (LiverX) and the Peroxisome Proliferator-Activated Receptor (PPAR) γ. Afterwards the software quantifies by QSAR models the individual binding affinities for each NR and gives therefrom an estimation of what it's called toxic potential. It ranges from 0 (none) to 1 (extreme) and may be interpreted as a toxic alert. The calculated 3D structures of all protein-ligand complexes are available to the user via a friendly interface. VirtualToxLab is freely accessible to universities, governmental or regulatory bodies and non-profit organizations [7].

Establishing Level of Safety Concern of Chemicals in Food without the Need for Toxicity Testing

To respond to the increasing demand for methodologies to establish levels of safety concern associated with dietary exposures to chemicals for which no toxicological data are available, the European branch of the International Life Science Institute (ILSI) convened an expert group to define the process in order to better exploit data from computational toxicology in food safety. In this frame, computational chemists and toxicologists developed a

Decision Tree (Figure 2) to aid integrating exposure information and predicted toxicological reference points obtained with (Q)SAR and read across approaches. The DT keeps the basic logic of classical risk assessment encompassing the four main steps (exposure assessment, hazard identification, hazard characterization and risk characterization) and leave the flexibility to feed each of this step with all available information coming from both classical toxicological studies (if any available) and *in silico* predictions. To make a safety statement requires quantitative predictions of toxicological reference points such as chronic toxicity (e.g., LOAEL) and carcinogenic potency (TD_{50}) for DNA-reacting chemicals [21].

In this framework the DT can be summarized as followed:

1. **Exposure Assessment** (box 1, Figure 2). This is the first and mandatory step to apply the DT. Indeed a quantitative estimate of the actual or anticipated exposure is a prerequisite for conducting risk characterization (last step of the process) where exposure is compared with toxicological information through a margin of exposure (MoE) approach. In this context all information regarding potential exposure to the substance must be collected, including occurrence of the substance in relevant contaminated foods and their intake, the population potentially affected and the duration of exposure. It is expected that in general, only fragmentary and theoretical information will be available, and in this case the conservative assumptions may be applied to define worst-case exposure scenarios. The strength of the evidence and the uncertainties regarding all aspects of exposure should be thoroughly reviewed and described.
2. **Hazard identification** (boxes 2, 3, 4, 5, Figure 2). Classically it is defined as the intrinsic property of a chemical agent to produce adverse health effects and it uses results from toxicity studies in animal models. In the context of the DT application, little direct toxicological data on the substance of interest is expected and therefore hazard identification focuses on the use of predicted values obtained with (Q)SAR and read across methods. Box 2 concerns the collection of all the information available for the targeted compound. It includes structure characterization, identification of potentially degradates (if the chemical is not stable in food), the similar compounds (analogs) and the collection of relevant toxicological, metabolic and mechanistic data available (e.g., Molecular Initiator Event or affinity for endocrine receptors) for both target chemical and analogs. The homogeneity of this data will provide increased confidence for the prediction of the target chemical. In box 3 the genotoxic potential through DNA reactivity is assessed. Indeed in risk assessment a key distinction is made between those effects that exhibit a biological threshold in their dose-response relationship and those that do not and in practice the only mechanism for which it is assumed that there is no threshold is genotoxicity via direct DNA reactivity, with the consequence of genetic mutations. Hence, in the absence of toxicological information, an important early consideration is whether a substance maybe a DNA-reactive mutagen or not. The Ames/Salmonella test is a relative reliable predictor of such potential and in absence of data, several software packages, both commercially and freely available, can assist in such prediction. Alert for carcinogenicity through DNA-reactivity requires proceeding to box 4, but at the same time to process chemicals with alert for genotoxicity through box 5 may also be advisable because carcinogenicity may be related to other toxic effects. On the

contrary the absence of any evidence for genotoxicity through DNA reactivity allows proceeding directly to box 5 dealing with threshold toxicity. In box 4 the carcinogenic potency (TD_{50}) and in box 5 the chronic toxicity (LOAEL) can be predicted using either (or better both) QSAR or read-across approaches.

3. **Hazard characterization** (box 6, Figure 2). It may be defined as the determination of the relationship between the magnitude of exposure to a chemical agent and the severity and/or frequency of adverse effects. Classically, hazard characterization is the process where the most appropriate toxicological data is selected in order to either establish a safe level of human exposure or to provide exposure levels associated with a pre-determined level of risk. Because the present work deals with situations where specific experimental toxicological data are absent or insufficient, the establishment of safe levels is not considered appropriate. In this context, hazard characterization therefore refers to the selection of the most adequate prediction values to be used in the calculation of the margin of exposure. Therefore in box 6 both qualitative information (e.g., mechanistic) and quantitative values (e.g., LOAELs, TD50) obtained previously from box 2 to box 5 are listed and described in detail. The main information analyzed are the applicability domain, the uncertainty of the models used, the similarity of the analogs used for read-across and the quality of the toxicological data found for them, together with consideration about the exposure time/duration (e.g., chronic, short term). The main meaningful values will be selected and brought to the next step on risk characterization.

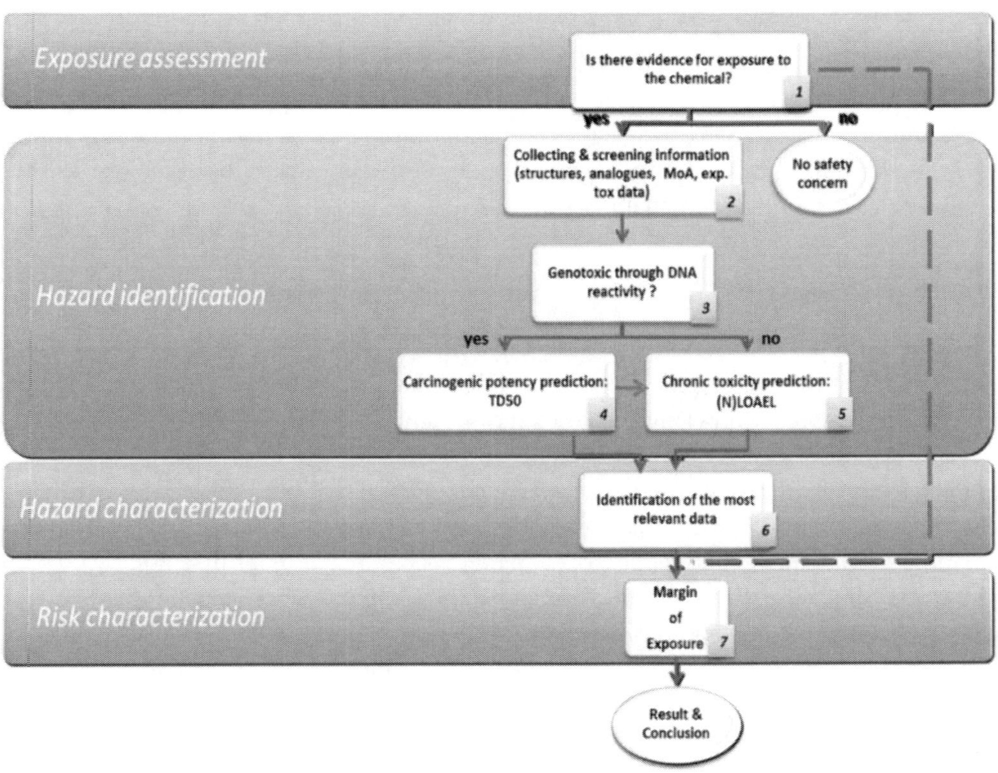

Figure 2. Decision Tree reprinted from Schilter et al. [21] with permission from Elsevier.

4. **Risk characterization** (box 7, Figure 2). It is classically defined as the quantitative or semi-quantitative estimate of the probability of occurrence and severity of adverse effects in a given population under defined exposure conditions based on hazard identification, hazard characterization and exposure assessment. In the present context, risk characterization refers to the calculation of Margins of Exposure (MoE) corresponding to the ratio between relevant predicted toxicological values with exposure estimated. In practice it is done by dividing the individual toxicological values selected at box 6 with the exposure estimate obtained at box 1. Even if MoE corresponds to the distance between a dose expected to produce toxicity and actual or estimated human exposure, it is not a measure of risk per se, but its size is considered to reflect the level of concern associated with a specified exposure. Although it appears obvious that the larger the MoE, the lower the level of concern, the interpretation of MoEs may not be straight forward and should be done on a case by case basis. Final interpretation will have to consider and compare all independently calculated MoEs. In absence of specific reasons to exclude one of the MoEs, as a precaution the most conservative may be considered for decision making. The size of the MoE defines the level of safety concern and should account for a number of uncertainties such as the classical interspecies and inter-individual variability as well as others determined on a case by case basis. In general these uncertainties can be adequately dealt by factors already widely applied in classical risk assessment. Proper interpretation of the size of the calculated MoEs will provide insight on the level of safety concern. For example in classical hazard characterization based on animal data, a safe level of exposure for human is usually derived from the application of a default Uncertainty Factor (UF) of 100 to the No Observed Adverse Effect Level (NOAEL) in the most relevant animal study (often a chronic toxicity study). This UF takes into account potential differences between the test animal and human (interspecies, UF =10), and the variability in sensitivity to the test substance within the human population (intraspecies, UF =10). Therefore, in this case to conclude on a low safety concern, a MoE based on predicted chronic animal data should be > 100, in order to account for potential inter-species and inter-individual differences.

In general it is important to keep in mind that, like in any scientific field, data interpretation requires knowledge and expertise and using *in silico* models is not an exception. All the results obtained should always be reviewed considering as well the belonging to the applicability domain, in order to identify errors, chance correlations and results that contradict with current knowledge and discard them if necessary. The proper application of the methodology proposed requires significant inter-disciplinary expertise, including competences in structural chemistry, computational modeling, toxicology, mechanism/mode of action and risk assessment.

The application of the DT should be seen as a practical aid to provide toxicological advice within a relatively fast timeframe. An analysis of the uncertainties of *in silico* approaches together with results from case studies suggest that establishing safety concern based on application of the DT is unlikely to be significantly more uncertain than based on experimental data. Indeed the DT makes a full use of all data available, ensuring an adequate degree of conservatism and it can be used when fast decision making is required and to decide on the need for further toxicological studies. The outputs of the DT are expected to be communicated in a pragmatic format containing quantitative elements of risk together with levels of confidence based on an analysis of uncertainties and the quality of the data used.

Identification of Chemicals of Concern in Packaging

The main role of food packaging is to contain the food but also to protect it from the outside. Therefore food packaging is generally recognized as contributing directly to product safety through prevention of biological and chemical contaminations. Packaging impacts food safety also indirectly by providing consumers with health related information on ingredients, mode of preparation and shelf-life. Beside these well documented beneficial roles in food safety, there has been an increasing scientific, media and public attention regarding packaging as a source of food contaminations. Indeed packaging materials may contain a number of chemicals which then can potentially migrate into food products. Some of them, called Non Intentionally Added Substances (NIAS), originate from various sources such as impurities in starting materials as well as from chemical reactions and degradations over packaging product manufacturing and use [22]. Often they may neither be known nor expected and in general, they are not toxicologically characterized. Currently, it is acknowledged that NIAS of concern (e.g., Carcinogenic, Mutagenic and Reprotoxic, CMR, compounds) should be assessed from a risk perspective, but there is no consensus on how to identify them. To address this challenge, efforts are increasingly focused on non-targeted chemical screening. In addition, the use of alternative toxicology methods, providing insight on the occurrence of chemicals acting through mechanisms considered of concern at low level of exposure, mainly receptor mediated and mutagenicity, is increasingly seen as important. Up to now, mainly *in vitro* bioassays have been applied for that purpose. Recently, we also proposed to use *in silico* tools [23].

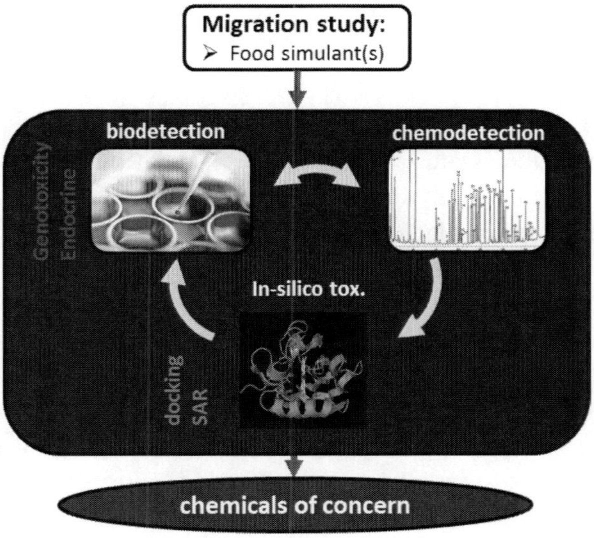

Figure 3. Interdisciplinary approach applied internally for the identification of chemical(s) of concern in packaging.

In brief, the identification of NIAS of concern is thought to benefit from the application of *in silico* methods as summarized on Figure 3. Samples resulting from laboratory migration studies are analysed using both untargeted chemical analysis (chemodetection, using GC and LC –MS techniques) and biological screening (biodetection, using cell culture systems). The list of chemically identified substances is then screened with *in silico* methods, mainly QSAR for genotoxicity assessment and docking for receptor-mediated effects. The *in silico* data not only allow to identify in the list of detected compounds the ones potentially active, but also provide a strong opportunity to identify substances potentially responsible of *in vitro* bioactivity if any. For example, in case where a genotoxic activity is observed in a migration sample, chemically identified substances can be screened for genotoxicity alert using *in silico* models. The chemical(s) associated with such a structural alert can then be tested experimentally in order to verify its (their) activity and determine its (their) actual responsibility in the overall genotoxicity of the sample. The concordance between the intensity of the peak (quantitative estimation of the substance) and the *in vitro* activity found for the single compound tested provides the origin of the biological activity. Obviously, the identified chemicals of concern can then be assessed from a risk perspective, using either actual experimental toxicological data if available, or predicted data following the DT scheme presented in the previous section. It is important to keep in mind that such an approach is generic and can also be applied to other fields requiring to identify substances of concern in complex mixtures.

Conclusion

There is an increasing pressure to reduce animal experimentation for safety assessment of food components, and in this context *in silico* predictive methods have obvious advantages and roles to play. In the field of medicinal chemistry, the *in silico* tools are much more used and developed and the big challenge is now to learn how and when to apply them in the field of food safety. This requires a mind shift and an understanding of food safety issues. Indeed to make safety statements often requires the comparison of human intake with levels of exposure either considered safe, or associated with a defined level of risk. Therefore, quantitative predictions of toxicological reference points are necessary, including for examples LOAEL, or carcinogenic potency, in case of chemicals with an alert of DNA reactive genotoxicity. Unfortunately endpoints such as chronic toxicity and carcinogenicity are often considered too heterogeneous and complex to be adequately modeled and quantitatively predicted. However, new evidences have indicated that adequate and sufficiently reliable methods can be developed and applied, such as read-across and QSAR. Interestingly, comparing the error of prediction with both the variability of experimental data and with the uncertainties inherently present in the classical risk assessment (e.g., on exposure), it seems that *in silico* approaches, if integrated and interpreted correctly, can be used to establish levels of safety concern. Moreover computational approaches, such as docking, can be useful tools to integrate mechanistic information such a MIEs. Their applications to endocrine active substances within the AOP framework are considered very promising.

It is always important to remember that *in silico* toxicology is a powerful discipline but, like in any other scientific field, the proper interpretation of the results is of key importance. This requires both computational and toxicology knowledge to be able to discard results coming from well identified computational errors and that contradict with current toxicological knowledge. Different from pharma, in food safety the transparency and availability of the models used is strongly requested by regulatory because it will give the possibility to better review the results obtained. The acceptance of these models will be possible only if the analysis is full transparent and therefore the promotion of validated freely available tools based on open-source codes should be strongly promoted.

References

[1] Lo Piparo, E., Worth, A., Manibusan, M., Yang, C., et al. (2011). Use of computational tools in the field of food safety. *Reg. Tox. Pharm.*, *60*, 354-362.

[2] EFSA Journal. (2012). Scientific Opinion on Exploring options for providing advice about possible human health risks based on the concept of Threshold of Toxicological Concern (TTC), *10*(7), 2750.

[3] Cramer, G. M., Ford, R. A. & Hall, R. L. (1978). Estimation of toxic hazard--a decision tree approach. *Food and Cosmetics Toxicology*, *16*, 255-276.

[4] Kroes, R., Renwick, A. G., Cheeseman, M., Kleiner, J., Mangelsdorf, I., Piersma, A., Schilter, B., Schlatter, J., van Schothorst, F., Vos, J. G. & Wurtzen, G. (2004). Structure-based thresholds of toxicological concern (TTC): guidance for application to substances present at low levels in the diet. *Food and Chemical Toxicology*, *42*, 65-83.

[5] Sushko, I., Salmina, E., Potemkin, V. A., Poda, G. & Tetko, IV. (2012), ToxAlerts: a Web server of structural alerts for toxic chemicals and compounds with potential adverse reactions, *J Chem Inf Model.*, *52*(8), 2310-6.

[6] OECD (Organisation for Economic Co-Operation and Development Environment). (2013). Health and Safety Publication Series on Testing and Assessment No 184. *Guidance Document on Developing and Assessing Adverse Outome Pathways*. Paris, 2013.

[7] Vedani, A., Dobler, M., Hu, Z. & Smieško, M. (2014). OpenVirtualToxLab-A platform for generating and exchanging in silico toxicity data, *Toxicol Lett.*, *232*(2), 519-532.

[8] Ehrlich, V. A., Dellafiora, L., Mollergues, J., Dall'Asta, C., Serrant, P., Marin-Kuan, M., Lo Piparo, E., Schilter, B., Cozzini, P. (2015). Hazard Assessment Through Hybrid In Vitro / In Silico Approach: The Case of Zearalenone. *ALTEX,* May 18.

[9] Wu, S., Blackburn, K., Amburgey, J., Jaworska, J. & Federle, T. (2010). A framework for using structural, reactivity, metabolic and physicochemical similarity to evaluate the suitability of analogs for SAR-based toxicological assessments. *Reg. Tox. Pharm.*, *56*, 67-81.

[10] Blackburn, K., Bjerke, D., Daston, G., Felter, S., Mahony, C., Naciff, J., Robison, S. & Wu, S. (2011). Case studies to test: a framework for using structural, reactivity, metabolic and physicochemical similarity to evaluate the suitability of analogs for SAR-based toxicological assessments. *Reg. Tox. Pharm.*, *60*, 120-135.

[11] OECD (Organisation for Economic Co-Operation and Development Environment) 2014, Guidance on grouping of chemicals, second edition.

[12] Blackburn, K1., Bjerke, D., Daston, G., Felter, S., Mahony, C., Naciff, J., Robison, S. & Wu, S. (2011). Case studies to test: A framework for using structural, reactivity, metabolic and physicochemical similarity to evaluate the suitability of analogs for SAR-based toxicological assessments. *Regul Toxicol Pharmacol.*, 2011, *60*(1), 120-35.

[13] Mazzatorta, P., Estevez, M. D., Coulet, M. & Schilter, B. (2008). Modeling oral rat chronic toxicity. *J Chem Inf Model.*, *48*, 10, 1949-54.

[14] USFDA. (2008). Guidance for Industry: Genotoxic and Carcinogenic Impurities in Drug Substances and Products: Recommended Approaches. Available at: http://www.fda.gov/downloads/ Drugs/ .../Guidances/ucm079235.pdf

[15] Gold, L. S., Slone, T. H., Ames, B. N. & Manley, N. B. (2001) Pesticide Residues in Food and Cancer Risk: A Critical Analysis. Handbook of Pesticide Toxicology, Second Edition (R. Krieger, ed.), San Diego, CA, Academic Press, 799-843.

[16] Bercu, J. P., Morton, S. M., Deahl, J. T., Gombar, V. K., Callis, C. M. & van Lier, R. B. (2010). In silico approaches to predicting cancer potency for risk assessment of genotoxic impurities in drug substances. *Regul Toxicol Pharmacol.*, *57*, 2, 300-6.

[17] Contrera, J. F. (2011). Improved in silico prediction of carcinogenic potency (TD50) and the risk specific dose (RSD) adjusted Threshold of Toxicological Concern (TTC) for genotoxic chemicals and pharmaceutical impurities. *Regul Toxicol Pharmacol.*, *59*, 1, 133-41.

[18] Lo Piparo, E., Maunz, A., Helma, C., Vorgrimmler, D. & Schilter, B. (2014). Automated and reproducible read-across like models for predicting carcinogenic potency, *Regul Toxicol Pharmacol.*, *70*(1), 370-8

[19] JRC. (2010). Applicability of QSAR analysis to the evaluation of the toxicological relevance of metabolites and degradates of pesticide active substances for dietary risk assessment. Report from the European Commission's Joint Research Centre (JRC) to the European Food Safety Authority (EFSA). Available at: http://www.efsa.europa.eu/ en/ scdocs/ scdoc/50e.htm

[20] Mazzatorta, P., Tran, L. A., Schilter, B. & Grigorov, M. (2007). Integration of structure-activity relationship and artificial intelligence systems to improve in silico prediction of Ames tetast mutagenicity. *J. Chem. Model.*, *47*, 34-38.

[21] Schilter, B., Benigni A., Boobis, A., Chiodini, A., Cockburn, A., Cronin, M.T.D, Lo Piparo, E., Modi, S., Thiel, A., Worth, A. (2014). Establishing the level of safety concern for chemicals in food without the need for toxicity testing. *Reg. Tox. Pharm.* 68, 275-296.

[22] Nerin, C., Alfaro, P., Aznar, M. & Domeno, C. (2013). The challenge of identifying non-intentionally added substances (NIAS) from food packaging materials: a review. *Analytica Chimica Acta.*, http://dx.doi.org/doi:10.1016/j.aca.2013.02.028.

[23] Schilter, B., Simian, H., Spack, L., Varela, J., Veyrand, J., Marin-Kuan, M., Lo Piparo, E., Guerin, V. & Ehrlich, V. A. (2013). Assessing and managing Non Intentionally Added Substances (NIAS): a food manufacturing perspective, Mitteilungen aus lebensmittel- und Umweltchemie (Travaux de chimie alimentaire et environmentale). Volume *102*, 60-68.

In: From Medicinal Chemistry to Food Science
Editor: Pietro Cozzini

ISBN: 978-1-63483-587-9

Chapter 2

IN VIVO TEST FOR FOOD CONTAMINANTS AND TOXINS EVALUATION

***N. Rizzi*[1,2] *and S. Della Torre*[1,*]**

[1]University of Milan, Center of Excellence on Neurodegenerative Diseases and Department of Pharmacological and Biomolecular Sciences, Milan, Italy

[2]University of Milan, Polo Veterinario di Lodi, Lodi, Italy

Abstract

Nutrients *per se* can have a physiological activity and both naturally-occurring or industrial-derived food contaminants and toxins can exert negative effects by interfering with some physiological pathways and causing dangerous consequences for health. The negative properties of such as food contaminants could be due to a specific compound or could be ascribable to (and maybe enhanced by) the combinatorial effects of different compounds, matrix effects and different metabolites of specific compounds. This complexity of action and the relative systemic consequences of specific food-associated contaminants or toxins emphasizes the need of an *in vivo* evaluation and underlines the increasing requirement of new animal tools for testing the presence/deposit of these substances and their influence on specific pathways. Although animal studies are routinely involved in testing food-associated contaminants and toxins [1-3], the potential multifactorial effects of these compounds in different tissues with time and in relation to the physiological/pathological context requires a *spatio*-temporal analysis that can not be extensively accomplished by using wild-type animals.

To overcome these limitations, in this study we propose three new reporter mice as tools for the evaluation of food-associated contaminants and toxins and their related consequences on the activation of the estrogen receptor-dependent signalling and on the development of oxidative stress and inflammation, two negative factors common to numerous metabolic pathologies.

*Corresponding Author. Email: sara.dellatorre@unimi.it. Tel: +39-02-50318405. Fax: +39-02-50318284

1. Introduction

Increasing literature points on the relevance of food-associated effects on different physiological processes: nutrients *per se* can have a physiological activity, as it is well known that some classes of nutrients can act as modulators of nuclear receptors activity, by modulating the signal transduction of the respective regulated pathways. For example, it is well known that fatty acids [4-7] are the naturally inducers of PPARs (Peroxisome Proliferator-Activated Receptors alpha, beta, and gamma), nuclear receptors that regulate the transcription of genes involved in the regulation of cellular differentiation, development, metabolism (carbohydrate, lipid, protein), and tumorigenesis. Like fatty acids for PPARs, also other classes of nutrients may act as agonists on other transcription factors, such as glucose for Liver X Receptor alpha (LXRα) [8] or aminoacids for Estrogen Receptor alpha (ERα) [9]. Furthermore, some pathways are greatly inter-connected each other, as demonstrated by different studies highlighting the relevance of the cross-talk between LXRα and PPARα, crucial in regulating hepatic lipogenesis [10] or between LXRα and Farnesoid X Receptor (FXR), which in turn regulates bile acids, sterols and fatty acids, that are the activators of FXR, LXRα and PPARα, respectively [11]. The food-dependent modulation of nuclear receptors activity and the related consequences of this regulation have deep roots in the strong inter-relation between nutrients and nuclear receptors that has been established and improved with evolution. The ability to sense nutrients and adapt a physiological response to nutrient availability and to their specific characteristic has allowed to organisms to increase their opportunities to survive and to expand their progenies: in one word, to evolute. This ability became of particular relevance for female mammals, for whose, with the placenta advent, energy metabolism and fertility started to be tightly connected and reciprocally regulated to limit fertility in a nutritionally poor environment and to adapt energy production and storage to the requirements of reproduction [12].

In this view, nutrients exert a bioactive role and can not be merely considered a source of energy, but represent the first/main modulators/signals for an organism, which life expectancy can not disregard from the environment by taking advantage of available nutritional resources.

The understanding of the properties of principles present in some nutrients [13-16] lead humans to appreciate the role of bioactive food components in the prevention of metabolic diseases [17-22], cancer [23-27], inflammation [28, 29] or aging-associated neuro-degenerative diseases [30, 31].

If the proofs of the potential benefic effects of food components become every day more evident, on the other hand others naturally-occurring or industrial-derived food contaminants and toxins can have adverse effects *per se* or interfere with signalling regulated by nuclear receptors and cause dangerous consequences for health. This is the case of genistein, a phytoestrogen known to be naturally present in food, in particular in soy milk. Even if low dose of phytoestrogens seem to have beneficial effects, high doses of these compounds could have adverse effects [32], in particular for children health; in fact, has been reported an increased morbidity, mainly due to infections of upper respiratory tract, in infants fed with soymilk compared to those fed with cow's-milk [33].

Similarly, endocrine disrupting chemicals (EDCs) present in food as additives or contaminants [34] have been suspected to be associated with altered reproductive function in males and females [35-40] and increased incidence of breast cancer [41, 42], abnormal growth patterns and neuro-developmental delays in children [43], as well as changes in immune function [44].

In addition to their action on nuclear receptors, food contaminants could affect the composition of gut microbiota [45, 46], an integral factor that impacts host metabolism and plays a vital role in metabolic diseases such as obesity [47], insulin resistance [48], type 2 diabetes [49, 50], cardiovascular diseases [51, 52] and immune-related disorders [53].

Some substances, for their chemical structures, could accumulate in fat tissue and then could be mobilized under energetic imbalance (e.g., in weight-loss inducing diseases or diets) and, once mobilized, could exert their action on whole organism. For example, it has been reported that genistein accumulates in body depots and is mobilized during fasting, reaching estrogenic levels in serum that are able to counteract the hormonal actions of endogenous estradiol [54].

Also metals could accumulate in some tissues and alter both the local and whole metabolism by interfering with some signalling or by causing a damage that could progressively degenerate or could be extended to the whole organism. For example, iron overload can affect major tissues involved in glucose and lipid metabolism (pancreatic β cells, liver, muscle, and adipose tissue) and organs affected by chronic diabetic complications and it is associated to the clinical course of several chronic metabolic diseases such as type 2 diabetes, obesity, atherosclerosis and non-alcoholic fatty liver disease [55]. In particular, in the liver iron overload is associated with hepatic oxidative damage to DNA in nonalcoholic steatohepatitis [56].

In addition, the nature and the power of the effects caused by contaminants, additives and toxins present in food could be attributable to (or enhanced by) the combinatorial effects of different compounds [57-60], the matrix effects [61] and different metabolites of specific compounds [62, 63].

Furthermore, the relevance of food-associated contaminants or toxins could be linked to the physiological state and changes depending on particular physiological or pathological conditions. In fact, some compounds could have major effects on children or in pregnant women by affecting also the health of the unborn. For example, EDCs could be transferred from the pregnant woman to the developing fetus or child through the placenta and breast milk [64] and affect the development and increase the susceptibility to non-communicable diseases that can become evident until later in life [65-68]. An unbalanced alimentary regimen [69, 70] and/or pathological conditions can aggravate the power of negative effects of food-associated contaminants or toxins, contributing to the progression of the pathologies themselves [71, 72]. This aspect is particularly relevant in respect to the increasing, epidemic incidence of obesity and its related metabolic disorders: projections of the World Health Organization estimate that by 2015 there will be 2.3 billion adults overweight with more than 700 million obese adults [WHO 2006; www.who.int]. The World Health Organization has declared obesity as one of the major global health risks, as excessive body weight is associated with increased risk for several diseases, such as coronary heart disease, hypertension, stroke, hyperinsulinemia, insulin resistance, reproductive problems, and cancer and a number of other issues, including social problems [73]. The obesity epidemic is not attributable solely to overeating or inactivity, although these are obvious factors. One possibility is that the environment and food-associated contaminants and toxins also play a role [67, 74]; furthermore, few epidemiological and experimental studies have suggested a connection between the environment and obesity or indicated the existence of an obesogenic environment that may have worsened in recent decades [75, 76]. The rise in obesity coincides with an exponential increase in the use of industrial chemicals over the last 40 years and chemical compounds have attracted attention for their potential contribution to the increased obesity rate [77, 78]. Different studies have indicated that exposure during fetal development to chemicals capable of hormone-like action may be one contributing factor to weight gain during adulthood by altering body [65-67].

The complexity underlying the effects exerted by food components and food-associated contaminants and toxins emphasizes the need of the *in vivo* evaluation of these dietary compounds and highlights the increasing requirement of new animal tools for testing the presence/deposit of these substances and their consequences for health.

In fact, unlike *in vitro* analysis, the studies done with animal models could be greatly useful for understanding the related consequences of food components and food-associated contaminants and toxins on whole physiology [79], an analysis that could not be exhaustively reached with *in silico* or *in vitro* studies.

Although animal studies are routinely involved in testing food-associated contaminants and toxins [1-3], the potential multifactorial effects of these compounds in different tissues with time and in relation to the physiological/pathological context requires a *spatio*-temporal analysis that can not be extensively accomplished by using wild-type animals.

To overcome these limitations, in this study we assess the relevance of new reporter mice as tools for the evaluation of food-associated contaminants and toxins and their related consequences on the activation of the estrogen receptor-dependent signalling and on the development of oxidative stress and inflammation, two negative factors common to numerous metabolic pathologies [80-89]. The reporter mice are animals that were genetically modified to respond to a specific stimulus with the production of proteins, named "reporters", that are easily detected and quantitated by *in vivo* and *ex vivo* imaging. As well as all the advantages using reporter mice already described, for example the possibility to follow in *spatio*-temporal dimension the effects of compound *in vivo*, another main benefit is the reduction of animals number used for each experiments with very high attention to the animal wellness and in accordance with the 3R philosophy (Reduction, Replacement, Refine).

2. The ERE-Luc Mouse, a Tool for the Study of Food-Associated Estrogenic Compounds

As estrogens, estrogenic compounds may exert their physiological effects acting through two estrogen receptors, named ERα and ERβ, belonging to a large family of the nuclear receptor family of ligand-activated transcription factors and being known for their relevance in reproductive and metabolic functions in mammals [90]. ERα is mainly expressed in reproductive and metabolic tissues, whereas ERβ is expressed in the ovary, prostate, lung, gastrointestinal tract, bladder, hematopoietic cells, and the central nervous system [90].

Once activated by estrogens or estrogenic compounds, ERs dissociate from the inhibitory binding of heat shock proteins (HSPs) and undergo to structural modifications enabling the receptor protein to directly bind to specific DNA sequences (named Estrogen Responsive Elements or ERE) by regulating gene transcription or to bind to other signalling molecules by interfering with their activities. In addition to estrogenic compounds, ERs intracellular activity may be induced *via* a so-called unliganded-mechanism by other molecules such as growth factors, neurotransmitters, cytokines, and a variety of steroidal and non-steroidal compounds [90].

By such a complex mechanism of action ERs may modulate a large number of physiologically relevant processes and the presence of food-associated estrogenic compounds could interfere with this finely tuned regulation causing an impairment of ERs action with consequences for health.

In the last years has becoming even more evident the involvement of ERs in the regulation of different metabolic processes [91, 92]. In particular, the relevance of ERα action in the liver (the metabolic organ most responsive to estrogen action) has been highlighted by the use of the ERE-Luc, a reporter mouse that expresses a construct containing two ERE sequences driving the gene encoding the firefly luciferase flanked by insulator sequences that protects the construct from the influences of the surrounding chromatin and ensures that the transgene is ubiquitously expressed in all cells [93]. This bioluminescent reporter model allows the *spatio*-temporal study of ERs activity *in vivo* and the screening and identification of compounds acting on ERs, including estrogenic compounds present in nutrients that can act as EDCs.

By using this model, it has been demonstrated that in females hepatic ER activity (mainly due to ERα, the most expressed ER in the liver) is associated both to the hormonal and nutritional status and is greatly associated to dietary aminoacid intake, allowing the progression of the reproductive cycle only in case of nutrient availability [9, 94, 95]. The studies done *in vivo* with ERE-Luc mice allowed to evaluate the changes in the activation of estrogen receptor exerted by genistein as pure isoflavone, by genistein and daidzein as a mixture of pure isoflavones or by a soy-based diet (soy milk) [61]. Similarly, it has been demonstrated that estrogen-like effects of diet-derived cadmium differ from those of orally administered $CdCl_2$ [96, 97]. The results described represent a proof of the effects of a given compound, which power could be enhanced by the combination with other compounds or by a matrix-effect and further underline the relevance of the use of mice as complex organisms for the evaluation of the effects of food-associated compounds altering the endocrine balance. In addition, the opportunity to use the ERE-Luc model allowed to understand the differential effect exerted by cadmium on ER activation in a physiological context changes during time and affect differentially various tissues. This aspect is of particular relevance and further give prominence to the use of mice as complex organisms for the evaluation of the effects of food-associated compounds altering the endocrine balance.

2.1. Case Study: Imaging of Sexually Immature ERE-Luc Reporter Mice to Study the Estrogen-Like Effects of Dietary Cadmium (Cd)

In the recent study [97] our group described an *in vivo* bioluminescent imaging methodology using sexually immature male ERE-Luc reporter mice to test the estrogen-like (EL) effects of Cd as a natural component in wheat and flax bread based diets.

Emerging evidences show that Cd is clearly a potent non-steroidal estrogen [96, 98]. A large body of evidence points to a major endocrine disruptive action of this heavy metal due to its ability to bind with high affinity to ERα independently of estrogen binding and to activate this receptor. In most *in vivo* studies, Cd effects were analysed by administering $CdCl_2$ and this type of treatment does not really mimic the effects of dietary exposure to this metal. Food accounts for approximately 90% of human exposure to Cd in the non-smoking general population and the major foodstuffs that contribute to this exposure are cereal products, vegetables, nuts and starchy roots. Using sexually immature male ERE-Lucreporter mice [93] we show a non-invasive bioluminescent in *vivo* imaging methodology to measure the changes in estrogenic activity in response to two cereal-based diets and $CdCl_2$for a period of 21 days.

Study Design

Treatment	Animals	Mice number	*In vivo* imaging (time, days)	*Ex vivo* imaging and enzymatic assay
control group (standard estrogen-free diet)	2-3 weeks male ERE-Luc	5	0, 1, 7, 14, 21	Small Intestine Brain Liver
$CdCl_2$ (1ug/kg *per* day by gavage)	2-3 weeks male ERE-Luc	5	0, 1, 7, 14, 21	
wheat bread diet (low Cd content 17.57 μg/kg)	2-3 weeks male ERE-Luc	5	0, 1, 7, 14, 21	
flax bread diet (high Cd content 49.22 μg/kg)	2-3 weeks male ERE-Luc	5	0, 1, 7, 14, 21	

Four different experimental groups: (1) controls group (standard estrogen-free diet); (2) $CdCl_2$ (1μg/kg *per* day by gavage); (3) wheat bread diet (low Cd content, 17.57 μg/kg); and (4) flax bread diet (high Cd content, 49.22 μg/kg) were fed for 21 days with the indicated diets while for the $CdCl_2$ group, $CdCl_2$ dissolved in saline was administered as a single 100 μl bolus every 24 hours for 21 days. Bioluminescence *in vivo* imaging sessions were carried out on days 0, 1, 7, 14 and 21.On the last day of the experiment, the animals were sacrificed and selected organs were excised for the *ex vivo* measurement of tissue specific signals.

Results and Discussion

Whole body *in vivo* analysis of ER activity, done by measuring photon emission at 0, 1, 7, 14 and 21 days of treatment did not show major changes associated to the treatments (Figure 1A), yet at the end of the study mice were euthanized, organs dissected and photon emission was measured by CCD camera in small intestine, brain and liver (Figure 1B). Significant changes were observed in brain: as indicated by photon emission quantification, while Cd did not affect ER activity in liver and small intestine (Figure 1C).

The study for the first time demonstrates that a cereal-based diet or $CdCl_2$ may affect ER signalling in specific organs (brain but also prostate and WAT, data do not shown), in sexually developing immature male mice. It also shows that the observed Cd related effects occur only at the molecular level and not at the anatomo-pathological level. In this study, ERE-Luc reporter mice coupled bioluminescent imaging technology was shown as an experimental system to study the effects of low doses of contaminants in a complex food on estrogen receptor activity. The concept of receptor mediated toxicity has driven the field of toxicology to carry out tests that are more predictable, leading to a thorough understanding of the mechanism based toxicity.

In conclusion, this study undertakes to analyse the Cd related EL effects and shows an imaging-based methodology to investigate the effect of extremely low concentrations of Cd embedded in food matrix and emphasizes the novel possibility to carry out long-term studies of suspected chemicals in their inherent presence in the food chain or environment.

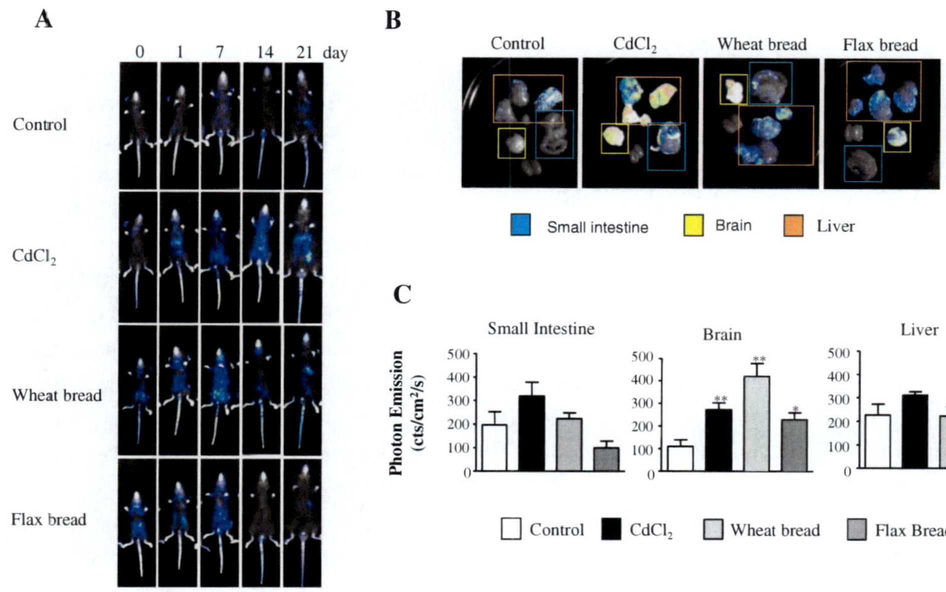

Figure 1. In vivo and ex vivo bioluminescence imaging of ERE-Luc male mice fed with Cd-based diets. A. Representative images of bioluminescent in vivo imaging of male ERE-Luc mice at 0, 1, 7, 14 and 21 days after indicated diet-treatment. B. Representative images of bioluminescent ex vivo imaging of freshly excised small intestine, brain and liver dissected from euthanized mice at the end of 21 day of indicated diet-treatment. C. Photon emission from isolated tissues was measured and expressed as photon emission (cts/cm2/s).Data represents MEAN ± SEM (n=5); $^*p<0.05$ and $^{**}p<0.01$ versus control by one-way ANOVA followed by Bonferroni post hoctest.

Methods

In vivo imaging: for the semi-quantitative analysis of photon emission, animals were injected i.p. with 80 mg/kg of luciferin (Beetle Luciferin Potassium Salt; Promega, Madison, WI, USA) 15 min prior the imaging session. For the imaging, mice were anaesthetized using Isofluorane (Isofluorane-Vet; Merial, Lyon, France) and kept under anesthesia during the 5 minutes of the session carried out with a CCD-camera (Xenogen IVIS Lumina System; Caliper, PerkinElmer or Night Owl imaging unit; Berthold Technologies, Bad Wildbad, Germany). Photon emission in selected body areas was measured, respectively, using the Living Image Software (Caliper, PerkinElmer) or using WinLight32 imaging software (Berthold Technologies). *Ex vivo imaging:* the selected organs were dissected and immediately subjected to *ex vivo* imaging session of 5 minutes. Photon emission was quantified with the Living Image Software (Caliper, PerkinElmer) or using WinLight32 imaging software (Berthold Technologies).

3. The Oxidative Stress Reporter ARE-Luc2 Mouse, a Tool for the Evaluation of Food-Associated Oxidative Stress *In Vivo*

In recent years, oxidative stress has been implicated in a wide variety of degenerative processes, diseases and syndromes, including: mutagenesis, cell transformation and cancer; heart attacks, strokes, atherosclerosis, and ischemia/reperfusion injury; chronic inflammatory diseases; acute inflammatory problems; neurological disorders, such as Parkinson's and Alzheimer's diseases; and other age-related disorders, perhaps even including factors underlying the aging process itself [99]. Oxidative damage is a result of the intrinsic and extrinsic reactive oxygen species (ROS) formation factors. The most important endogenous sources of oxidants are mitochondrial electron transport chain and nitric oxide synthase reaction, and the non-mitochondrial sources: Fenton reaction, reactions involving cytochromes P450 in microsomes, peroxisomal beta-oxidation and respiratory burst of phagocytic cells [100]. The main exogenous causes of oxidative stress can be summarize in: decreased intake of exogenous antioxidants from food; increased metal ion intake (e.g., Fe, Cu, Cr) also as food contaminants; ROS from ionizing radiation, air pollution, smoking; chronic inflammation. The reduction of oxidative stress can be achieved on two levels: i) by lowering exposure to oxidative agents; ii) by increasing the levels of endogenous and exogenous antioxidants in order to scavenge ROS before they can cause any damage. In fact, if on the one hand the food is one of the main sources of exogenous oxidative stress, on the other hand the compounds derived from the diet represent the important antioxidant defenses. Therefore, the identification of the actual ingredient in a specific diet responsible for the antioxidant effect remains an important bottleneck for translating observational epidemiology to the development of functional food ingredients. A large number of different techniques have been developed to monitor oxidative damage and its consequences; besides, these approaches can be used to identify dietary antioxidants. We are exploring this topic with recent developed reporter mouse designed to measure the response to oxidative stress by optical imaging. Similarly to ERE-Luc also this new model, called ARE-Luc2 is a tool to perform *in vivo* imaging, allowing to monitor the oxidative stress response in length of time (temporal dimension) and provide a global view of the potential target organs of toxicity in all spectrum of body action of whole mouse (spatial dimension). After an extensive bibliographic research and a consequent bioinformatics analysis, we have designed a promoter sequence composed by multimerized anti-oxidant responsive elements (AREs), sequences inducible by NF-E2-related factor 2 (Nrf2), a transcription factor that controls the basal and induced expression of an array of antioxidant response element–dependent genes [101]. The new generated synthetic promoter sequence is composed by 8 minimal ARE derived from four genes (Hmox1, Nqo1, GstTa, GSR) optimized to have only binding sites for the Nrf-2 factor. This promoter was tested in transient transfection assays in several cell lined treated with different oxidative stress inducers (tert-butylhydroquinone; 2-tert-butyl-1,4-benzoquinone; diethyl maleate; DL-sulforaphane; Cystamine; Curcumin; Quercetin) and with unrelated stimuli (Ethidium bromide; Acetylsalicylic acid; Dithiothreitol; 2-Mercaptoethanol) as negative controls.

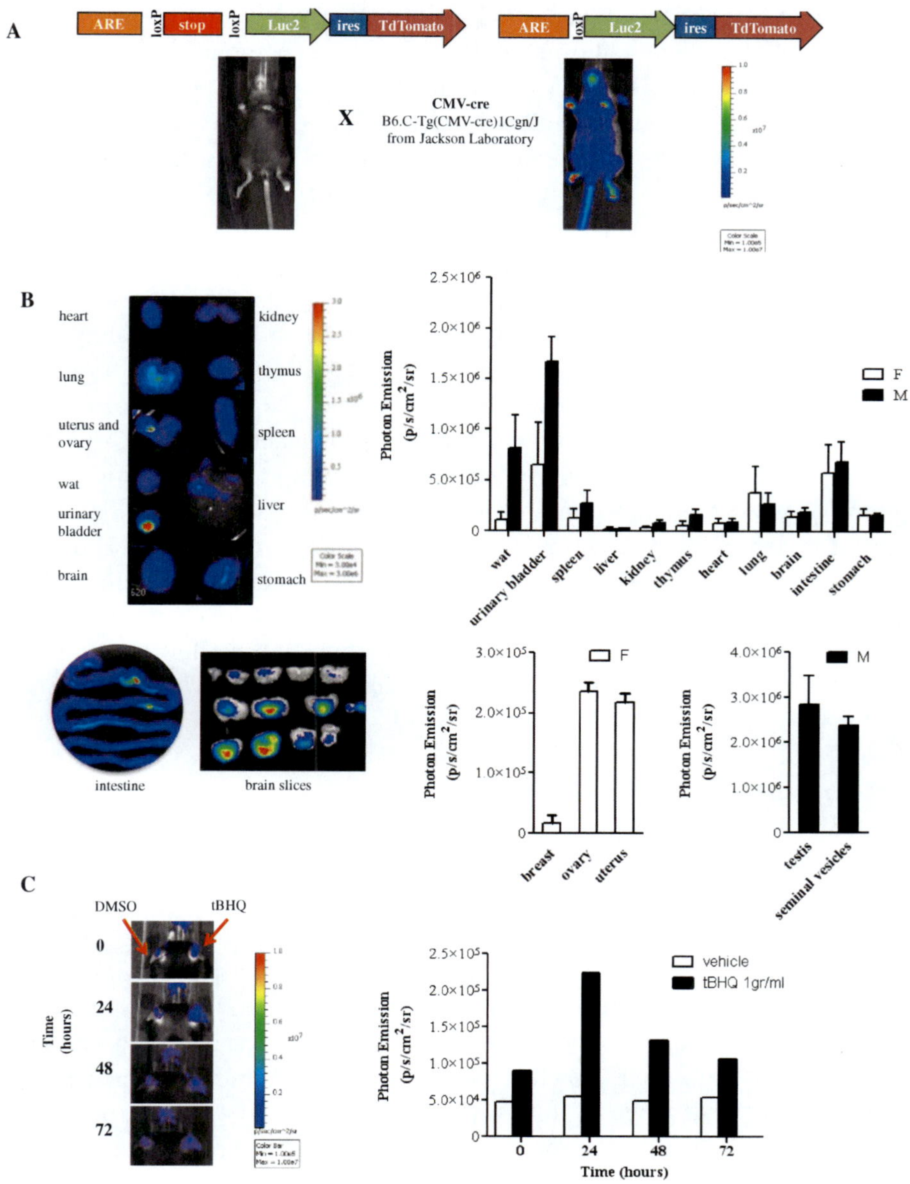

Figure 2. *In vivo* and *ex vivo* bioluminescence imaging of the oxidative stress reporter ARE-Luc2 mouse. A. *In vivo* bioluminescence imaging of ARE mice: photos portray of ARE-STOP-Luc2-ires-TdTomato and ARE-Luc2-ires-TdTomato mice (ventral view) in a pseudocolor image processed to indicate the intensity of photon emission in whole body. *In vivo* imaging reveals that the presence of STOP sequence blocks the expression of reporter gene, while after deletion of STOP sequence by breeding with CRE transgenic mouse (B6.C-Tg(CMV-cre)1Cgn/J the Luc2 reporter gene is ubiquitous expressed. B. *Ex vivo* bioluminescence imaging of 12 dissected organs from ARE-Luc2-ires-TdTomato mice. The quantification of photon emission from the indicated organs confirms the expression of Luc2 protein in all analyzed tissues at basal level. Each bars correspond to MEAN ±SEM (n=2M; n=4F) and they are expressed as photon emission (p/s/cm^2/sr). C. The ears of ARE-Luc2 reporter mice are treated topically with tBHQ (right ear) or with vehicle (left ear). *In vivo* imaging analysis is carried out at the indicated time points after treatments. Pseudocolor images indicate the intensity of photon emission from the ear skin. Quantification of photon emissions are plotted in the inserted graph and expressed as photon emission (p/s/cm^2/sr).

The selectivity and the magnitude of the reporter response were evaluated by luciferase enzymatic assay and univocally allowed us to choose the appropriate promoter arrangement for ARE reporter system. However, in addition this reporter system is suitable for multimodality imaging, because it allows *in vivo* detection *via* both fluorescent and bioluminescent imaging. The binding of Nrf2 to AREs induces the transcription of a bicistronic mRNA coding for a fluorescent (tdTomato) and a bioluminescent (Firefly Luciferase 2) reporters. The tdTomato is a red fluorescent protein, an evolution of the monomeric DsRed fluorescent protein, that shows excellent photostability and accumulates rapidly (one hour) into the cytoplasm [102]. TdTOMATO's emission wavelength and brightness make it ideal for live animal imaging. Further the protein TdTomato will enable *ex vivo* 2-photon and 3D microscopy or immunohistochemistry analysis. The synthetic firefly Luc2 gene has been extensively optimized for more efficient expression in mammals and the Luc2 gene will guarantee the *in vivo* visualization and quantification of oxidative stress response in specific cells types. Moreover, the reporter system has a conditional expression by the insertion between promoter and reporters genes of the cassette loxP-STOP-loxP. The presence of a floxed RNA PolII termination signal allows to restrict the expression of the reporter genes to any specific cell lineage after breeding with transgenic Cre mice (Figure 2A) [103]. Similarly to ERE-Luc mouse, the promoter and the reporter cassette were inserted between MAR insulator sequences into the final targeting vector allowing the site specific integration of the reporter system in the mouse genome. This technology based on the combined use of:

- Mar insulator sequences (European patent No. EP 1298988B1; US patent No. 7,943,815), able to insulate the reporter gene from the influence of the host genome;
- appropriate integration site in the murine genome which ensures the expression of the reporter in all cells;

allows the production of functional reporter mice in which the reporter system is pervasively expressed in every cell of the genetically modified mouse (Figure 2B).As a proof of principle to the faculties of ARE-Luc2 reporter mouse to monitor *in vivo* the oxidative stress dynamics we tested the response to a well know antioxidant agent tert-Butylhydroquinone (t-BHQ). t-BHQ is added to a wide range of foods as a preservative [104]. As expected, this test demonstrated that the Nrf2-inducing agent tBHQ increases the luminescent emission from the ARE-Luc2 reporter mouse and, interestingly, the system is also able to monitor the decrease of photon emission that corresponds to the recovery of the damage (Figure 2C).

4. NFkB-Luc2 Mouse, a Tool for Study Food Associated Inflammation

As previously said, the overeating and the obesity is often associated to high level of inflammation and in fact many efforts are involved to find the best diet to decrease the inflammation level and increase the anti-inflammatory response.

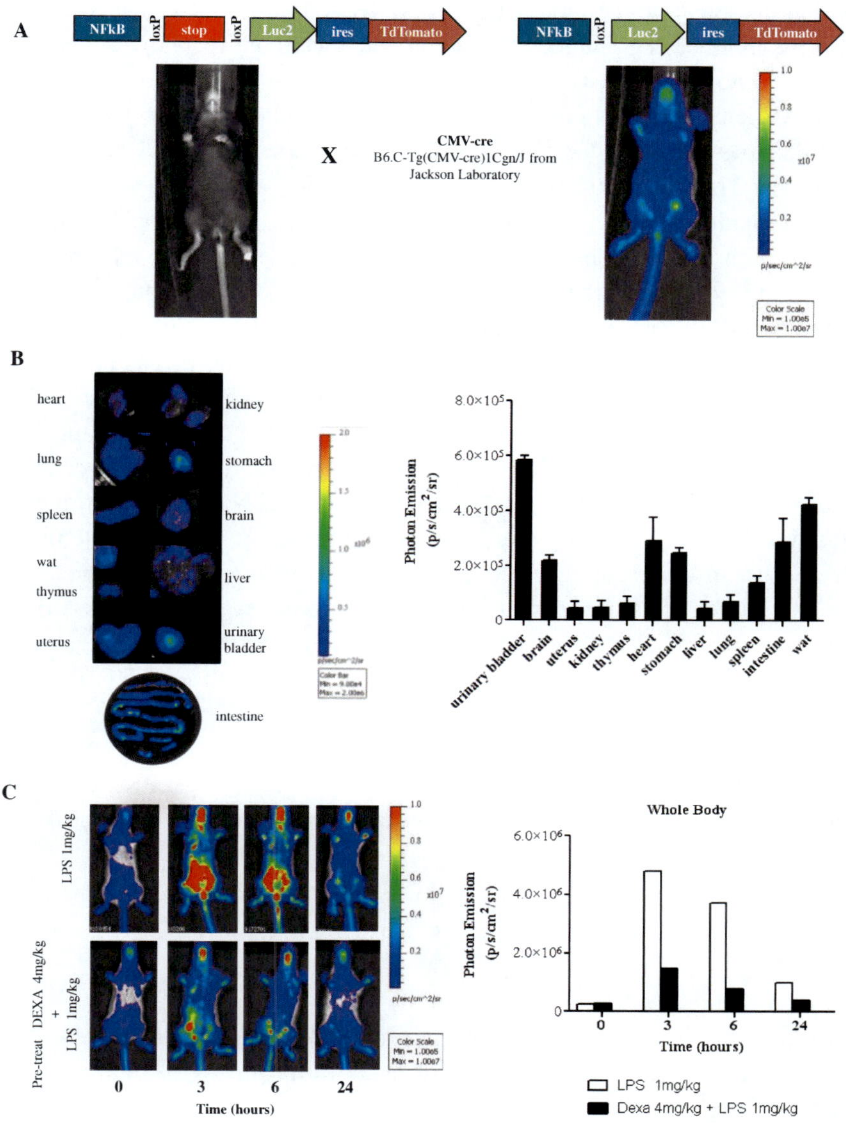

Figure 3. *In vivo* and *ex vivo* bioluminescence imaging of NFkB mice. A. *In vivo* bioluminescence imaging of NFkB mice: photos portray of NFkB-STOP-Luc2-ires-TdTomato and NFkB-Luc2-ires-TdTomato mice (ventral view) in a pseudocolor image processed to indicate the intensity of photon emission in whole body. *In vivo* imaging reveals that the presence of STOP sequence blocks the expression of reporter gene, while after deletion of STOP sequence by breeding with CRE transgenic mouse (B6.C-Tg(CMV-cre)1Cgn/J the Luc2 reporter gene is ubiquitous expressed. B. *Ex vivo* bioluminescence imaging of 12 dissected organs from NFkB-Luc2-ires-TdTomato mice, the quantification of photon emission from the indicated organs confirms the expression of Luc2 protein in all analyzed tissues at basal level, each bars correspond to MEAN ±SEM (n=2) and they are expressed as photon emission (p/s/cm^2/sr). C. *In vivo* bioluminescence imaging of NFkB mice treated with Dexamethasone (Dexa) and/or LPS. Both Dexa (4 mg/kg) and LPS (1 mg/kg) are injected i.p. at time 0, the mice are then imaged after indicated time (0, 3, 6, 24 hours post compounds injection). Photon emission from whole body (ventral view) is measured and expressed as photon emission (p/s/cm^2/sr).

To further demonstrate the modulation of NFkB activity in the transgenic mice by the *in vivo* imaging technique, LPS-induced transgenic mice were pretreated with dexamethasone (4 mg/kg), a well known suppressor of NFkB activity, one hour before the LPS induction (Figure 3C). As demonstrated in Figure 3C, whole body imaging of living mice revealed that 4 mg/kg of dexamethasone reduced the total luminescence from the whole body animal by 60% at 3 hours and by 75% at 6 hours after LPS induction.

As for the oxidative stress reporter mouse, similarly, we develop a reporter mouse to monitor the inflammatory response *in vivo*. For the generation of inflammation reporter system we proposed to use the Nuclear factor kB (NFkB) promoter, a synthetic DNA sequence composed of 4 kB-binding motifs each repeated twice; this promoter arrangement was experimentally proven to be selective for the immune response. The promoter was tested *in vitro* in several cells lines: as macrophage (RAW 264.7), fibroblast (NIH-3T3) and microglia (BV-2) and it was confirmed as activated by LPS stimulation and by other pro-inflammatory agent and it is not activated by unrelated stimuli (e.g.: serum) or by specific M2 activator such as interleukin 4(IL-4) [105]. We generated the NFkB reporter mouse following the previously described technology used for oxidative stress reporter system: the insulator sequences able to insulate the reporter gene from the influence of the host genome *plus* appropriate integration sites in the mouse genome that ensures the expression of the reporter in all cells; the loxP-STOP-loxP sequence for the conditional tissue specific expression and the multimodality (Luc2 and TdTomato reporter genes) for the use of NFkB reporter mouse for several applications (Figure 3 A-B).

As a proof of principle, we performed also the assessment of luminescence after induction by pro-inflammatory agent: in mice treated with 1mg/kg LPS we observed a great increase of the luminescence in a time specific manner and, after 24 hours, the bioluminescent signals go back to the basal level.

Conclusion

The *in silico* and *in vitro* studies represent useful tools to predict and assess the potential negative effects of thousands and thousands of naturally-occurring or industrial-derived food contaminants and toxins.

Although the usefulness of these high-throughput approaches, however the *in silico* and *in vitro* studies limit the ability to evaluate the consequences of a potential food contaminant or toxin in a physiological context. Conversely, *in vivo* assays have received much interest since they appear to mimic better the complexity of a physiological organism in a more reliable way. Furthermore, as the potential toxicity of a food component could be due to (or enhanced by) combinatorial effects, matrix effects or the derived metabolites, the *in vivo* analysis leads to better analyze the effects of food-associated contaminants or toxins by allowing to study the integrative consequences of a certain compound on health.

The *in vivo* analysis requires to take in account that many negative effects of a food-associated contaminant could be latent in a short-term assay and could be conversely relevant with time or in particular physiological (e.g.: pregnancy, pre-puberty, ageing) or pathological (e.g.: metabolic diseases) conditions. This aspect highlights the relevance of developing long-term *in vivo* studies that allow to monitor the potential interfering action of a contaminant on a specific signalling in different districts and at different time points. Such a *spatio*-temporal analysis could be achieved by the use of reporter mice specifically generated to evaluate a specific signal/pathway.

In this view, we propose three kind of reporter mice able to monitor the effects of a particular substance on ER signalling (the ERE-Luc mouse), on oxidative stress response (the ARE-Luc2 mouse), and on inflammatory response (the NFkB -Luc2 mouse).

The well characterized ERE-Luc mouse leads to evaluate whether a specific compound or a food-associated component shows an estrogenic activity, at which dose and in which district (organ) may act and how this activity could interfere with the physiological signalling of ER by potentially leading to endocrine unbalance and to pathological consequences. By the use of this mouse model we were able for example to discriminate between the action of a pure compound ($CdCl_2$) and of a food-associated Cd (wheat and flax bread) and demonstrate how this food-associated metal could interfere with estrogenic signalling in immature male mice. The ERE-Luc mouse represents once again a useful tool to evaluate in a physiological setting the estrogenic activity of a unkown food-associated contaminant (such as EDCs) previously predicted or assessed by *in silico* or *in vitro* studies.

As for the ERE-Luc for ER signalling, the ARE-Luc2 and the NFkB-Luc2 mouse represent two novel and invaluable tools to assess whether a food-associated contaminant could affect (by enhancing or counteracting) an oxidative stress or inflammatory response, respectively.

All these reporter mice could furthermore represent valid instruments to evaluate *in vivo* the potential estrogenic, oxidative/anti-oxidative or pro-/anti-inflammatory properties of a certain food-associated component, contaminant or toxin. These reporter models and the bioluminescence *in vivo* imaging analysis, by enhancing the ability to study in a physiological setting the properties of food-associated components, represent new technological facilities that will lead to increase our knowledge and will lead the industrial companies to improve the food processing, by limiting the negative effects and improving the beneficial properties of food for health.

Acknowledgment

This work was supported by grants from the European Community (ERC-Advanced Grant 322977; FP7 INMiND 278850).

References

[1] Brendler-Schwaab, S., et al., The in vivo comet assay: use and status in genotoxicity testing. *Mutagenesis,* 2005. 20(4): p. 245-54.

[2] Stokes, W.S., Selecting appropriate animal models and experimental designs for endocrine disruptor research and testing studies. *ILAR journal / National Research Council, Institute of Laboratory Animal Resources*, 2004. 45(4): p. 387-93.

[3] Uzal, F.A. and B.A. McClane, Animal models to study the pathogenesis of enterotoxigenic Clostridium perfringens infections. *Microbes andinfection / Institut Pasteur*, 2012. 14(12): p. 1009-16.

[4] Keller, H., et al., Fatty acids and retinoids control lipid metabolism through activation of peroxisome proliferator-activated receptor-retinoid X receptor heterodimers. *Proceedings of the National Academy ofSciences of the United States of America*, 1993. 90(6): p. 2160-4.

[5] Forman, B.M., J. Chen, and R.M. Evans, Hypolipidemic drugs, polyunsaturated fatty acids, and eicosanoids are ligands for peroxisome proliferator-activated receptors alpha and delta. *Proceedings of theNational Academy of Sciences of the United States of America*, 1997. 94(9): p. 4312-7.

[6] Sanderson, L.M., et al., Effect of synthetic dietary triglycerides: a novel research paradigm for nutrigenomics. *PloS one,* 2008. 3(2): p. e1681.

[7] Bunger, M., et al., Exploration of PPAR functions by microarray technology--a paradigm for nutrigenomics. *Biochimica et biophysica acta*, 2007. 1771(8): p. 1046-64.

[8] Mitro, N., et al., The nuclear receptor LXR is a glucose sensor. *Nature,* 2007. 445(7124): p. 219-23.

[9] Della Torre, S., et al., Amino acid-dependent activation of liver estrogen receptor alpha integrates metabolic and reproductive functions via IGF-1. *Cell metabolism*, 2011. 13(2): p. 205-14.

[10] Ide, T., et al., Cross-talk between peroxisome proliferator-activated receptor (PPAR) alpha and liver X receptor (LXR) in nutritional regulation of fatty acid metabolism. II. LXRs suppress lipid degradation gene promoters through inhibition of PPAR signaling. *Molecular endocrinology,* 2003. 17(7): p. 1255-67.

[11] Calkin, A.C. and P. Tontonoz, Transcriptional integration of metabolism by the nuclear sterol-activated receptors LXR and FXR. Nature reviews. *Molecular cell biology*, 2012. 13(4): p. 213-24.

[12] Della Torre, S., et al., Energy metabolism and fertility--a balance preserved for female health. Nature reviews. *Endocrinology*, 2014. 10(1): p. 13-23.

[13] Jew, S., S.S. AbuMweis, and P.J. Jones, Evolution of the human diet: linking our ancestral diet to modern functional foods as a means of chronic disease prevention. *Journal of medicinal food*, 2009. 12(5): p. 925-34.

[14] Milner, J.A., Nutrition in the 'omics' era. *Forum of nutrition,* 2007. 60: p. 1-24.

[15] Milner, J.A., Molecular targets for bioactive food components. *TheJournal of nutrition*, 2004. 134(9): p. 2492S-2498S.

[16] Milner, J.A., Diet and cancer: facts and controversies. *Nutrition andcancer*, 2006. 56(2): p. 216-24.

[17] Sirtori, C.R., et al., Nutritional and nutraceutical approaches to dyslipidemia and atherosclerosis prevention: Focus on dietary proteins. *Atherosclerosis,* 2009. 203(1): p. 8-17.

[18] Lattimer, J.M. and M.D. Haub, Effects of dietary fiber and its components on metabolic health. *Nutrients,* 2010. 2(12): p. 1266-89.

[19] Martinez-Augustin, O., et al., Bioactive anti-obesity food components. International journal for vitamin and nutrition research. Internationale Zeitschrift fur Vitamin- und Ernahrungsforschung. *Journal internationalde vitaminologie et de nutrition*, 2012. 82(3): p. 148-56.

[20] Ku, C.S., et al., Health benefits of blue-green algae: prevention of cardiovascular disease and nonalcoholic fatty liver disease. *Journal of medicinal food*, 2013. 16(2): p. 103-11.

[21] Oh, Y.S. and H.S. Jun, Role of bioactive food components in diabetes prevention: effects on Beta-cell function and preservation. *Nutrition andmetabolic insights*, 2014. 7: p. 51-9.

[22] Soory, M., Nutritional antioxidants and their applications in cardiometabolic diseases. *Infectious disorders drug targets*, 2012. 12(5): p. 388-401.

[23] Stefanska, B., et al., Epigenetic mechanisms in anti-cancer actions of bioactive food components--the implications in cancer prevention. *British journal of pharmacology*, 2012. 167(2): p. 279-97.

[24] Milner, J.A., Nutrition and cancer: essential elements for a roadmap. *Cancer letters*, 2008. 269(2): p. 189-98.

[25] Kim, Y.S., et al., Bioactive food components, inflammatory targets, and cancer prevention. *Cancer prevention research*, 2009. 2(3): p. 200-8.

[26] Stan, S.D., et al., Bioactive food components and cancer risk reduction. *Journal of cellular biochemistry*, 2008. 104(1): p. 339-56.

[27] Davis, C.D. and J.A. Milner, Biomarkers for diet and cancer prevention research: potentials and challenges. *Acta pharmacologica Sinica*, 2007. 28(9): p. 1262-73.

[28] Szarc vel Szic, K., et al., Nature or nurture: let food be your epigenetic medicine in chronic inflammatory disorders. *Biochemical pharmacology*, 2010. 80(12): p. 1816-32.

[29] Kris-Etherton, P.M., et al., Bioactive compounds in nutrition and health-research methodologies for establishing biological function: the antioxidant and anti-inflammatory effects of flavonoids on atherosclerosis. *Annual review of nutrition*, 2004. 24: p. 511-38.

[30] Aruoma, O.I., T. Bahorun, and L.S. Jen, Neuroprotection by bioactive components in medicinal and food plant extracts. *Mutation research*, 2003. 544(2-3): p. 203-15.

[31] Bender, A., et al., Creatine improves health and survival of mice. *Neurobiology of aging*, 2008. 29(9): p. 1404-11.

[32] Patisaul, H.B. and W. Jefferson, The pros and cons of phytoestrogens. *Frontiers in neuroendocrinology*, 2010. 31(4): p. 400-19.

[33] Zoppi, G., et al., Immunocompetence and dietary protein intake in early infancy. *Journal of pediatric gastroenterology and nutrition*, 1982. 1(2): p. 175-82.

[34] Diamanti-Kandarakis, E., et al., Endocrine-disrupting chemicals: an Endocrine Society scientific statement. *Endocrine reviews*, 2009. 30(4): p. 293-342.

[35] Matsuura, I., et al., Evaluation of a two-generation reproduction toxicity study adding endpoints to detect endocrine disrupting activity using lindane. *The Journal of toxicological sciences*, 2005. 30 Spec No.: p. 135-161.

[36] Fusani, L., et al., Altered reproductive success in rat pairs after environmental-like exposure to xenoestrogen. Proceedings. *Biological sciences / The Royal Society*, 2007. 274(1618): p. 1631-6.

[37] Maranghi, F., et al., The food contaminant semicarbazide acts as an endocrine disrupter: Evidence from an integrated in vivo/in vitro approach. *Chemico-biological interactions*, 2010. 183(1): p. 40-8.

[38] Zawatski, W. and M.M. Lee, Male pubertal development: are endocrine-disrupting compounds shifting the norms? *The Journal of endocrinology*, 2013. 218(2): p. R1-12.

[39] Schiffer, C., et al., Direct action of endocrine disrupting chemicals on human sperm. *EMBO reports*, 2014. 15(7): p. 758-65.

[40] Schug, T.T., et al., Endocrine disrupting chemicals and disease susceptibility. *The Journal of steroid biochemistry and molecularbiology,* 2011. 127(3-5): p. 204-15.

[41] Recchia, A.G., et al., Xenoestrogens and the induction of proliferative effects in breast cancer cells via direct activation of oestrogen receptor alpha. *Food additives and contaminants*, 2004. 21(2): p. 134-44.

[42] Maffini, M.V., et al., Endocrine disruptors and reproductive health: the case of bisphenol-A. *Molecular and cellular endocrinology*, 2006. 254-255: p. 179-86.

[43] Meeker, J.D., Exposure to environmental endocrine disruptors and child development. *Archives of pediatrics & adolescent medicine*, 2012. 166(10): p. 952-8.

[44] Pistol, G.C., et al., Natural feed contaminant zearalenone decreases the expressions of important pro- and anti-inflammatory mediators and mitogen-activated protein kinase/NF-kappaB signalling molecules in pigs. *The British journal of nutrition*, 2014. 111(3): p. 452-64.

[45] Saint-Cyr, M.J., et al., Evaluation of an oral subchronic exposure of deoxynivalenol on the composition of human gut microbiota in a model of human microbiota-associated rats. *PloS one*, 2013. 8(11): p. e80578.

[46] Grenier, B. and T.J. Applegate, Modulation of intestinal functions following mycotoxin ingestion: meta-analysis of published experiments in animals. *Toxins*, 2013. 5(2): p. 396-430.

[47] Cox, A.J., N.P. West, and A.W. Cripps, Obesity, inflammation, and the gut microbiota. The lancet. *Diabetes & endocrinology*, 2014.

[48] Khan, M.T., M. Nieuwdorp, and F. Backhed, Microbial Modulation of Insulin Sensitivity. *Cell metabolism*, 2014.

[49] Karlsson, F., et al., Assessing the human gut microbiota in metabolic diseases. *Diabetes,* 2013. 62(10): p. 3341-9.

[50] Breen, D.M., et al., Nutrient-sensing mechanisms in the gut as therapeutic targets for diabetes. *Diabetes,* 2013. 62(9): p. 3005-13.

[51] Wong, J.M., et al., Gut microbiota, diet, and heart disease. *Journal ofAOAC International*, 2012. 95(1): p. 24-30.

[52] Manco, M., L. Putignani, and G.F. Bottazzo, Gut microbiota, lipopolysaccharides, and innate immunity in the pathogenesis of obesity and cardiovascular risk. *Endocrine reviews*, 2010. 31(6): p. 817-44.

[53] Jeffery, I.B. and P.W. O'Toole, Diet-microbiota interactions and their implications for healthy living. *Nutrients,* 2013. 5(1): p. 234-52.

[54] Penza, M., et al., Genistein accumulates in body depots and is mobilized during fasting, reaching estrogenic levels in serum that counter the hormonal actions of estradiol and organochlorines. *Toxicological sciences : an official journal of the Society of Toxicology*, 2007. 97(2): p. 299-307.

[55] Fernandez-Real, J.M. and M. Manco, Effects of iron overload on chronic metabolic diseases. The lancet. *Diabetes & endocrinology*, 2014. 2(6): p. 513-26.

[56] Fujita, N. and Y. Takei, Iron overload in nonalcoholic steatohepatitis. *Advances in clinical chemistry*, 2011. 55: p. 105-32.

[57] van Meeuwen, J.A., et al., Mixture effects of estrogenic compounds on proliferation and pS2 expression of MCF-7 human breast cancer cells. Food and chemical toxicology

: an international journal published for the British Industrial *Biological Research Association*, 2007. 45(11): p. 2319-30.

[58] Taxvig, C., et al., Effects of nutrition relevant mixtures of phytoestrogens on steroidogenesis, aromatase, estrogen, and androgen activity. *Nutrition and cancer*, 2010. 62(1): p. 122-31.

[59] Katchy, A., et al., Coexposure to phytoestrogens and bisphenol a mimics estrogenic effects in an additive manner. *Toxicological sciences : anofficial journal of the Society of Toxicology*, 2014. 138(1): p. 21-35.

[60] Boberg, J., et al., Endocrine disrupting effects in rats perinatally exposed to a dietary relevant mixture of phytoestrogens. *Reproductivetoxicology,* 2013. 40: p. 41-51.

[61] Rando, G., et al., Differential effect of pure isoflavones and soymilk on estrogen receptor activity in mice. *Toxicology and applied pharmacology*, 2009. 237(3): p. 288-97.

[62] Lee, W.J. and K. Hase, Gut microbiota-generated metabolites in animal health and disease. *Nature chemical biology*, 2014. 10(6): p. 416-24.

[63] Simarro Doorten, A.Y., et al., Metabolism-mediated cytotoxicity of ochratoxin A. *Toxicology in vitro : an international journal published in association with BIBRA*, 2004. 18(3): p. 271-7.

[64] Aravindakshan, J., et al., Consumption of xenoestrogen-contaminated fish during lactation alters adult male reproductive function. *Toxicological sciences : an official journal of the Society of Toxicology,* 2004. 81(1): p. 179-89.

[65] Tang-Peronard, J.L., et al., Association between prenatal polychlorinated biphenyl exposure and obesity development at ages 5 and 7 y: a prospective cohort study of 656 children from the Faroe Islands. *TheAmerican journal of clinical nutrition*, 2014. 99(1): p. 5-13.

[66] Fall, C.H., Evidence for the intra-uterine programming of adiposity in later life. *Annals of human biology*, 2011. 38(4): p. 410-28.

[67] Legler, J., et al., The OBELIX project: early life exposure to endocrine disruptors and obesity. *The American journal of clinical nutrition*, 2011. 94(6 Suppl): p. 1933S-1938S.

[68] Biro, F.M., M.S. Wolff, and L.H. Kushi, Impact of yesterday's genes and today's diet and chemicals on tomorrow's women. *Journal of pediatricand adolescent gynecology*, 2009. 22(1): p. 3-6.

[69] Ding, S., et al., High-fat diet aggravates glucose homeostasis disorder caused by chronic exposure to bisphenol A. *The Journal of endocrinology*, 2014. 221(1): p. 167-79.

[70] Tsuchiya, H., et al., High-fat, high-fructose diet induces hepatic iron overload via a hepcidin-independent mechanism prior to the onset of liver steatosis and insulin resistance in mice. *Metabolism: clinical and experimental*, 2013. 62(1): p. 62-9.

[71] Leffel, E.K., et al., Drinking water exposure to cadmium, an environmental contaminant, results in the exacerbation of autoimmune disease in the murine model. *Toxicology*, 2003. 188(2-3): p. 233-50.

[72] Mailloux, R.J., et al., Exposure to a Northern Contaminant Mixture (NCM) Alters Hepatic Energy and Lipid Metabolism Exacerbating Hepatic Steatosis in Obese JCR Rats. *PloS one*, 2014. 9(9): p. e106832.

[73] Pi-Sunyer, F.X., The obesity epidemic: pathophysiology and consequences of obesity. *Obesity research*, 2002. 10 Suppl 2: p. 97S-104S.

[74] Girardet, C., et al., The food-contaminant deoxynivalenol modifies eating by targeting anorexigenic neurocircuitry. *PloS one*, 2011. 6(10): p. e26134.
[75] Pruller, F., et al., Activated protein C resistance assay and factor V Leiden. *The New England journal of medicine*, 2014. 371(7): p. 685-6.
[76] Mangge, H., et al., Laboratory medicine for molecular imaging of atherosclerosis. *Clinica chimica acta; international journal of clinical chemistry*, 2014. 437: p. 19-24.
[77] Baillie-Hamilton, P.F., Chemical toxins: a hypothesis to explain the global obesity epidemic. *Journal of alternative and complementary medicine,* 2002. 8(2): p. 185-92.
[78] Norman, R.E., et al., Environmental exposures: an underrecognized contribution to noncommunicable diseases. *Reviews on environmental health*, 2013. 28(1): p. 59-65.
[79] Mantovani, A., Risk assessment of endocrine disrupters: the role of toxicological studies. *Annals of the New York Academy of Sciences,* 2006. 1076: p. 239-52.
[80] Mangge, H., et al., Antioxidants, inflammation and cardiovascular disease. *World journal of cardiology*, 2014. 6(6): p. 462-77.
[81] Lugrin, J., et al., The role of oxidative stress during inflammatory processes. *Biological chemistry*, 2014. 395(2): p. 203-30.
[82] Bondia-Pons, I., L. Ryan, and J.A. Martinez, Oxidative stress and inflammation interactions in human obesity. *Journal of physiology and biochemistry*, 2012. 68(4): p. 701-11.
[83] Bryan, S., et al., Redox-inflammatory synergy in the metabolic syndrome. *Canadian journal of physiology and pharmacology*, 2013. 91(1): p. 22-30.
[84] Pitocco, D., et al., Oxidative stress in diabetes: implications for vascular and other complications. *International journal of molecular sciences,* 2013. 14(11): p. 21525-50.
[85] Hajjar, D.P. and A.M. Gotto, Jr., Biological relevance of inflammation and oxidative stress in the pathogenesis of arterial diseases. *TheAmerican journal of pathology*, 2013. 182(5): p. 1474-81.
[86] Crujeiras, A.B., et al., Oxidative stress associated to dysfunctional adipose tissue: a potential link between obesity, type 2 diabetes mellitus and breast cancer. *Free radical research*, 2013. 47(4): p. 243-56.
[87] Wang, Y., et al., Autophagy regulates inflammation following oxidative injury in diabetes. *Autophagy,* 2013. 9(3): p. 272-7.
[88] Dali-Youcef, N., et al., Metabolic inflammation: connecting obesity and insulin resistance. *Annals of medicine*, 2013. 45(3): p. 242-53.
[89] Cai, D. and T. Liu, Inflammatory cause of metabolic syndrome via brain stress and NF-kappaB. *Aging*, 2012. 4(2): p. 98-115.
[90] Dahlman-Wright, K., et al., International Union of Pharmacology. LXIV. Estrogen receptors. *Pharmacological reviews*, 2006. 58(4): p. 773-81.
[91] Barros, R.P. and J.A. Gustafsson, Estrogen receptors and the metabolic network. Cell metabolism, 2011. 14(3): p. 289-99.
[92] Faulds, M.H., et al., The diversity of sex steroid action: regulation of metabolism by estrogen signaling. *The Journal of endocrinology*, 2012. 212(1): p. 3-12.
[93] Ciana, P., et al., Engineering of a mouse for the in vivo profiling of estrogen receptor activity. *Molecular endocrinology*, 2001. 15(7): p. 1104-13.
[94] Ciana, P., et al., In vivo imaging of transcriptionally active estrogen receptors. *Nature medicine*, 2003. 9(1): p. 82-6.

[95] Della Torre, S., et al., The conundrum of estrogen receptor oscillatory activity in the search for an appropriate hormone replacement therapy. *Endocrinology,* 2011. 152(6): p. 2256-65.

[96] Ramachandran, B., et al., Estrogen-like effects of diet-derived cadmium differ from those of orally administered CdCl(2) in the ERE-luc estrogen reporter mouse model. *Toxicology letters*, 2011. 202(2): p. 75-84.

[97] Ramachandran, B., N. Rizzi, and A. Maggi, Sexually immature male ERE-Luc reporter mice to assess low dose estrogen-like effects of CdCl2 versus dietary Cd. *American journal of nuclear medicine and molecularimaging,* 2014. 4(3): p. 270-82.

[98] Johnson, M.D., et al., Cadmium mimics the in vivo effects of estrogen in the uterus and mammary gland. *Nature medicine*, 2003. 9(8): p. 1081-4.

[99] Davies, K.J., Oxidative stress: the paradox of aerobic life. *Biochemical Society symposium,* 1995. 61: p. 1-31.

[100] Poljsak, B.a.M., I., Aging, Oxidative Stress and Antioxidants in Oxidative Stress and Chronic Degenerative Diseases - A Role for Antioxidants2013.

[101] Johnson, J.A., et al., The Nrf2-ARE pathway: an indicator and modulator of oxidative stress in neurodegeneration. *Annals of the NewYork Academy of Sciences*, 2008. 1147: p. 61-9.

[102] Shaner, N.C., et al., Improved monomeric red, orange and yellow fluorescent proteins derived from Discosoma sp. red fluorescent protein. *Nature biotechnology*, 2004. 22(12): p. 1567-72.

[103] Tsien, J.Z., et al., Subregion- and cell type-restricted gene knockout in mouse brain. *Cell,* 1996. 87(7): p. 1317-26.

[104] Kropfl, J.M., et al., Exercise-induced norepinephrine decreases circulating hematopoietic stem and progenitor cell colony-forming capacity. *PloS one*, 2014. 9(9): p. e106120.

[105] Locati, M., A. Mantovani, and A. Sica, Macrophage activation and polarization as an adaptive component of innate immunity. *Advances in immunology,* 2013. 120: p. 163-84.

In: From Medicinal Chemistry to Food Science
Editor: Pietro Cozzini
ISBN: 978-1-63483-587-9

Chapter 3

IN VITRO AND *IN SILICO* MODELS OF DIGESTION: LIMITATIONS AND CHALLENGES

Thomas Moxon, Clifford Latty, Ourania Gouseti and Serafim Bakalis
School of Chemical Engineering,
The University of Birmingham, United Kingdom

Abstract

In this chapter, reports on *in-vitro* and *in-silico* models of digestion are reviewed, looking at the food & chyme properties and how these affect the bioaccessibility of nutrients, and the physiological responses. Currently, *in-silico* models are mainly empirical in nature, though some have attempted more mechanistic approaches, e.g., looking at mass transfer within intestinal lumen, absorption rate, effect of boli, or gastric emptying rate, but those are still simplistic in nature and further developments are needed. *In-vitro* models reproduce one or more events of food digestion, including mechanical and/or chemical processes. Models are typically application specific, and can be static or dynamic depending on whether temporal changes (e.g., segmentation and peristaltic forces) are reproduced. *In-vitro* and *in-silico* methods promise the potential of important uses in health claims, and possibly aid elucidating the reasons behind diet related diseases such as diabetes and obesity, offering understandings which will enable clinical or dietary changes to tackle the diseases. The trust of these models relies heavily on their validation, with effective correlation between *in-vitro*, *in-vivo* and *in-silico* systems, and this will be one of the challenges for future researchers within the field.

Introduction

The relationship between diet and health is gaining increasing attention within consumers and researchers in the last decades. The occurrence of diet-related diseases, such as obesity continue to threaten world population health, some of which are evidently life style related while others are metabolically linked [1]. It has been estimated that the United States spent over 123 billion dollars for diagnosis, direct medical cost and decreased productivity related to diabetes in 2007 [2, 3]. In 2007–2008, 33.8% of the adult population was clinically obese in the United States [4-6]. In the United kingdom in 2012 the obesity rate was 25% [7] the estimated cost of this on the economy was 5.1 billion pounds [8] and it has been estimated that if the current trends persist then around half the UK population will be classed as obese with a cost of £50 billion per annum on society by 2050 [9]. These implications have triggered the need to reduce the number of cases to ease the economic strain as well as to promote healthy lifestyles and extended lives [10]. But, who is responsible for such health management: Government, food manufacturer, nutritionists, or individual consumers? These challenges and responsibilities have already initiate detail studies.

In the past two decades, the industry and academia have teamed up to study and understand how these diseases may be controlled and prevented, by understanding the gastro-intestinal phenomena in the digestion of foods and food delivery systems [10, 11]. This has led to using human subjects for *in-vivo* assays in clinical trials, *in-vitro* models and *in-silico* models for experiments and predictions respectively. All these categories of experiment have their advantages and disadvantages. *In-vivo* research to date has employed mostly the use of various animal models and human trials based on several experimental designs [10]. While it might seem trustworthy to use and interpret the data obtained from *in-vivo* results to assess studies of interest, they can be costly, and notably differ due to the variability found between individuals [11].

This chapter will look at the *in-silico* and *in-vitro* approaches to modelling the digestion of food, and will highlight the current research and challenges, with the aim to set a basis for a better understanding of the physiological processes involved in digestion and a greater predictability and standardised approach to the study of nutrition and food related diseases.

In-Silico Approach to Digestion

In-silico modelling has been used widely in the field of pharmacokinetics, and there are many pieces of software available for this purpose; but the field of *in-silico* modelling of food digestion is less well developed, due in part to the increased complexity of the process for food, e.g., formulation/ microstructure of the meal, hormonal response to digestion, feedback mechanisms etc.

The idea of using *in-silico* modelling for digestion came from papers by Penry & Jumars, in the late 1980's, suggesting the mammalian digestive system can be modelled as a system of ideal reactors [12, 13]. With this idea we can break the digestive system into compartments, and the mathematical descriptions of the ideal reactors (batch, continuous stirred, and plug flow) can be used to describe each compartment.

In contrast to this method a non-compartmental or empirical approach can be used, where by the system is analysed as a whole from experimental data. This method is generally used for data description and interpolation and does not offer much in terms of predictive capabilities [14].

The two major areas of the digestive system that will be covered in the *in-silico* model are the stomach and the small intestine. The majority of nutrient absorption will take place in the small intestine but the stomach will play a key role in the release of nutrients into the duodenum making them available for digestion.

Stomach

Most mathematical studies of the stomach have been empirical in nature; looking at the gastric content after consumption of a liquid generally shows an exponential decay, which can be defined using the following equations:

$$V(t) = V_0 e^{-\gamma t} \quad (1)$$

$$\frac{dV}{dt} = -\gamma V(t) \quad (2)$$

Where V is the volume in the stomach at a specific time, V_0 is the initial volume in the stomach, t is the time from consumption, and γ is the decay constant. The emptying can be characterised with the half-life i.e., the time for half the content to leave the pylorus to the duodenum:

$$t_{1/2} = \frac{\ln(2)}{\gamma} \quad (3)$$

Calbet et al., looked at comparing the emptying rate to the caloric density, a clear trend was found between the gastric emptying rates in kcal/min to the caloric content of the food [15], as shown in figure 1.

Looking more closely at the inhibition of gastric emptying via duodenal calories, one study analysed the effect that glucose has on the emptying by infusing various concentrations straight into the duodenum, and inputting a saline solution into the stomach. Results indicated that calories sensed in the duodenum inhibit gastric emptying rates [16]. Multiple authors have measured the rate of emptying of glucose in humans and primates; for humans Calbet et al., found a high emptying rate of 3.4kcal/min [15], and Brener et al., found a value of 2.13kcal/min [17], whereas for monkeys the value found by McHugh was lower at 0.4kcal/min [16]. The methods used varied with Calbet et al., measuring how a glucose solution emptied from the stomach and adding a dye of known mass to stomach, then measuring the concentration of dye in the stomach to calculate the volume remaining. Whereas Brener et al., used a saline solution in the stomach, and infused glucose directly into the duodenum, the saline solution was then aspirated to measure the volume left after a certain time. Another investigation looked at the effect of high/low glucose concentrations with and without polymer on the gastric emptying rate of the carbohydrate, by measuring how the stomach volume changed with time [18]. It was found that the low glucose concentration solutions emptied faster than the higher concentration ones which is in agreement with figure 1(b), and had similar rates of emptying with and without the polymers. However, the high glucose polymer solution emptied at a faster rate than the high glucose solution. The author points to the change in osmolality of the two solutions as a cause for this [18], but others found no relationship between osmolality and emptying rate [15, 16]. Another explanation could be a reduction in mass transfer in the small intestine, due to the presence of the polymer, which could cause a reduction in the bioavailability of the glucose. A reduction in bioavailability in the intestine will reduce the amount of glucose which is sensed and hence will not signal for slower gastric emptying. Before the nutrients can enter the duodenum and feedback is initiated there is an initial high emptying rate of the nutrients. This initial emptying rate seems to be controlled by the volume of meal i.e., the distention of the stomach when the meal is consumed and viscosity of the meal [17-20].

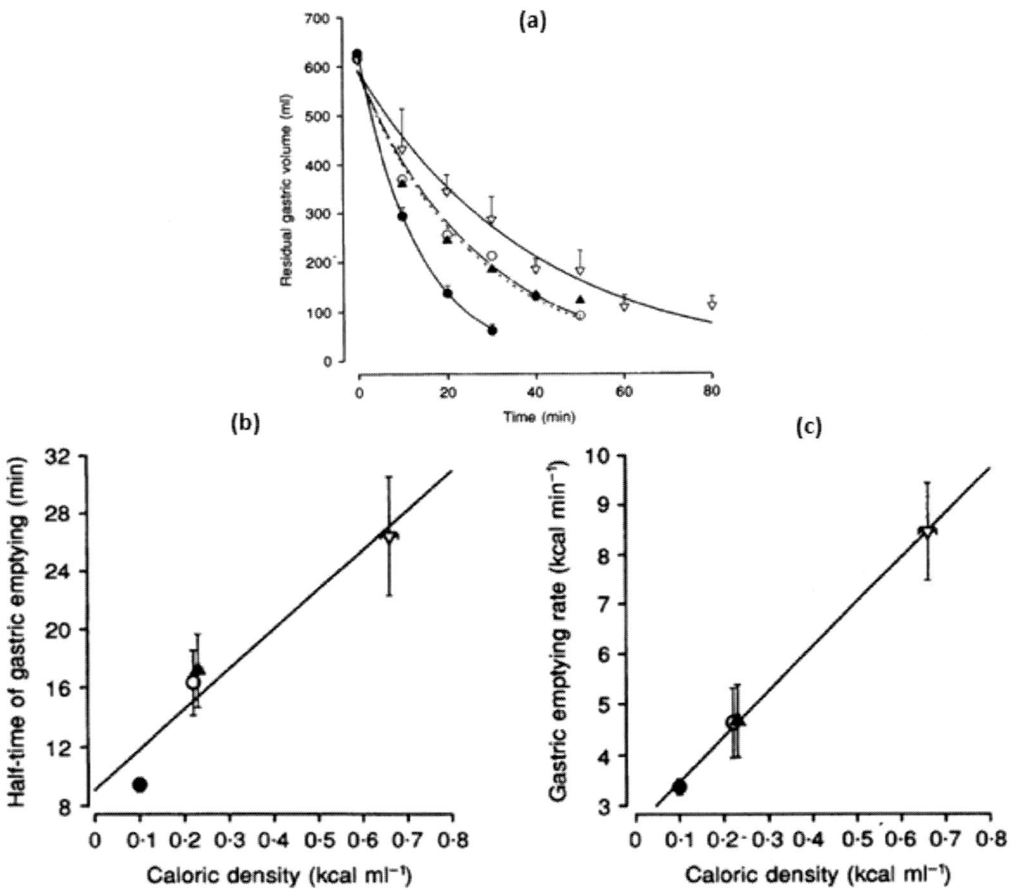

Figure 1. (a) Shows measurement of stomach volume against time for glucose (•) solution, milk protein (∇) solution , whey peptide hydrolysate(▲), pea peptide hydrolysate (○) , (b) shows the half-life of emptying for each solution against caloric density, and (c) shows the caloric emptying rate against the caloric density [15].

The emptying of solids from the stomach differs from that of liquids (figure 2); emptying of solids does not begin instantaneously but there is an initial lag phase, during which solids are broken down to particle sizes of < 1-2mm [21], small enough to allow them to pass through the pyloric sphincter and hence empty can occur. It is suggested that food will leave the stomach by a process of elution, where by more dilute parts of the food will empty first [20].

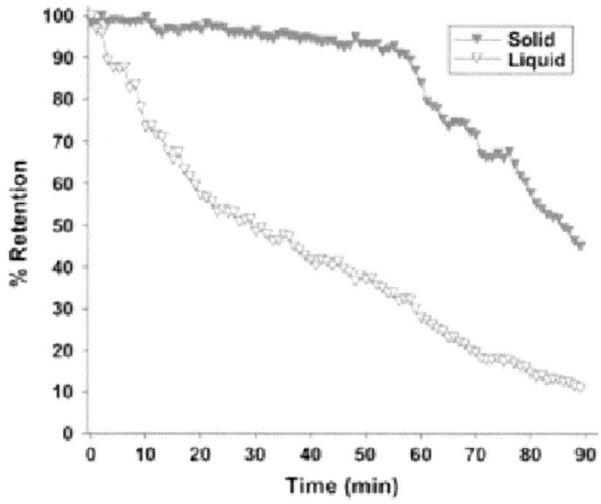

Figure 2. Comparison of liquid and solid gastric emptying [21].

In literature a modified exponential equation is generally used to describe the gastric emptying of solids [22-24]:

$$V(t) = 1-(1 - e^{-k.t})^{\beta} \tag{4}$$

Here k is the gastric emptying rate (min^{-1}) and β is used to define the lag phase. The curve can be defined by the lag time and half-life [24]:

$$t_{lag} = \frac{\ln(\beta)}{k} \tag{5}$$

$$t_{1/2} = \frac{\ln(1-0.5^{1/\beta})}{-k} \tag{6}$$

But to model gastric emptying in this fashion requires parameter k & β to be measured from *in-vivo* experiments and therefore offers no predictive capabilities.

To design a more physiologically relevant (with predictive possibilities) gastric emptying model there needs to be an understanding of the mechanisms involved in the breakdown of solid foods in the stomach and link this to a feedback response found for liquid emptying. The breakdown of foods in the stomach could be dependent on the size, density and hardness [23, 25].

A model by Dalla Man et al., looked at simulating an oral glucose tolerance test, as well as a test meal, with particular emphasis on the gastric emptying rates, by use of experimental data of glucose appearance in plasma by measuring labled glucose molecules, based on a multiple tracer technique [26]. The emptying rate can be linked to the stomach content of glucose and the difference between the current stomach mass and initial mass of input. The system was made up of 3 compartments, one for solids in the stomach, $\dot{q}_{sto1}(t)$, one for liquid in the stomach, $\dot{q}_{sto2}(t)$, and a gut (small intestine) compartment, $\dot{q}_{gut}(t)$, each represented by an ordinary differential equation. The rate of absorbed glucose was represented as Ra. Solid content is broken down and enters the liquid compartment of the stomach, from the liquid compartment of the stomach the glucose can empty and move to the small intestine (gut compartment) and be absorbed:

$$\dot{q}_{sto1}(t) = -k_{21}.q_{sto1}(t) + D\delta(t) \tag{7}$$

$$\dot{q}_{sto2}(t) = -k_{empt}.q_{sto2}(t) + k_{21}.q_{sto1}(t) \tag{8}$$

$$\dot{q}_{gut}(t) = -k_{abs}.q_{gut}(t) + k_{empt}.q_{sto2}(t) \tag{9}$$

$$Ra(t) = f.k_{abs}.q_{gut}(t) \tag{10}$$

Here, k_{21} is the rate of grinding, k_{abs} is the rate of absorption, f is the fraction of absorbed glucose which appears in the plasma, and Dδ(t) is the input of mass in the stomach. The emptying rate was set as follows:

$$k_{empt}(q_{sto}) = k_{min} + \frac{k_{max}-k_{min}}{2}.\{\tanh[\alpha(q_{sto} - b.D)] - \tanh[\beta(q_{sto} - c.D)] + 2\} \tag{11}$$

where:

$$q_{sto} = q_{sto1} + q_{sto2}$$

$$\alpha = \frac{5}{2.D.(1-b)}$$

$$\beta = \frac{5}{2.D.c}$$

Here b and c are constants: where b corresponds to the decrease in rate of emptying and c to the increase in emptying rate and k_{min} & k_{max} represent the minimum and maximum emptying rate respectively [26].

Whilst the results show good agreement with experimental data, the model requires many of the parameters to be fit; only the input mass of glucose is known, therefore the model offers little in the way of predictive capability.

The model also takes a simplistic approach in the description of the small intestine and does not take into account any of the chyme properties (e.g., viscosity), intestinal motility- which may have large effects on the absorption rates [27-31], or the formation of a bolus that enters the small intestine such as that modelled by Taghipoor [32]. In addition, the model does not take into account the properties of the (solid) foods in modelling breakdown and transformation into a 'liquid' form, which can be emptied.

In the Dalla man et al., model the emptying rate is quantified with the amount of glucose in the stomach (equation 11); whilst the consensus is that the rate of emptying is mainly dependent on the nutrient concentration [15-17, 19-21], this is likely sensed from within the duodenum [15-17, 19], and the emptying controlled by a feedback response. Therefore we must consider the bioavailability of the nutrients within the duodenum when considering the feedback response for the rate of emptying.

Small Intestine

In 1980 Ni et al., published a paper looking at the absorption and flow of drugs in the small intestine. A one dimensional plug flow model was used to describe the mass transfer in the lumen of the small intestine of the drug [33]:

$$\frac{\partial C}{\partial t} = \alpha \frac{\partial^2 C}{\partial x^2} - \beta \frac{\partial C}{\partial x} - \gamma C \tag{12}$$

Where:

C The concentration of the drug in the lumen at time, t, and distance along the lumen from the pylorus, x

$\alpha = D_e$ The diffusivity of the drug molecule

$\beta = Q/\pi r^2$ The mean velocity of flow through the intestinal tract

$\gamma = 2\, P_e/r$ The absorption rate through the epithelium, calculated as the permeability multiplied by the surface area to volume ratio (for cylinder this is equal to 2/radius).

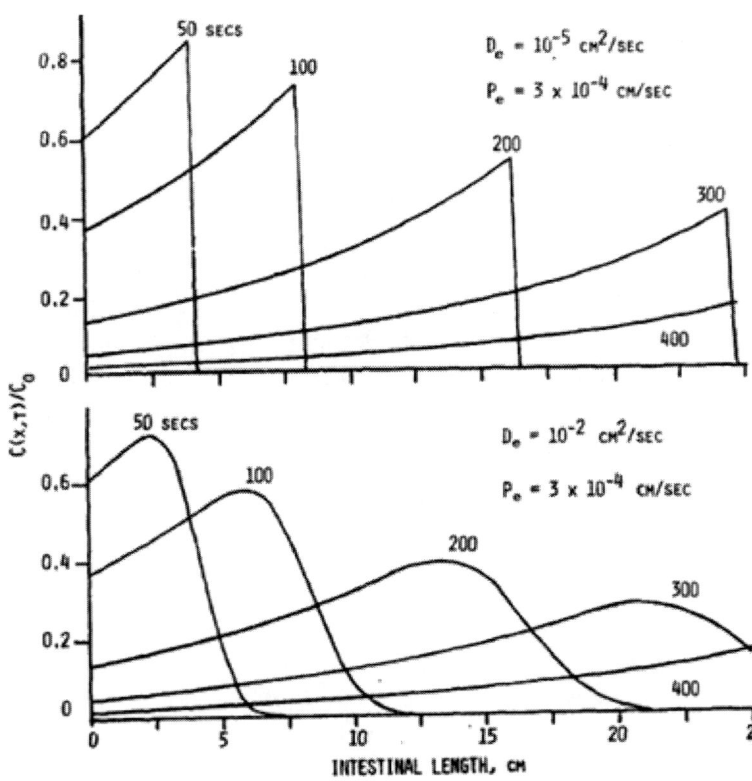

Figure 3. Concentration of drug in intestine along the length at different times, top shows the concentration when the diffusivity is small, bottom when there is higher diffusivity [33].

The authors used Laplace transforms to solve the partial differential equations for a variety of boundary conditions. Figure 3 shows the fraction of drug along the intestinal length at different time periods and diffusivities. The two graphs demonstrate diffusive spreading of the drug along the intestinal lumen. When diffusivity is higher we can see more spread of the drug along the length of the intestine but slower movement of the peak. If the ratio of advection to diffusivity is large the effect of diffusion will be negligible, and we will see a similar graph to the top of figure 3, with no noticeable spreading. This can be defined with the Peclet number (advective transport rate/ diffusive transport rate).

Stoll et al., [34] developed the use of a plug flow intestine for studying drug mass transfer and absorption. The model used a single step input of drug (representing a single 'bolus' of drug entering the intestine), and attempted to address the multi scale nature of the intestine, by characterising the absorption rate as a function of the folds and protrusions present at the wall of the intestine.

Here the absorption rate, κ, is defined as:

$$\kappa = \left(\frac{A^{pc}}{A^{ppc}}\right)\left(\frac{A^{v}}{A^{pv}}\right)K_p \tag{13}$$

Where the permeability, K_p is expressed as

$$K_p = \left(\frac{A^{mv}}{A^{pmv}}\right)\left(\frac{kD_L}{l} + k_t + k_p\right) \tag{14}$$

$\left(\frac{A^{pc}}{A^{ppc}}\right)$, $\left(\frac{A^{v}}{A^{pv}}\right)$ and $\left(\frac{A^{mv}}{A^{pmv}}\right)$ are the ratio of surface area to projected surface area for plicae circulares, villi, and microvilli respectably; and $\frac{kD_L}{l}, k_t, k_p$ represent the rate constants for different paths through the epithelium (transcellular, transcytotic, or paracellular).

Using the plug flow equation (generic form shown in equation 12) and looking at the different length scales present in the intestine, Stoll et al., showed the mass distribution of a drug in the intestine. Figure 4a shows how this drug concentration changes along the length of the intestine and with time. The plasma concentration of drug was modelled using a one-compartment model and assumed exponential decay for the removal of the drug in the body after it was absorbed [34]. The predicted drug plasma concentration was compared to experimental results for 4 drugs. Figure 4b shows the results for growth hormone releasing peptide (GHRP-1). The values of the absorption rate and luminal degradation rate of the drug where estimated so to match the peak concentration and peak time, respectively.

Looking at the different length scales Stoll et al., calculated the increase in surface area with ratios from Ganong, stating that the increase due to plicae circulares will be 3-fold, due to the villus 10-fold, and due to the microvilli 20-fold [35]. But it is speculated by some [36, 37] that rapidly absorbed molecules, such as glucose, will be absorbed at the tip of villi before they get a chance to diffuse down the length of the protrusion.

As such, it has been speculated that unless villi contraction offer a mixing effect within the unstirred layer then only 2% of the villus surface area will be involved in the absorption of molecules such as glucose [36, 37].

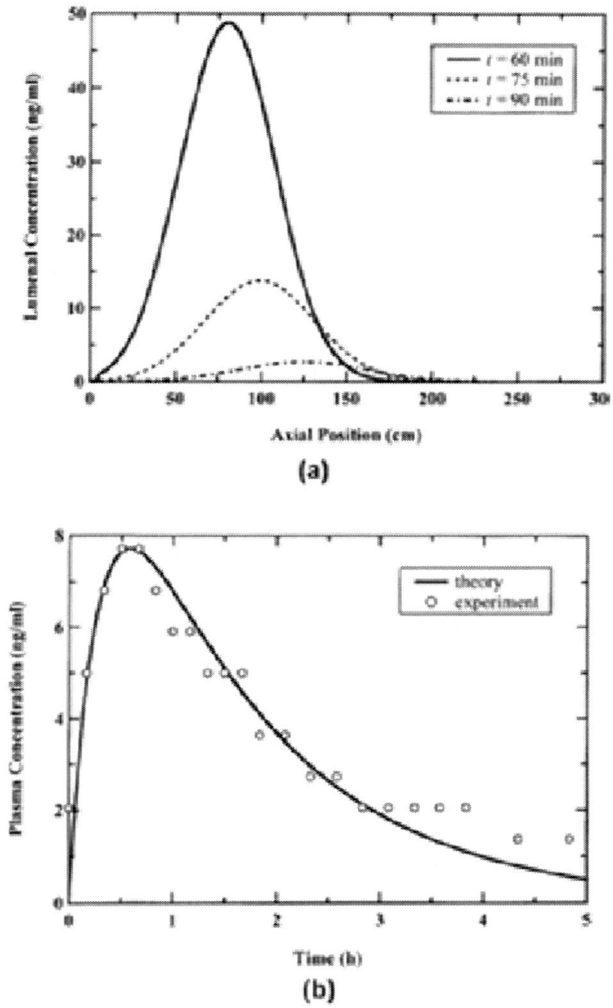

Figure 4. a) concentration of drug in the in the small intestine at different times b) change in plasma concentration of drug [34].

The mixing effect of the villi upon the unstirred water layer, and the subsequent effect on absorption, has been studied using a 2 dimensional lattice Boltzmann model by Wang et al., [38]. The authors argued that the motions of the villi produce a micro-mixing layer (MML) which couples with the macro-scale mixing produced from the contractions of the intestinal walls [38]. Results from the model indicated that while the macro-scale mixing increased absorption rate by a factor of 5, compared with diffusion alone, inclusion of the MML resulted in an 8-fold increase in absorption rate, signifying the importance of both length scales in digestion [38]. The model also showed that when the villi are stationary, 81-82% of absorption occurred through the tips, while at the highest villi motion frequency this reduced to 76-8% [38], backing up the concept that even with the motion the vast majority of nutrients are absorbed at the tips [36, 37].

We can assume that only 2% of the projected area of villi and microvilli are active in absorption, but in the case of Plicae circulares the nutrients are likely to be mixed between them and the full projected area will be involved in absorption. Therefore the total surface area avaliable for absorption will be defined by:

$$2\pi rL\left(\frac{A^{pc}}{A^{ppc}}\right)\times 0.02\left[\left(\frac{A^{v}}{A^{pv}}\right)\left(\frac{A^{mv}}{A^{pmv}}\right)\right] \tag{15}$$

Moving to food digestion, a model of 1996 looked at digestion of a complex meal in pigs [39]. The model incorporated 4 compartments: stomach, small intestine 1 (duodenum & proximal jejunum), small intestine 2, and large intestine; each compartment is modelled with an ordinary differential equation for each input of a variety of nutrient types into the compartments: protein, starch, sugar, fats, fibre, minerals etc. and it takes into account the endogenous secretions. Movement from each compartment is controlled by a fractional rate and absorption in the intestine is via Michaelis-Menten kinetics, with a maximum velocity and Michaelis parameter for each constituent.

Whilst the model tries to look at a whole picture- i.e., has a wide variety of nutrients likely to make up a meal, secretion of enzymes etc., it does not show how these nutrients are likely to interact with each other or affect gastric emptying, nor does it take into account the mass transfer within each compartment, the effect of scale in the intestine (e.g., presence of villi, & microvilli), or any motility effects. The author also point out the limitations of using CSTR like models, where the global concentration of nutrients does not represent the mean local concentration of nutrients this becomes significant when it is considered that the luminal concentration will affect the absorption [39].

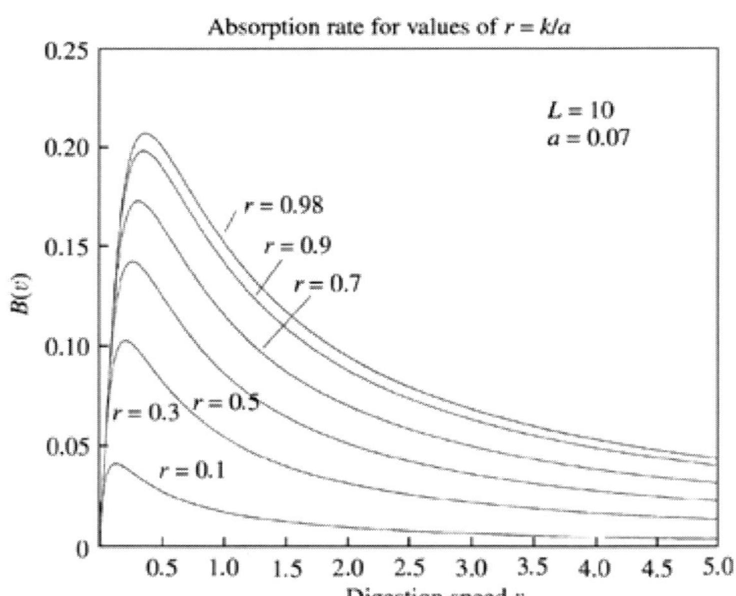

Figure 5. Graph shows absorption rate (B(v)) for different ratios of rate constants k/a (hydrolysis rate over absorption rate) and propulsion speeds (v) [41].

These effects need to be taken into account when trying to model a multi-competent system, but will lead to vast increases in complexity. Consideration of the relative impact on each component must be taken to allow the production of accurate but simplistic models e.g., the presence of fibre/protein microstructure in the chyme will have a greater effect on the mass transfer within the gut than other components [28, 29, 31, 40], or the effect on stomach emptying of sugars, or fatty acids [15, 16, 21] over other components.

When modelling the small intestine we also need to take into account such things as the residence time: how long the chyme is in the gut will affect the amount absorption and also any hydrolysis. Logan et al. looked at a variety of different reaction kinetics for a plug flow small intestine model, from the prospective of an arthropod gut [41]. Optimal gut speeds for maximum absorption at a variety of ratios of rate constants (rate of hydrolysis over rate of absorption) were considered, and are shown in figure 5. Figure 5 demonstrates that varying the digestion speed has a large effect on the absorption rate within the PFR model and highlights the importance of accurate measurements of properties of the gut, i.e., length, radius and speed of propulsion.

To look more closely at the mass transfer of nutrients in the small intestine we can link the absorption of glucose to the mass transfer through the lumen, with the overall mass transfer coefficient been defined as [31]:

$$\frac{1}{K} = \frac{1}{K_{lm}} + R_{sys} \tag{16}$$

Where R_{sys} is the resistance to mass transfer from the wall of the intestine and the recipient side: the rate of movement of the nutrients into the blood plasma from the intestinal wall. We can then define the K_{lm} as the convective mass transfer coefficient of the nutrients in the chyme (luminal side), which will be related to the diffusivity and therefore the viscosity. Assuming the resistance at the wall is small (i.e., absorption of nutrients is not rate limiting) [39, 42], the overall mass transfer coefficient can be assumed to equal the convective mass transfer coefficient of the chyme.

Putting the mass transfer coefficient into the advection-reaction equation:

$$\frac{\partial G(z,t)}{\partial t} = -u\frac{\partial G(z,t)}{\partial z} - \frac{2.f}{r_m}.K.G(z,t) \tag{17}$$

Where r_m is the radius of the intestine, f is the increase in surface area due to protrusions (=12 fold increase), u is the mean velocity of flow within the intestine, G(z,t) is the mass of glucose in the intestine at time distance, z, along the length and time, t.

Figure 6 shows that the viscosity of chyme will have a clear effect in reducing the amount of glucose which will be absorbed. Increasing the viscosity from 1mPa.s to 10mPa.s seems to halve the amount of glucose absorbed within 3 hours, this idea agrees with *in-vivo* studies [28, 29, 31, 40], though it has been shown that the difference in viscosity is reduced between consumption and reaching the jejunum for pigs, but will still have an effect on the absorption rate [29]. So an important consideration when defining the viscosity of chyme is to take into account the effect of secretions in the mouth, stomach and intestine, and effects of any shear forces.

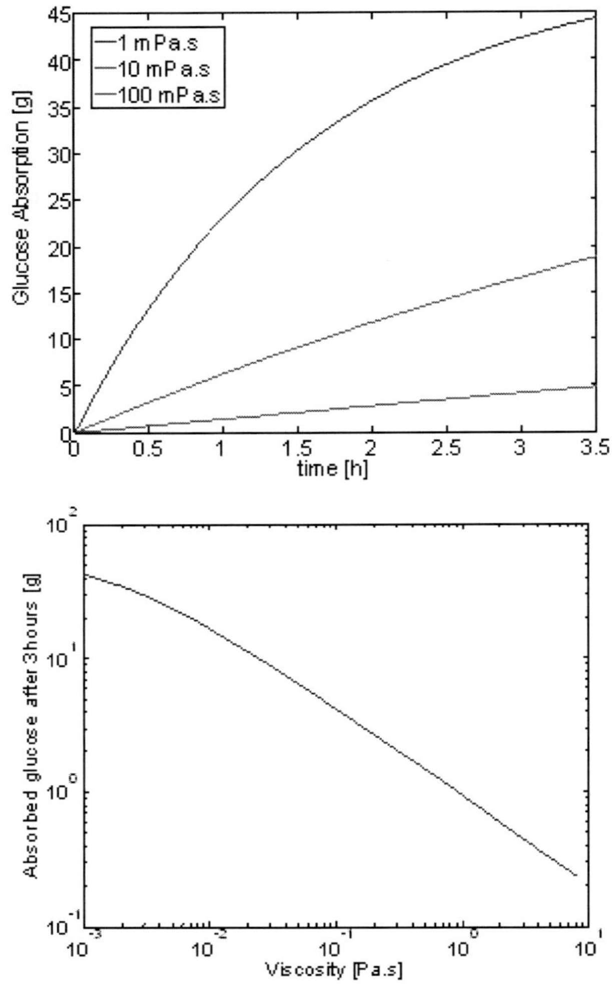

Figure 6. effect of changing viscosity of chyme on absorption of glucose solution: left absorption curves for different viscosities, right, total absorbed glucose after 3 hours against the chyme viscosity (from eq. 17).

A model by Taghipoor et al. looked at the movement and degradation of a bolus through the small intestine of pigs [32]. The model tries to address the idea of velocity changes along the length of the intestine. The model has a bolus that is solubilised in the presence of water, and will contain 3 components: A_s, A_{ns}, and A_{nd}, representing solubilised A, non-solubilised A, and non-degradable A components respectively. Non-solubilised A can be converted to solubilised A by the presence of water, solubilised A can be hydrolysed in the presence of enzymes to B_{abs} which can be absorbed or B_{int} which must be hydrolysed by brush boarder enzymes to give B_{abs} [32].

The transport of the bolus is described by the following equations:

$$\frac{d^2x}{dt^2}(t) = \bar{a}(t)\frac{c_0 + c_1V(t)}{a + bx(t)} - \frac{\dot{K}}{[W](t)}\frac{dx}{dt}(t) \tag{18}$$

$$\frac{dx}{dt}(0) = v_0, \quad x(0) = 0$$

$$\bar{a}(t) = \tau\left(1 - \frac{1}{c}\frac{dx}{dt}(t)\right)$$

Results are shown in figure 7, where the non solubilised A is quickley solubilised, and a spike is seen between 1-2m of the intestine. However, the conversion to absorbable B is slower, showning a peak around the 12m point.

In including a step for the solubilisation of A_{ns} to A_s which can then be hydrolysed, the model takes a realistic step ahead of other models whereby a molecule must be made bioavailable within the intestine before it can be either hydolysed and/or absorbed. This is important in solid meals whereby the chyme will consist of solid boli where molecule must be freed before hydrolysis can occur.

But the model suffers the limitation of only observing one bolus, when in reality boli will be entering the small intestine over a period of hours from the stomach. To address this issue, changes must be made to the small intestine model and a comprehensive mechanistic model of gastric emptying must be produced.

Future models need to develop a more mechanistic approach, linking the gastric emptying to food properties and physiological process e.g., nutrient feedback mechanism from duodenum. Models also need to take into account the properties of the ingested food e.g., the breakdown and disintegration of solids in the stomach, although studied by some [22-25], it has yet to be full described in a realistic mathematical model. Steps have been taken to look at the disintegration of a bolus in the small intestine [32, 43] but need to be developed further to take into account the appearance of multiple boli in the small intestine and the input into the intestine – gastric emptying.

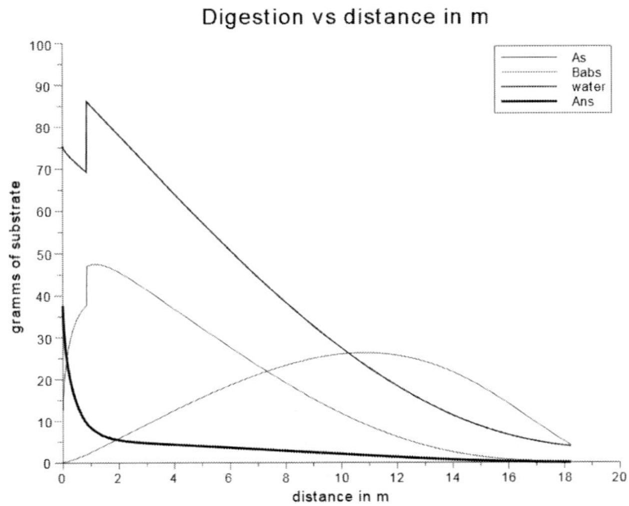

Figure 7. Change in mass of substances along length of intestine during digestion [32].

Along with these there are many other things that may affect the digestion and absorption of nutrients: the secretion of enzymes and other fluids, effect of diffusion through the unstirred water layer, the use of nutrients within the gut wall, interactions between different nutrients, effect of pH on Diffusion and mass transfer, or the effect of mixing [44-46].

In-Vitro Study of Digestion

Mimicking food digestion outside the body, albeit fascinating, poses serious challenges generated by the inherent complexities and variability associated with both the digestive system as well as the food matrices. Unlike most pharmacological preparations, foods are multi-component, multi-phase materials with complex structures, and dimensions extending between 8 orders of magnitudes from cm (e.g., first bite) to mm (e.g., sugar granule), μm (e.g., emulsion droplets, starch granules, ice crystals), nm (e.g., plant cell walls), angstrom (e.g., single molecules of glucose, water) [47, 48]. The different food formulations and food structures significantly affect physiological responses during digestion, e.g., enzymatic secretions, feedback mechanisms, satiety perception, residence times, etc., which are therefore difficult to simulate *in-vitro*.

The digestive tract is a multi-compartmental organisation with complex anatomy, structures and functions, and with diverse physiological operating conditions and feedback mechanisms. For example, control of satiety is regulated through a complex, neurohormonal brain-gut axis, where signals such as stomach distension or energy detected in the small intestine or ileum (ileal break) are communicated to and stimulate the satiety centres of the brain [49-51]. Similarly, control of gut emptying rates depends on a number of factors, e.g., calorific intake, composition of the ingested food, etc. [15-17]. Such control mechanisms are complex, not always well defined and understood, and definitely challenging to reproduce outside the body [11]. Adding to this complexity, postprandial responses also vary substantially between humans as well as within a human, depending on the mood, time of the day, quantity and quality of ingested food, etc. For example, a study looking at duodenal pH of controls and patients with functional dyspepsia during eating has reported values ranging between 4 and 7 for the 'control' group, indicating the prevalence of inter-human variability [52]. The human gut also involves many length scales. Indicatively, the small intestine is the region where nutrients (molecular size) are absorbed at specific sites in the wall membrane, which is lined with villus protrusions (μm size) and folded into larger villi organisations (mm), all embedded in the tube-like organ (about 3cm diameter). All length scales, as well as their interrelations, are important in food digestion and pose challenges in simulating them. For example mass transfer in the small intestine is difficult to imitate *in-vitro*, as it has been reported to combine macro- (contractions of the tube) and micro- (movements of the villi) scale mixing events resulting from intestinal wall contractions [38, 53].

An ideal, complete *in-vitro* model would realistically simulate all digestive processes, it would be easy to use, and it would produce fast (high throughput) and accurate results [54]. It would also be flexible in its functions to allow for any sensitivity analysis or variations in the digestive conditions desired to be studied. It would also be affordable to operate. Today, some reagents are expensive and/or they are biological and very sensitive. Mucins, for example, are known to significantly lose their rheological properties fast and should therefore be used fresh

[55]. Overall, *in-vitro* models for food digestion have become increasingly popular since 1990s and since then they have seen substantial progress [11, 54, 56, 57]. However, *in-vitro* digestion models simulate model conditions and therefore care should be taken in data interpretation.

In-Vitro Models Used to Simulate Digestive Processes That Occur In-Vivo

In-vitro models for food digestion are typically application specific, and should be principally considered for the application that they were designed. For example, there are mouth models designed to study specifically the effect of biting [58], mastication [59], mixing [60], shearing [61], compression [62, 63], or the action of the tongue [64] on different aspects of oral digestion, such as taste perception, texture perception, or bolus formation. Such models are useful to isolate and offer insight on specific angles of food digestion, while indirectly they highlight the complexities involved and the associated challenges in replicating it. It is important to carefully select the *in-vitro* models we use and to understand their advantages and limitations during data acquisition and analysis.

There are a number of physiological features and conditions of digestion that may be incorporated in *in-vitro* models. Those include, but are not limited to, the temperature, the pH and pH gradients, the enzyme types, concentrations and flow rates, the addition of other digestive secretions such as water, electrolytes or mucins, the volumes and flow rates of the foods and secretions, the residence times, the motility of and mechanical actions exerted by the digestive tract, the diffusion and mass transfer, or the different types of nutrient absorption. It is typical to find temperature, pH, secretions, and residence times among the controlled variables of an *in-vitro* model.

Models may simulate one or more digestive processes occurring after the food enters in the alimentary tract. The entry point is the mouth, which signals the beginning of digestion, and from there, the food passes through to the stomach, small intestine, and large intestine. A model may consider one or more parts of this digestive path and may simulate various digestive conditions.

In-vitro food digestion models have been broadly classed into two categories: dynamic and static, depending on whether temporal changes of digestion are incorporated or not, respectively, in the model [11, 65, 66]. Changes may refer to mechanical actions, gut wall contractions, fluid flow to, within, and out of the digestive part that is simulated, etc.

Similarly to the USP models for pharmaceuticals, static models typically consist of a series of static, batch vessels each replicating a different part of the digestive tract, such as oral, gastric, or small intestinal, with their different conditions (pH, temperature, enzymes, etc.). Those models do not account for time-depended processes of digestion and are generally considered advantageous when ease of use and/or high throughput are of importance. Mixing is typically performed homogeneously using mixing devices such as magnetic stirrers or shaking incubators (see for example Englyst, Veenstra & Hudson, 2007 [67]). Interestingly, the volumes that those models operate with can vary substantially from microliters [68] to few hundreds of milliliters [69]. Figure 8 shows two examples of static in-vitro models using different sample volumes for analysis. It should be noted that the different models are designed to investigate different aspects of digestion and therefore should be considered with care.

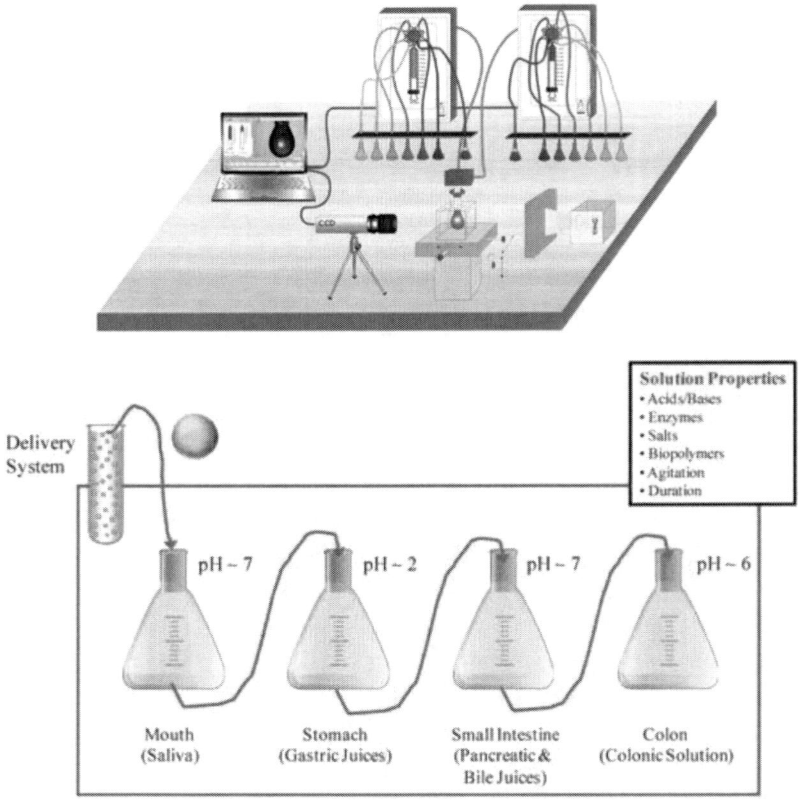

Figure 8. Examples of static in-vitro models used for food digestion: (a) octopus analyses samples of drop volume; (b) pH stat can analyse samples of few hundreds of mL volume [68, 69].

Static models also offer substantial flexibility in their operation, allowing for independent control of digestive conditions, such as pH and/or enzyme concentrations. This enables sensitivity analysis and systematic investigation of the effect of selected features on food digestion; however it causes difficulties in comparing results from different models that use different digestive conditions. Indicatively, a literature review published in 2013 by EFSA reported that protease concentrations in *in-vitro* models used for protein gastric digestion varied by 4 orders of magnitude, the range is huge [56]. Similarly, at least 36 different models have been reported in 2008 to study starch digestion [57]. Different models may vary not only in the levels of enzymes used, but also in the type and/or origin [10]. Enzymes extracted from different sources, e.g., human, animal, or plant origin, may substantially vary in their activities. Recently there have been attempts in standardising static *in-vitro* models used for food digestion, and a group of scientists working in a European (COST) project have published a suggested universal standardised model in an attempt to unify and facilitate comparison between results from different research groups [70]. Yet even when similar conditions are used, experimental errors between different laboratories may still be considerable. A paper looking at inter-laboratory variations of peanut protein gastric digestion using similar protocols, reported digestion of the protein in time durations ranging from 0 to 60min, and an overall 77% agreement between the different labs [71]. Studying digestion of specific target groups, such as infants or elderly people, also requires special attention, as the conditions may substantially vary [72].

Dynamic models offer the potential of incorporating elements of the temporal profile of digestion, such as mechanical forces and/or fluid flow, together with other physiological conditions (such as pH, type and concentration of digestive enzymes and other secretions, etc.) in the simulation [11, 66, 73]. As with static, they may replicate one or more parts of the digestive system. Table 1 shows examples of dynamic oral, gastric, and intestinal models developed for food digestion, which replicate mechanical action of the digestive tract (with references). It should be noted that the list of table 1 is not exhaustive, but rather indicative.

Depending on the desired application, oral digestion has been simulated using commercial meat mincers [74, 75] or other, more sophisticated mouth analogues that have been specifically developed to study oral digestive processes [58-64]. In some cases, the main aim of oral digestion is the production of a bolus that will then be introduced in a gastric and/or intestinal model for further investigations [74].

A number of gastric dynamic models has been developed that incorporate mechanical forces by using different techniques. For example, the model introduced by Chen and co-workers uses the action of a vertically moving spherical bead positioned in a jacketed tube to replicate gastric fluid flow [76]. The Human Gastric Simulator (HGS), introduced in 2008, mimics gastric mechanical stresses exerted on food digesta through the abrasive motion of small plastic beads in a vessel siting on a turntable [23]. The Dynamic Gastric Model (DGM) mimics gastric wall contractions by the squeezing movement of an elastic conical vessel that sits on a cylindrical container. Physical mixing is simulated at representative shear rates, and the model overall simulates food breakdown and transit [77-79]. Intestinal motility (segmentation and peristalsis) has been replicated by squeezing a porous cylindrical membrane that simulates the intestinal lumen [31, 66]. In this model, chyme and enzymes are introduced in the inner (luminal) side of the membrane, while small molecules such as glucose are allowed to pass to the outer (recipient) side for sampling.

In-vitro models may also replicate more than one parts of the digestive tract. A multi-compartmental model (TNO's Intestinal Model, TIM1) was introduced in the mid-1990s in the Netherlands, and replicates gastric, duodenal, jejunal, and ileal digestion in the different compartments [80-83]. A second model (TIM2) was later developed to study large intestinal digestion [73]. Those models control temperature, pH, and secretions, while replicating peristaltic contractions by a squeezing the walls of a cylindrical tube.

Dynamic models are a new generation of *in-vitro* models, whose importance had been forecasted long before their popularity [84]. Evidently, dynamic models offer a more accurate representation of physiologically relevant digestive processes, and they are therefore advantageous in understanding different phenomena occurring during digestion, such as the effects of fluid dynamics on digestibility. As with static, dynamic models also offer flexibility in operation and the potential for individual control of digestive conditions (e.g., pH, enzymes, etc.). However, they are typically more laborious and time consuming to operate and therefore are not yet recommended for high throughput, fast experiments. Like static, dynamic models are also application specific and should be considered with care.

Table 1. Dynamic gastrointestinal models from the literature that simulate mechanical action of the digestive tract (the list is by no means exhaustive)

Model	Schematic	Mechanism to produce mechanical action	References
In-vitro mouth model	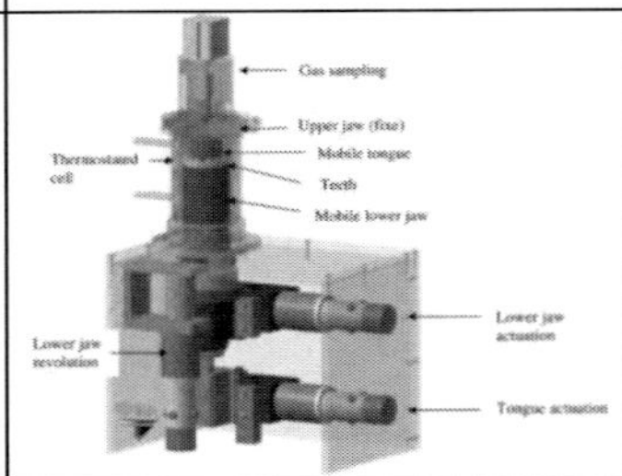	Mimics chewing by incorporating and controlling the movement of artificial teeth, jaws, and tongue	[59]
Dynamic Gastric Model (DGM)	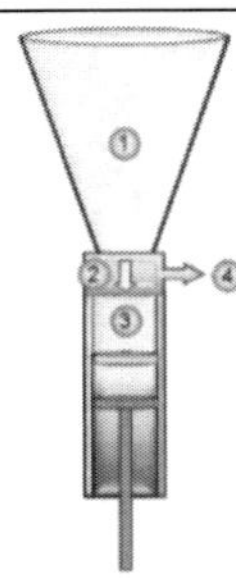	Mimic gastric wall contractions by complex wall motion; gastric motility simulated by squeezing of the conical shaped vessel	[77-79]
Human Gastric Simulator (HGS)	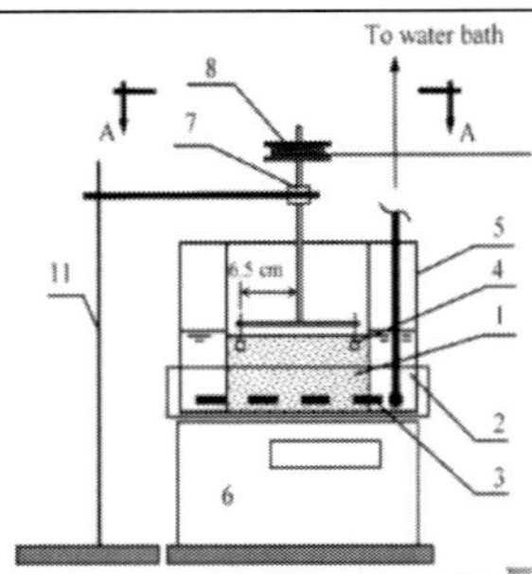	Mimics mechanical forces exerted by the stomach by using article-particle abrasion	[23]
In-vitro Gastric Model	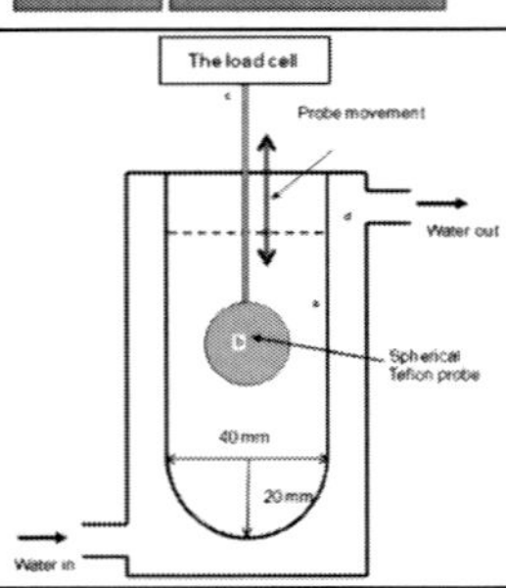	Mimics fluid flow in the stomach using compression action between spherical probe and cylindrical wall	[76]
Dynamic Duodenal Model (DDuo)	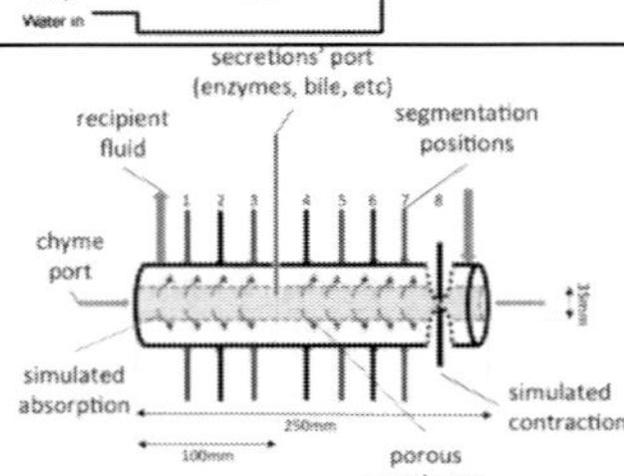	Mimics segmentation and peristaltic movements of the small intestine by squeezing the walls of a cylindrical shaped membrane at different locations by mechanical action	[31, 66]

Model	Schematic	Mechanism to produce mechanical action	References
TIM1 (gastric and small intestinal digestion model) (similar apparatus exists for large intestinal simulation)	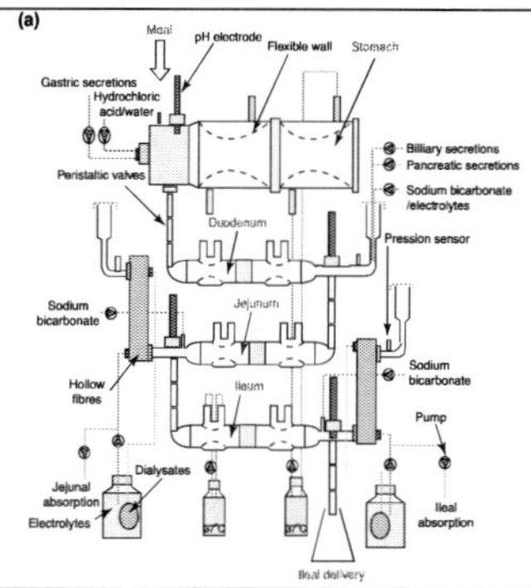	Mimics peristaltic mixing by squeezing the walls of a flexible tube using pressure	[73, 80-83]

Membrane permeation and nutrient transport through the gut wall are another important part of digestion and absorption, which determines nutrient bioaccessibility (i.e., how accessible are the nutrients in the luminal side of the digestive tract) and bioavailability (i.e., how available are the nutrients in the bloodstream). *In-vitro* models typically assess nutrient bioaccessibility by studying behaviour in the luminal side. However, there are models that incorporate nutrient transport through the gut wall membrane as part of their simulation. These typically use Caco-2 cell lines from the human colon and may be static [85, 86] or dynamic [87] in nature.

Overall, replicating food digestion is challenging due to the complex nature of both the human digestive system and the food matrices. A vast variety of *in-vitro* food digestion models exist, and have seen increasing popularity over the last decades. They may represent one or more digestive processes in a very rudimentary or very sophisticated way, and they are highly considered as application specific. Static models are easy to operate and relatively fast, compared to the laborious and time consuming dynamic models. However, dynamic models represent more accurately physiological processes occurring *in-vivo*. When selecting the appropriate *in-vitro* model we should understand the limitations and advantages of each choice and consider the model that addresses our question.

Artificial Digestion: The Possible Future

Although *in-vitro* models have seen significant progress over the last decades, they still have a number of limitations and challenges [10, 11]. Critical to *in-vitro* models is their biological significance, and their correlation with *in-vivo* processes (*in-vitro in-vivo* correlation, IVIVC), and therefore it is at the core of our interests to establish methodologies for model validation. Currently, models have been reported that agree reasonably well [88] or less accurately [89] with *in-vivo* data. In addition, models are needed that will link the various length scales involved in food digestion and identify the controlling mechanisms of digestibility and nutrient bioaccessibility.

Harmonisation of in-vitro models and standardisation of digestive conditions may be beneficial, as they will allow comparison between the different laboratories. For this, it is crucial to first understand the experimental conditions that we want to replicate. For example, the *in-vivo* biochemistry is still not well understood and agreed, as are the enzyme to food concentrations. There is a need to produce robust models that can be used beyond academic institutions, at a sustainable cost, and modellers of the future are likely to work towards this direction.

Conclusion

Simulating food digestion using *in-silico* and *in-vitro* techniques has gained increased popularity over the last decades. Simulation models offer advantages and limitations, which should be considered when selecting the appropriate model for a specific application. Replicating food digestion outside the body poses an attractive alternative to *in-vivo* methods, which are often time and labour consuming, complex, costly, and ethically compromised. However, currently *in-silico* models are often empirical, which limits their predictive ability, and *in-vitro* models are typically application specific (and often expensive), and therefore work well only within predefined frameworks.

Better understanding of the digestive processes, and increased quality and predictability of the models, may be achieved by combining data from *in-silico* and *in-vitro* models in a systematic way. This lies with the design and optimisation of experimental techniques that will fast and effectively feed the required parameters of *in-silico* simulations to produce validated predictions. In such a future, models will be used to efficiently scan different food formulations and predict postprandial responses. Undoubtedly, models will never be able to predict *in-vivo* processes with complete certainty; however optimised models can serve as a tool to health claims and to strategically design *in-vivo* experiments, for the benefit of public health.

References

[1] Jamison, D.T., et al., Global health 2035: a world converging within a generation. *The Lancet*. 382(9908): p. 1898-1955.
[2] Ogden, C.L., et al., Prevalence of overweight and obesity in the United States, 1999-2004. *Journal of the American Medical Association*. 295(13): p. 1549-1555.
[3] Association, A.D., Economic Costs of Diabetes in the U.S. in 2007. *Diabetes Care* 2008. 31: p. 1-20.
[4] McCann, S.J.H., Personality and American State Differences in Obesity Prevalence. *Journal of Psychology*, 2011. 145(5): p. 419-433.
[5] Cawley, J. and C. Meyerhoefer, The medical care costs of obesity: An instrumental variables approach. *Journal of Health Economics*, 2012. 31(1): p. 219-230.
[6] Flegal, K.M., et al., Prevalence and trends in obesity among US adults, 1999-2008. JAMA - *Journal of the American Medical Association*, 2010. 303(3): p. 235-241.
[7] HSCIC, Statistics on Obesity, physical activity and diet: England. 2014.
[8] Scarborough, P., et al., The economic burden of ill health due to diet, physical inactivity, smoking, alcohol and obesity in the UK: an update to 2006-07 NHS costs. *Journal of public health* (Oxford, England). 33(4): p. 527.

[9] McPherson, K., T. Marsh, and M. Brown, Modelling future trends in obesity and the impact on health. *Foresight tackling obesity: future choices*, 2007.

[10] Benshitrit, R.C., et al., Development of oral food-grade delivery systems: Current knowledge and future challenges. *Food Funct.*, 2012. 3(1): p. 10-21.

[11] Guerra, A., et al., Relevance and challenges in modeling human gastric and small intestinal digestion. *Trends Biotechnol*, 2012. 30(11): p. 591-600.

[12] Penry, D.L. and P.A. Jumars, Chemical reactor analysis and optimal digestion. *BioScience,* 1986. 36(5): p. 310-315.

[13] Penry, D.L. and P.A. Jumars, Modeling animal guts as chemical reactors. *The American Naturalist*, 1987. 129(1): p. 69-96.

[14] Peng, H. and B. Cheung, A review on pharmacokinetic modeling and the effects of environmental stressors on pharmacokinetics for operational medicine. 2009, Defence R&D Canada: Toronto.

[15] Calbet, J.A.L. and D.A. MacLean, Role of caloric content on gastric emptying in humans. *Journal of Physiology*, 1997. 498(2): p. 553-559.

[16] McHugh, P.R., The control of gastric emptying. *Journal of the autonomic nervous system*, 1983. 9: p. 221-231.

[17] Brener, W., T.R. Hendrix, and P.R. McHugh, Regulation of the gastric emptying of glucose. *Gastroenterology,* 1983. 85: p. 76-82.

[18] Vist, G.E. and R.J. Maughan, The effect of osmolality and carbohydrate content on the rate of gastric emptying of liquids in man. *Journal of Physiology,* 1995. 486(2): p. 523-531.

[19] Moran , T.H., et al., Interactions between gastric volume and duodenal nutrients in the control of liquid gastric emptying. *American Journal of Physiology*, 1999. 276(4): p. R997-R1002.

[20] Marciani, L., et al., Effect of meal viscosity and nutrients on satiety, intragastric dilution, and emptying assessed by MRI. *American Journal of Physiology Gastrointestinal and liver physiology*, 2001. 280: p. G1227-G1233.

[21] Hellstrom, P.M., P. Gryback, and H. Jacobsson, The physiology of gastric emptying. *Best practice & research clinical anaesthesiology,* 2006. 20(3): p. 397-407.

[22] Siegel, J.A., et al., Biphasic nature of gastric emptying. *Gut,* 1988. 29: p. 85-89.

[23] Kong, F. and R.P. Singh, Disintegration of solid foods in human stomach. *journal of food science*, 2008. 73(5): p. R67-R80.

[24] Bornhorst, G.M., et al., Gastric emptying rate and chyme characteristics for cooked brown and white rice meals in vivo. *Journal of the science of food and agriculture*, 2013. 93(12): p. 2900-2908.

[25] Bornhorst, G.M., K. Kostlan, and R.P. Singh, Particle size distribution of brown and white rice during gastric digestion measured by image analysis. *Journal of food science*, 2013. 78(9): p. E1383- E1391.

[26] Dalla Man, C., M. Camilleri, and C. Cobelli, A System Model of Oral Glucose Absorption: Validation on Gold Standard Data. *IEEE transactions on biomedical engineering*, 2006. 53(12): p. 2472-2478.

[27] Singh, J., A. Dartois, and L. Kaur, Starch digestibility in food matrix: a review. *Trends in food science & technology*, 2010. 21: p. 168-180.

[28] Takahashi, T., et al., Hydrolyzed guar gum decreases postprondial blood glucose and glucose absorption in the rat small intestine. *Nutrition research*, 2009. 29: p. 419-425.

[29] Ellis, P.R., et al., The effect of high-molecular-weight guar gum on net apparent glucose absorption and net apparent insulin and gastric inhibitory polypeptide production in the growing pig: relationship to rheological changes in jejunal digesta. *British journal of nutrition,* 1995. 74: p. 539-556.

[30] Slaughter, S.L., et al., the effect of guar galactomannan and water availability during hydrothermal processing on the hydrolysis of starch catalysed by pancreatic alpha-amylase. *Biochimica et biophysica acta,* 2002. 1571: p. 55-63.

[31] Tharakan, A., et al., Mass Transfer and Nutrient Absorption in a Simulated Model of Small Intestine. *Journal of Food Science*, 2010. 75(6): p. E339-E346.

[32] Taghipoor, M., et al., Mathematical modeling of transport and degradation of feedstuffs in the small intestine. *journal of theoretical biology,* 2012. 294: p. 114-121.

[33] Ni, P.F., et al., Theoretical model studies of intestinal drug absorption V. Non-steady-state fluid flow and absorption. *International Journal of Pharmaceutics*, 1980. 5: p. 33-47.

[34] Stoll, B.R., et al., A theory of molecular absorption from the small intestine. *Chemical engineering science*, 2000. 55: p. 473-489.

[35] Ganong, W., gastrointestinal physiology, in ganong's review of medical physiology. 2005, The McGraw-Hill Companies. p. 429-486.

[36] Strocchi, A. and M.D. Levitt, Role of Villous Surface Area in Absorption: Science Versus Religion. *Digestive Diseases and Sciences,* 1993. 38(3): p. 385-387.

[37] Lentle, R.G. and P.W.M. Janssen, *The Physical processes of digestion.* 2011: Springer.

[38] Wang, Y., et al., A multiscale lattice boltzmann model of macro- to micro-scale transport, with applications to gut function. *Philosophical transactions of the royal society*, 2010. 368: p. 2863-2880.

[39] Bastianelli, D., D. Sauvant, and A. Rerat, Mathematical modeling of digestion and nutrient absoprtion in pigs. *journal of animal science,* 1996. 74: p. 1873-1887.

[40] Sasaki, T. and K. Kohyama, Influence of non-starch polysaccharides on the in vitro digestibility and viscosity of starch suspensions. *Food Chemistry,* 2012. 133: p. 1420-1426.

[41] Logan, D.J., A. Joern, and W. Wolesensky, Location,Time and Temperature Dependence of Digestion in Simple Animal Tracts. *journal of theoretical biology*, 2002. 216: p. 5-18.

[42] Stumpel, F., et al., Normal kinetics of intestinal glucose absorption in the absence of GLUT2: evidence for a transport pathway requiring glucose phosphorylation and transfer into the endoplasmic reticulum. *Proceeding of the national academy of sciences,* 2001. 98(20): p. 11330-11335.

[43] Taghipoor , M., et al., Digestion modelling in the small intestine: Impact of dietary fibre. *Cornell University Library*: arXiv:1211.7025, 2012.

[44] Sugano, K., Introduction to computational oral absorption simulation. *Expert opinion on drug metabolism and toxicology*, 2009. 5(3): p. 259-293.

[45] Van Wey, A.S., et al., A mathematical model of the effect of pH and food matrix composition on fluid transport into foods: An application in gastric digestion and cheese binding. *Food Research International*, 2014. 57: p. 34-43.

[46] Lentle, R.G. and P.W.M. Janssen, Physical characteristics of digesta and their influence on flow and mixing in hte mammalian intestine: a review. *J. Comp. Physiol. B*, 2008. 178: p. 673-690.

[47] Aguilera, J.M., Why food microstructure? *journal of food engineering,* 2005. 67(1-2): p. 3-11.

[48] James, B., Food microstructure analysis. 4th ed. Engineering Properties of Foods, ed. M.A. Rao, Rizvi, S.S.H., Datta, A.K., *Ahmed, J.* . 2014, US: CRC press.

[49] Wilde, P.J., & Chu, B. S., Interfacial & colloidal aspects of lipid digestion. *Advances in Colloid and Interface Science*, 2011. 165(1): p. 14–22.

[50] Fiszman, S. and P. Varela, The role of gums in satiety/satiation. A review. *Food Hydrocolloids*, 2013. 32(1): p. 147-154.

[51] Gidley, M.J., Hydrocolloids in the digestive tract and related health implications. *Current Opinion in Colloid & Interface Science*, 2013. 18(4): p. 371-378.

[52] Bratten, J. and M.P. Jones, Prolonged recording of duodenal acid exposure in patients with functional dyspepsia and controls using a radiotelemetry pH monitoring system. *Journal of Clinical Gastroenterology*, 2009. 43(6): p. 527-33.

[53] Lentle, R.G., et al., Mucosal microfolds augment mixing at the wall of the distal ileum of the brushtail possum. Neurogastroenterology and Motility : *The Official Journal of the European Gastrointestinal Motility Societ,* 2013. 25(11): p. 881-e700.

[54] Hur, S.J., et al., In vitro human digestion models for food applications. *Food Chemistry*, 2011. 125(1): p. 1-12.

[55] Bongaerts, J.H.H., D. Rossetti, and J.R. Stokes, The Lubricating Properties of Human Whole Saliva. *Tribology letters*, 2007. 27(3): p. 277-287.

[56] Mills, C.E.N., Marsh, J.T., Boyle, R., Hoffmann-Sommergruber, K., DuPont, D., Bartra, J., Bakalis, S., McLaughlin, J., Shewry, P.R., Literature review: "In-vitro digestibility tests for allergenicity assessment", in European Food Safety Authority (EFSA) External Scientific Report, EN-529. 2013.

[57] Woolnough, J.W., Monro, J. A., Brennan, C. S., & Bird, A. R., Simulating human carbohydrate digestion in vitro : a review of methods and the need for standardisation. *International Journal of Food Science & Technology*, 2008. 43(12): p. 2245–2256.

[58] Meullenet, J.F. and R.K. Gandhapuneni, Development of the BITE Master II and its application to the study of cheese hardness. *Physiology & Behavior,* 2006. 89(1): p. 39-43.

[59] Salles, C., et al., Development of a chewing simulator for food breakdown and the analysis of in vitro flavor compound release in a mouth environment. *J. Food Eng.*, 2007. 82(2): p. 189-198.

[60] De Wijk, R.A., A.M. Janssen, and J.F. Prinz, Oral movements and the perception of semi-solid foods. *Physiology & Behavior*, 2011. 104(3): p. 423-428.

[61] Lvova, L., et al., Salt release monitoring with specific sensors in in vitro oral and digestive environments from soft cheeses. *Talanta,* 2012. 97: p. 171-180.

[62] De Loubens, C., et al., Mechanistic model of in vitro salt release from model dairy gels based on standardized breakdown test simulating mastication. *J. Food Eng.*, 2011. 105(1): p. 161-168.

[63] Mills, T., et al., Development of an in-vitro mouth model to quantify salt release from gels. *Food Hydrocolloids*, 2011. 25(1): p. 107-113.

[64] Benjamin, O., et al., Development of a model mouth containing an artificial tongue to measure the release of volatile compounds. *Innovative Food Science & Emerging Technologies*, 2012. 15(0): p. 96-103.

[65] Vieira, M.L.T., Kirby, B., Ragueneau-Majlessi, I., Galetin, A., Chien, J. Y. L., Einolf, H. J., … Huang, S.-M. , Evaluation of various static in vitro-in vivo extrapolation models for risk assessment of the CYP3A inhibition potential of an investigational drug. *Clinical Pharmacology and Therapeutics,* 2014. 95(2): p. 189-98.

[66] Gouseti, O., et al., Hydrocolloids in human digestion: Dynamic in-vitro assessment of the effect of food formulation on mass transfer. *Food Hydrocolloids,* 2014.

[67] Englyst, H.N., J. Veenstra, and G.J. Hudson, Measurement of rapidly available glucose (RAG) in plant foods: a potential in vitro predictor of the glycaemic response. *British Journal of Nutrition*, 2007. 75(3): p. 327.

[68] Maldonado-Valderrama, J., et al., In vitro digestion of interfacial protein structures. *Soft Matter*, 2013. 9(4): p. 1043.

[69] McClements, D.J. and Y. Li, Review of in vitro digestion models for rapid screening of emulsion-based systems. *Food & Function*, 2010. 1(1): p. 32-59.

[70] Minekus, M., et al., A standardised static in vitro digestion method suitable for food - an international consensus. *Food & Function*, 2014. 5(6): p. 1113-24.

[71] Thomas, K., et al., A multi-laboratory evaluation of a common in vitro pepsin digestion assay protocol used in assessing the safety of novel proteins. *Regulatory Toxicology and Pharmacology : RTP*, 2004. 39(2): p. 87-98.

[72] Bourlieu, C., et al., Specificity of infant digestive conditions: some clues for developing relevant in vitro models. *Critical Reviews in Food Science and Nutrition,* 2014. 54(11): p. 1427-57.

[73] Minekus, M., et al., A computer-controlled system to simulate conditions of the large intestine with peristaltic mixing, water absorption and absorption of fermentation products. *Applied Microbiology and Biotechnology,* 1999. 53(1): p. 108-114.

[74] Bornhorst, G.M. and R.P. Singh, Kinetics of in Vitro Bread Bolus Digestion with Varying Oral and Gastric Digestion Parameters. Fo*od Biophys.*, 2013. 8(1): p. 50-59.

[75] Hoebler, C., et al., Development of an in vitro system simulating bucco-gastric digestion to assess the physical and chemical changes of food. *International Journal of Food Sciences and Nutrition*, 2002. 53(5): p. 389-402.

[76] Chen, J., et al., Development of a simple model device for in vitro gastric digestion investigation. *Food & Function*, 2011. 2(3-4): p. 174-182.

[77] Wickham, M.S.J.F., R. , *Dynamic gastric model.* 2013.

[78] Mercuri, A., et al., Dynamic gastric model (DGM): a novel in vitro apparatus to assess the impact of gastric digestion on the droplet size of self-emulsifying drug-delivery systems. *J. Pharm. Pharmacol.*, 2008. 60: p. A2-A2.

[79] Lo Curto, A., et al., Survival of probiotic lactobacilli in the upper gastrointestinal tract using an in vitro gastric model of digestion. *Food Microbiology*, 2011. 28(7): p. 1359-1366.

[80] Minekus, M., et al., A Multicompartmental Dynamic Computer-Controlled Model Simulating the Stomach and Small-intestine. *Alternatives to Laboratory Animals*, 1995. 2: p. 197-209.

[81] Marteau, P., et al., Survival of Lactic Acid Bacteria in a Dynamic Model of the Stomach and Small Intestine: Validation and the Effects of Bile. *Journal of Dairy Science*, 1997. 80(6): p. 1031-1037.

[82] Blanquet, S., et al., *The 'biodrug' concept: An innovative approach to therapy.* 2001. p. 393-400.

[83] Krul, C., et al., Application of a dynamic in vitro gastrointestinal tract model to study the availability of food mutagens, using heterocyclic aromatic amines as model compounds. *Food Chem. Toxicol.*, 2000. 38(9): p. 783-792.

[84] Lea, A.S., A Comparative Study of Natural and Artificial Digestions. (Preliminary Account). *Proceedings of the Royal Society of London,* 1889. 47: p. 192-197.

[85] Tenore, G.C., et al., Simulated gastrointestinal digestion, intestinal permeation and plasma protein interaction of white, green, and black tea polyphenols. *Food Chemistry*, 2015. 169: p. 320-326.

[86] Etcheverry, P., et al., Calcium, zinc, and iron bioavailabilities from a commercial human milk fortifier: a comparison study. *Journal of Dairy Science,* 2004. 87(11): p. 3629-37.

[87] Pusch, J., et al., The physiological performance of a three-dimensional model that mimics the microenvironment of the small intestine. *Biomaterials,* 2011. 32(30): p. 7469-7478.

[88] Hettiaratchi, U.P.K., S. Ekanayake, and J. Welihinda, Prediction of glycaemic indices (GI) of meals by starch hydrolysis indices. *International Food Research Journal*, 2012. 19: p. 1153-1159.

[89] Ferrer-Mairal, A., et al., In vitro and in vivo assessment of the glycemic index of bakery products: influence of the reformulation of ingredients. *European Journal of Nutrition*, 2012. 51(8): p. 947-54.

In: From Medicinal Chemistry to Food Science
Editor: Pietro Cozzini
ISBN: 978-1-63483-587-9

Chapter 4

POLYPHENOLS: FROM COMPUTER SIMULATIONS TO WORKBENCH ANALYSIS

***Luca Dellafiora*[1,2], *Luca Calani*[1], *Daniele Del Rio*[1] and *Pietro Cozzini*[1,2,•]**

[1]Department of Food Science, University of Parma, Parma, Italy
[2]Molecular Modelling Laboratory, Department of Food Science, University of Parma, Parma, Italy

Abstract

Phytochemicals are non-essential and naturally occurring plant compounds found in a wide number of products intended for human consumption. Polyphenolic compounds are perhaps the most widely studied class of phytochemicals, and produce a series of health-protective effects on humans as a consequence of dietary intake. In spite of the number of studies found in the literature so far – the most of which addressed the phytoestrogenic activity – the wide-spectrum analysis of biological activities is intrinsically challenging. In this frame, the use of *in silico* analysis gains a substantial step forward. In this chapter, it has been presented the metabolic fate of polyphenols in humans, and the phytoestrogenic activity of this sub-class of compounds has been particularly emphasized. Moreover, some of the most relevant advances gained by using computational approaches have been reviewed.

1. Introduction

Phytochemicals are non-essential and naturally occurring plant compounds found in a wealth of products intended for human consumption, which produce a series of health-protective effects on humans as a consequence of dietary intake. Indeed, an increased consumption of fruits and vegetables is strongly promoted in public health programs, according to the latest EU policy on healthier food. The theoretical-minimum-risk distribution

• E-mail: pietro.cozzini@unipr.it; Phone: +39 0521 905669.

for fruits and vegetables intake for adults was estimated to be 600 g/day, suggesting a protective effect from ischemic heart disease, stroke and cancer of the lung and digestive tract [1]. The epidemiological evidence for the benefit of consuming a diet that is high in fruits and vegetables is quite compelling, and the probable health-related factors were the presence of fibre and antioxidant vitamins (C, E and β-carotene), besides other chemical compounds, including phytochemicals. According to Harborne classification [2], the class of phytochemicals mainly includes polyphenols, terpenoids and alkaloids. The polyphenol sub-class is perhaps the most widely investigated. Even if the dietary intake of polyphenols widely exceeds that of all other classes of phytochemicals and known dietary antioxidants [3, 4], only in the last two decades researchers and food manufacturers focused the public's attention on (poly)phenolic compounds as potential candidates in the prevention of several chronic-degenerative diseases. This fact has led to several epidemiological studies evaluating if protective effects of vegetables and fruits versus several aging-related disorders could be explained by regular intake of phenolic and (poly)phenolic compounds [5-7]. Furthermore, since they act through different mechanisms and by targeting a wide array of biological targets, they are a case study of great pharmacological relevance. However, in spite of the current wealth of knowledge, some aspects of polyphenol action are vague so far and the understanding from systemic point of view is still scarce and unripe. Undoubtedly, their pleiotropic action contributes to encrypt the comprehensive view in terms of molecular mechanisms in living organisms. Nevertheless, regardless to the endpoints of the research, deciphering the complex network of interactions is paramount to rationally use polyphenols as instrumental for human health and wellbeing. In recent years, such a wide research field has extensively taken advantage of computational approaches. The so defined *in silico* techniques are well-known tools in medicinal chemistry. They are widely used in drug discovery and development mainly to design and optimize pharmacologically active compounds or to decipher the mechanisms of action of bioactive chemicals. Likewise, they can be a cost-effective analytical choice to try to find a solution to the current lack of knowledge in the matter of bioactivity assessment of polyphenols.

In the present chapter, the classification and the metabolic fate of dietary polyphenols is discussed and, furthermore, insights on the phytoestrogenic activity – which is perhaps the most widely investigated – are detailed. Moreover, a brief overview on the possible endpoints reached by using *in silico* techniques is also reported.

2. Classification of Polyphenols and Their Metabolic Fate in Humans

The plant-derived polyphenols are characterised by having at least two aromatic rings with one or more hydroxyl groups attached. Several thousand molecules with a polyphenol structure have been identified in plants, of which many are food plants. Polyphenols range from simple, low molecular weight, single aromatic-ring substances (phenolic compounds) to large and complex tannins and derived polyphenols [8].

An easier classification divides the (poly)phenolic class in two main groups: the flavonoids and the non-flavonoids [8]. The first group comprises a large family of compounds that share a common chemical structure, which consists of 2 aromatic rings (A and B rings) linked by a 3-carbon chain that forms an oxygenated heterocyclic ring (C ring) [9]. The flavonoids may then be further divided in several subclasses as a function of the type of heterocycles involved. Six subclasses are those that have a greater dietary relevance, and these are flavonols, flavanones, flavones, isoflavones, anthocyanidins and flavanols (catechins and proanthocyanidins). Among non-flavonoids, the phenolic acids are the compounds that have dietary significance, in particular hydroxybenzoic and hydroxycinnamic acids. Among the first group, gallic acid is the most widespread in the food plants and can be involved in the biosynthesis of gallotannins. In regard to hydroxycinnamates, the most common are caffeic, ferulic and *p*-coumaric acids, although the first is usually the most common [8]. Ellagic acid and ellagitannins are also considered as non-flavonoids. The latter compounds are formed by esterified linkages, which can be hydrolysed, releasing ellagic acid (dilactone form of hexahydroxydiphenic acid). Instead, stilbenoid compounds are less widespread in plants and their daily intake is lower than flavonoids and phenolic acids. The main dietary source of stilbenes is resveratrol (3,5,4'-trihydroxystilbene) that occurs as *trans* and *cis* isomers, but primarily as *trans*-resveratrol-3-*O*-glucoside [10].

After ingestion, (poly)phenol-glycosides can be modified immediately in the oral cavity by the hydrolysing activity of saliva. Although glycosides can also be hydrolysed prior to reaching the small intestine, the absorption at this site is mostly dependent by hydrolysing activity towards (poly)phenol-glycosides. The release of aglycone is mediated through the action of lactase phloridzin hydrolase (LPH) in the brush-border of the small intestine epithelial cells [11]. LPH exhibits broad specificity for flavonoid-*O*-*β*-D-glucosides and the released aglycone may then enter the epithelial cells by passive diffusion due to its increased lipophilicity [12]. An alternative hydrolytic step is mediated by a cytosolic *β*-glucosidase (CBG). For this process, the active sodium-dependent glucose transporter, SGLT-1, seems to be involved in the transport of flavonoid glycosides into epithelial cells, a necessary step to permit the action of *β*-glucosidase [13].

Once absorbed, flavonoids follow the common metabolic pathway of exogenous organic substances and, like drugs and most xenobiotics, undergo phase II enzymatic detoxification. They can be conjugated with glucuronic acid, sulfate and methyl groups, through UDP-glucuronosyltransferases (UGTs), sulphotransferases (SULTs) and catechol-*O*-methyltransferase (COMT) respectively [11]. Phase II metabolism first occurs in the small intestine, and metabolites subsequently undergo further conversions in the liver, before entering the systemic circulation, recycling back to the small intestine through bile excretion and finally being excreted in urine. However, the fraction of flavonoids unabsorbed in the small intestine will travel further down the gastrointestinal tract (GI), reaching the large intestine where it is exposed to the resident microflora and is rapidly degraded [11]. In some cases, an entero-hepatic circulation may ensure a longer residence time in plasma for some metabolites, thus extending their bioavailability if compared to that of their parent compounds. Gut bacteria can hydrolyse glycosides, glucuronides, sulfates, amides, esters and lactones, and are able to break down the flavonoidic skeleton as well as carry out reactions of reduction, decarboxylation, demethylation and dehydroxylation [14].

These complex modifications generate several low molecular weight metabolites that can be efficiently absorbed *in situ*, subsequently undergoing further phase II metabolism locally and/or in liver before entering the blood compartment and finally being excreted in urine in substantial amounts, largely exceeding the excretion of flavonoid metabolites formed in the upper GI tract [13].

3. Polyphenol Metabolites as Phytoestrogenic Compounds

The dietary polyphenols are associated with beneficial health effects in humans on the basis of a wealth of *in vitro* studies. In the literature, the large numbers of *in vitro* investigations performed are focused on the evaluation of the biological effects of both polyphenol-rich foods, their extracts and on the molecules present *in planta* [15]. As a matter of fact, the *in vivo* feeding studies conducted in recent years suggest that the molecules circulating in the organism are different from the molecules found *in planta* [13].

Phytoestrogens are a diverse group of plant-derived chemicals that structurally or functionally mimic estrogens and have been associated with potential benefits for human health, in particular in the context of breast and prostate cancers, cardiovascular diseases, menopausal symptoms, and osteoporosis [16]. Polyphenols are the most dietary phytoestrogens, and the human gut microbiota has been reported to strongly affect the estrogenic effects of these phytochemicals through metabolic transformations that have been widely described in the case of isoflavones and lignans [17].

Enterolactone and hydroxymatairesinol as well as estradiol concentration-dependently increased the proliferation of MCF-7 human breast cancer cells. Enterolactone exerted its activity in the same concentration range as that of estradiol (in the pM concentration range), but with lower efficacy (41% of the estradiol). Instead, hydroxymatairesinol exhibited a slightly lower efficacy (71% of estradiol) at μM concentration.The effects of both lignans were significantly reduced in the presence of tamoxifen, an antagonist of estrogen receptor (ER) [18].

Penttinen and colleagues tested the effects of enterolactone on ER transcriptional activation in a noncarcinoma cell line. A dose-dependent activation of the related gene expression started at μM concentration, suggesting that enterolactone can directly enhance ER-dependent transcription. However, some differences emerged between the two ER isoforms. ERα was activated at nM concentration, which is physiologically relevant for enterolactone, while activation through ERβ required a μM concentration [19]. Moreover, the same authors confirmed that enterolactone can activate the transcriptional activity of the ERs through direct binding to the Ligand Binding Domains of the both receptor isoforms. The estrogenic effect of enterolactone was further demonstrated *in vivo* in adult ovariectomized mice treated with a single dose of this mammalian lignan at 10 mg/kg. Interestingly, the treatment on mice induced tissue-specific estrogen-responsive reporter gene expression at the level of uterus and vagina, while in the bone tissue activation of the related gene was not observed [19]. The sesame lignans, sesamin and sesamolin, are metabolized by gut microflora to form enterolactone and enterodiol as end-products, as usually occurs for the most dietary lignans. The activation of ER was assayed by using T47D-KBluc breast cancer cells, observing as all tested compounds had estrogenic activity except enterodiol. Enterolactone showed estrogenic activity starting from 10 μmol/L, even if a nanomolar concentration usually occurs in systemic circulation [20]. The enterolignans, enterodiol and enterolactone, showed different behaviour for transactivation of ERα and ERβ in MCF cells. Cell incubation with enterodiol at 10 and 100 nM did not induce a significant increase in transcriptional activity for both ERs. Instead a significant increase in transcriptional activity is obtained with 1 and 10 μM compared to control, while for ERβ it was required a concentration equal to 10 μM to have the same effect [21]. In like manner, enterolactone induced ERα transcriptional activity in a dose-dependent mode with a significant effect obtained in the range 1-10 μM, while this enterolignans had no effect on ERβ transcriptional activity at concentration ranged between 10 nM and 10 μM. The same authors investigated the ER transcriptional activities of both mammalian lignans in presence of 17-β-estradiol. Both enterolignans showed neither stimulatory nor inhibitory effects on the E2-induced ERα transcriptional activation. Conversely, both enterolactone and enterodiol inhibited the ERβ transcriptional activation [21].

The colon microbiota can boost the bioactivity of soy-derived isoflavones, as well lignans, leading to the biotransformation of daidzein into the equol metabolite. This microbial flavonoid exists in two forms: S-(-)-equol and R-(+)-equol, but only the S form occurs naturally, as a results of gut microflora activity of 30–50% of the population [22]. Kelly and colleagues [23] incubated for 24 h HepG2 cells and Hep89 cells (expressing ERα) with 50 nmol/L 17β-estradiol or equol and measured tissue plasminogen activator (tPA), plasminogen activator inhibitor-1 (PAI-1), Factor VII, fibrinogen γ, protein C and protein S mRNA expression. Equol increased tPA and PAI-1 expression in Hep89 cells with fold changes greater than those observed for estradiol. In HepG2 cells (which do not express ERα), PAI-1 and tPA expression were unchanged. Factor VII was increased by equol in Hep89 cells but not in similarly treated HepG2s. Prothrombin gene expression was increased by equol exposure in HepG2 cells in the absence of the classical estrogen receptors. Taken together, these results suggest that physiological levels of equol can modulate coagulation and fibrinolytic processes in a human hepatocyte cell line. Zhang and colleagues [24] identified beneficial effects of S-(-)-equol with respect to cytoprotective antioxidant gene activation, ideally defining a novel strategy to prevent and treat cardiovascular diseases. They reported that S-(-)-equol, at physiological concentrations ranging between 10 and 250 nmol/L, was able to increase nuclear factor-erythroid 2-related factor 2 (Nrf2) as well as gene products of Nrf2 target genes hemeoxygenase-1 (HO-1) and NAD(P)H quinone oxidoreductase 1 (NQO1). The authors also performed a wide array of confirmatory experiments, clarifying that equol estrogenic activity was involved in this process. Interestingly, pre-incubation of cells with S-(-)-equol significantly improved cell survival in response to H_2O_2 or *tert*-Butyl hydroperoxide and this effect was reduced in cells exposed to a siRNA against Nrf2. The effects of genistein and equol (in racemic form) on cell proliferation were examined in MCF-7 cells after exposition up to 72 h to different concentrations of the related isoflavones ranging from 0.01 to 50 μM [25]. Both isoflavones stimulated cell proliferation in a dose-dependent manner and treating the cells at concentrations <10 μM which could be reached at physiological level. However, the daidzein-derived metabolite equol exerted more efficiently a growth-stimulatory effect than genistein. The maximum stimulatory growth exhibited at 10 μM of both isoflavones was inhibited by the addition of an ERα antagonist (ICI or 4-hydroxytamoxifen). Interesting to note as the mitogen-activated protein extracellular kinase (MEK)/extracellular-signal-regulated kinase (ERK) blocked genistein- and equol-mediated cell proliferation when the specific inhibitor U0126 was added. Thus, these findings pinpointed as the cell proliferation mediated by both isoflavones involves an ERα-mediated mechanism and the MEK/ERK pathway [25]. Upon ingestion, soy-derived isoflavones circulated mainly as conjugated forms with glucuronic acid and sulfate groups. Thus, Pugazhendhi and colleagues [26] investigated the effects of daidzein-, genistein- and equol-*O*-sulfates on the estrogen agonist properties of these phase II metabolites in MCF-7 cells. Moreover, the authors evaluated the ability of these sulfate metabolites to compete with estradiol for binding to ER, and whether they are implicated in the cell growth-stimulation. In details, 4'-sulfation of genistein reduced estrogenic agonist activity to a small extent in whole-cell assays but increased the relative binding affinity to ER. The sulfation of genistein and equol in position 7, reduced oestrogen agonist activity substantially in all assays. On the contrary, the position of sulfate group of daidzein acted in an opposing manner on estrogen agonist activity. Daidzein-7-*O*-sulfate led an increase in estrogenic agonist activity, while sulfation at the 4'-position of daidzein led a modest reduction in estrogenic agonist activity [26]. Zhao and colleagues investigated the ERα/β binding profile of genistein, daidzein and its metabolite equol. It was evaluated the relative binding affinity of the tested compounds by comparing the percentage binding affinity of 17-β-estradiol, that was equal to 100%. Genistein exhibited a higher affinity for ERβ with an IC_{50} (concentration required to achieve 50% inhibition in the binding estradiol) close to 80 nM, as well daidzein that showed an IC_{50} value of 1.7 μM for ERβ. As observed for daidzein, its microbial metabolite equol showed more affinity for ERβ with a IC_{50} at 580 nM. Interesting to note as the combination of three isoflavones led to a greatest binding selectivity for ERβ and an overall improved efficacy/safety profile when compared with single or other combined formulations, observing an increase in ERβ-binding selectivity close to 30%. Moreover, neuroprotective effect of three isoflavones was tested in rat primary hippocampal neuronal cultures, pointing out as the combination of all three compounds showed the maximal estrogenic activity in this model. The improvement of neuronal cell viability against toxic insults in primary neurons was also observed [26]. Anthocyanidins are flavonoids widely dispersed in several edible plants, being particularly evident in fruits and vegetables where they are responsible for red, blue and purple colours. These coloured flavonoids are esterified with sugars or other compounds as organic and hydroxycinnamic acids, which are known as anthocyanins. Since anthocyanins are scarcely absorbed in the proximal GI tract, and thus reaching the colon, interact with resident microbiota which leads the formation of smaller phenolics that are then well absorbed *in situ*. Thus, the healthy properties related to anthocyanin intake should be assessed by using their microbial metabolites in experimental *in vitro* models. Protocatechuic, syringic, gallic, and vanillic acids, which are the main colon-derived anthocyanin catabolites, have been evaluated for their ability to perturb different parameters involved in atherosclerosis, including inflammation, cell adhesion, chemotaxis, endothelial function, estrogenic/anti-estrogenic activity and angiotensin-converting enzyme (ACE) inhibitory activity. Only protocatechuic acid exhibited a slight inhibitory effect on NO production and TNF-α secretion in LPS induced macrophages, whereas gallic acid caused a decrease in the secretion of MCP-1, ICAM-1, and VCAM-1 in endothelial cells [27]. It must be pointed out that the applied concentrations were always higher than 10 μmol/L, making these experiments of scarce physiological relevance.

Ellagitannins are polyphenols belonging to hydrolyzable tannins because they are readily broken down by dilute acid releasing gallic acid and/or ellagic acid. These tannins are widespread in *Rubus* berries, pomegranate and some nuts, especially walnuts and chestnuts [28]. Evidence suggests that these tannins, may decrease the risk of developing some estrogen-dependent diseases [13, 29]. Estrogenic/antiestrogenic activity via ER subtypes ERα and ERβ has been also proposed for ellagic acid [30]. However, dietary ellagitannins are generally not absorbed in the proximal GI tract. They release ellagic acid in the gut and this is poorly absorbed in the stomach and small intestine, and largely metabolized by bacteria in the intestinal lumen to produce urolithins. Microbial metabolism starts in the small intestine forming urolithins having three and four hydroxyls, which are further metabolized along the intestinal tract to remove hydroxyl units leading to urolithin A (two hydroxyls) and B (one hydroxyl) in the distal colon. The absorbed urolithins get into the liver through the portal system, where interact with phase II enzymes, thus leading the formation of conjugated metabolites, mainly *O*-glucuronides, which are the main ellagic acid-derived circulating metabolites [13]. Thus, considering the putative estrogenic and/or antiestrogenic activity of urolithins based on their chemical structure, Larrosa and colleagues assayed these microbial metabolites by using MCF-7 human breast cancer cells [31]. The potential estrogenic/activities of urolithins were compared to those of genistein, daidzein, resveratrol and enterolactone – which are well-known xenoestrogens. Genistein showed a comparable affinity for both ER isoforms, and a higher affinity for ERβ in respect to daidzein was observed. Instead, a strong difference in the affinity for both receptors emerged among urolithin A and B. Urolithin A had a IC_{50} values less than 1 μM unlike to urolithin B which had a IC_{50} values greater than 10 μM, thus revealing as the additional hydroxyl group owned by urolithin A affect considerably the affinity for both ERs. Moreover, the same authors measured the estrogenic activity of urolithins A and B, as the capacity to prevent or decrease the proliferation of MCF-7 cells in the presence of 1 pM estradiol. The lowest urolithin concentration that led a significant induction of cell proliferation was 1 μM. However, both urolithins exerted high estrogenic response at greater concentrations, starting from 10 μM and up to 40 μM. Instead, about antiestrogenic activity, that should be involved the prevention of the estradiol-induced cell proliferation, both urolithins showed slightly higher antiestrogenic activities than the other tested phytoestrogens polyphenols [31]. Since most of ERs are located inside the cells at cytosolic and nuclear level, Larrosa and colleagues investigated also the uptake and metabolism of urolithins by using the same cell lines. Besides aglycone, urolithin A was metabolized by MCF cells leading to the formation of urolithin A-*O*-sulfate, urolithin A-*O*-glucuronide and urolithin A-*O*-disulfate. Instead, six metabolites were detected after incubations with urolithin B. It has been metabolized into urolithin B-*O*-sulfate and urolithin B-*O*-glucuronide. Free urolithin B also undergoes hydroxylation forming urolithin A and one of its structural isomer, and both are then conjugated with one sulfate group [31]. *In vitro* studies should be carry out to unravel if the conjugation step would affect the agonist activity of the urolithins, even if glucuronidation and sulfation seemed to prevent the interaction with the closed pockets of ER by using *in silico* approach and likely deprived compounds of potential agonistic behaviour [32].

Contrary to urolithins, Ruotolo and colleagues revealed recently as the glucuronidation of quercetin boosts the agonist activity of this flavonol [33]. The authors evaluated the estrogenic activity of the main quercetin derivatives present *in planta* (isoquercitrin, rutin and free quercetin), along with that of its two major phase II circulating metabolites (quercetin-3-*O*-glucuronide and quercetin-3'-*O*-sulfate). Free quercetin did not exhibit ERα agonistic activity in yeast cells, but showed a weak agonistic activity with ERβ. Quercetin-3-*O*-glucuronide acted as an equally potent agonist for both receptors, at 100 μM for both ERs. Isoquercitrin displayed a ERβ agonist activity similar to that of quercetin-3-*O*-glucuronide, whereas rutin and quercetin-3'-*O*-sulfate showed no estrogenic activity with either receptor. All quercetin derivatives and metabolites were tested for their ability to inhibit E2-induced, ER-Tif2-dependent transactivation activity in yeast cells, but none of them exhibited any appreciable antiestrogenic activity. A weak estrogenic activity of quercetin-3-*O*-glucuronide was further confirmed in human MCF-7 cell lines as well those of aglycone, whereas no ER antagonistic activity of the tested compounds was observed by parallel antiestrogen activity assays conducted in the same cell line [33].

A recent study carried out by the same research groups, allowed to reveal as resveratrol-3-*O*-sulfate, which is one of the major phase II metabolites that appeared in bloodstream after intake of resveratrol, exhibited an antiestrogenic activity in an *in vitro* yeast model [34]. The sulfate metabolite exerted a pronounced antioestrogenic activity on both receptors, with a marked preference for ERα. In contrast, free resveratrol proved to be an extremely weak and apparently non ER selective antagonist. Resveratrol-3-*O*-glucoside, the main dietary resveratrol glycoside, had no estrogenic and antiestrogenic activity, as observed for resveratrol-3-*O*-glucuronide and resveratrol-4'-*O*-glucuronide, the main *in vivo* circulating glucuronides. As observed in the yeast assay, resveratrol-3-*O*-sulfate behaved as antagonist of estradiol in MCF-7 cells. The estradiol-dependent luciferase expression was reduced of the 40% using an external concentration of 50 μM. Instead, free resveratrol at 10 μM produced in MCF-7 cells a maximal level of agonistic activity two-fold higher, although much less potent, than that obtained at the maximal doses of estradiol. Moreover, at concentrations >1 μM, the aglycone increased the estradiol-dependent transcriptional activation in MCF-7 cells. Although the functional implications of the newly discovered and physiologically relevant anti-estrogenic activity of resveratrol-3-*O*-sulfate remain to be determined, it is interesting to note that estrogens have been reported to stimulate cellular proliferation by enhancing the production of various growth factors (e.g., transforming growth factor α (TGFα), insulin-like growth factor (IGF) and epidermal growth factor (EGF)), while down-regulating the expression of growth-inhibitory factors such as TGFβ [35].

4. Computer Simulations

In the past years, computer-driven investigation have been widely used to investigate polyphenolic compounds in many ways, thus revealing the versatility of such methods in investigating such a challenging class of bioactives. A brief overview on the possible endpoints reached by using *in silico* techniques is reported below.

4.1. Large-Scale Virtual Screening

The term virtual screening is usually referred to the attempt to identify bioactive molecules by using computational strategies. Traditionally, virtual screening studies have been used for the search of active molecules toward a given biological target. Large-scale screening is referred instead to the simultaneous screening of a wide array of biological target *versus* multiple molecules of interest. The possibility to investigate the bioactivity of molecules toward a wide set of biological targets is a promising strategy to better characterize the network of protein-ligand interactions and pathways underlying the polyphenol action. Even though they have been under study for decades, the spectrum of biological targets of polyphenols is still concealed for the most part. As a consequence, the modes of action are not well understood yet. Due to the pleiotropic action and the great deal of molecular targets underlying their biological effects, the unbiased search of protein partners via experimental procedures may be extremely challenging. Within this frame, the computational approach can be an affordable strategy that may provide a rational criterion to design workbench experiments in a focused manner. Depending on the reliability of *in silico* predictions, a number of abortive experimental trials can be avoided, thus saving time and money. Further, the current advances in hardware technology and in algorithm development make affordable the simultaneous analysis of thousands of queries.

For instance, Zheng and coworkers [36] focused on the antineoplastic activity of epigallocatechin-3-*O*-gallate, epigallocatechin, epicatechin-3-*O*-gallate and epicatechin which are part of the main functional components of the green tea. Albeit antineoplastic activity for epigallocatechin-3-*O*-gallate has been found, the modes of action – intended as the network of molecular targets involved – still remain unclear. Moreover, data in the literature are conflicting (e.g., ref. [37-39]) thus concurring to blur the real scenario. Furthermore, the most of data available so far concern solely epigallocatechin-3-*O*-gallate whilst data on other catechins are quite scarce. Therefore, the comparative search of molecular targets for this class of compounds is a goal of great relevance, even if challenging. In this framework the use of *in silico* screening is a straightforward choice. In particular, while canonical virtual screening protocols – intended as the computational screening of huge libraries of queries – involve the search of active molecules toward a single target, the authors proposed a multi-target reverse-docking study. Reverse docking indicates docking protocols where one or more small-molecules are docked into the binding cavities (or catalytic sites) of a broad set of clinically relevant macromolecular targets. Reverse docking already proved to be a powerful tool for drug rescue and repositioning [40] and it may concur to investigate the complex network of polyphenols' targets as well. In particular, Zheng used a comprehensive network database that contains more than 1100 3D protein structures and covers 15 therapeutic areas including tumor therapy (namely, the Potential Drug Target Database; http://www.dddc.ac.cn/pdtd; [41]). By using two separate workflows involving AutoDock software and TarFisDock web server, the authors identified a series of putative molecular targets, some of which were already known targets for the considered molecules (e.g., HIV protease, glutathione reductase and catalase). Molecular targets were ranked on the basis of the predicted binding affinity and the most relevant candidates were selected accordingly. Moreover the binding architecture of epigallocatechin-3-*O*-gallate in complex with Leukotriene A4 hydrolase – which is among the strongest targets – was detailed, bestowing on electrostatic interactions and hydrogen bonds a pivotal role in the binding event. Whit the aim to better understand the network of polyphenols interaction, the work appeared as a solid starting point to discover novel targets, albeit the impact from physiological point of view was limited. Indeed the metabolized forms, which are the truly circulating molecules in living organism, were neglected throughout.

Similarly, Chen and co-workers [42] developed a computational methodology to gain further understanding of anticancer and therapeutic effects exerted by flavonoids. Specifically, the authors showed how composite computational methods involving virtual screening, docking simulations and pharmacophore analysis attempt to search ligands for biological targets and vice versa. Notably, the possibility to put in act cross-search onto thousands of phytochemicals and proteins has been pointed out, thus providing a possible and affordable framework for simultaneously discovering novel relevant flavonoids and protein targets in cancer chemoprevention and therapy.

Besides these high-throughput studies, a great deal of less extensive investigations can be found in the literature. For instance, Phosrithong and Ungwitayatorn [43] addressed the computational search of biological targets involved in cell cycle, cell growth, and DNA replication for 29 plant derived compounds including several polyphenols and one synthetic structure. A total of 7 biological targets involved in tumor maintenance and growth were investigated, i.e., cyclin-dependent protein kinase 2 and 6, DNA topoisomerases I and II, B-cell lymphoma 2, vascular endothelial growth factor receptor 2 and the telomere: G-quadruplex. Even if limited in the number, the authors proposed a straightforward strategy to effectively identify new putative molecular targets by using a validated docking approach. The need for structural insights in order to better understand the mode of action of bioactive molecules was emphasized. Owing to the limited number of hits, it was possible to structurally detail the mode of interaction of the strongest predicted complexes. Although the fine description of ligand-target interaction is challenging when large-scale screening is performed, it should be kept in mind that it is paramount to better understand the structural features underlying the biological action.

4.2. Target-Specific Virtual Screening

Aside to the large-scale virtual screening, the interaction of sets of molecules with a unique target of interest can be searched by using target-specific virtual screening procedures. This kind of study is a paramount early stage for the discovery of novel active compounds toward a given pharmacological target of interest. In fact, owing to the broad spectrum of biological effects, in the last decades a huge number of polyphenolic compounds has been independently investigated for its pharmacological properties toward a great variety of molecular endpoints. For instance, Shan and co-workers [44] recently investigated the inhibition of polo-like kinase 1 (PLK1) by green tea catechins. PLK1 enzyme plays crucial functions in multiple stages of mitosis. Its overexpression is observed in cancers and it may correlate with poor prognosis [45]. Interestingly, the functional interferences mediated by small-molecule inhibitors selectively reduce the growth of cancer cells [46] with no effect on normal cell lines [47]. For this reason PLK1 is a target of great pharmacological relevance which can be considered for cancer therapy. The enzyme is organized into two domains with different functions, i.e., the N-terminal kinase domain (KD), which is similar to domain shared by kinase family, and the C-terminal polo-box (PBD), which is specific for PLK class. Due to the high degree of analogy, inhibitors targeting the KD can be affected by low specificity. Conversely, PBD diverges much more along the class and it can be accounted for a more specific inhibition. Over the past years a series of PDB inhibitors have been identified, but the modest selectivity and, even more, the scarce drug-likeness of these compounds highlighted the need for the discovery of novel and more effective PBD inhibitors. By combining virtual screening and fluorescence polarization screening, the authors screened a database containing almost 5 million of compounds. In particular, by using a modular virtual screening involving docking simulation, structures clustering and molecular dynamics optimizations, 50 lead compounds were identified and then subjected to experimental assay. Among them, (-)-epigallocatechin was found as a strong PBD binder. Next, some major green tea catechins were evaluated for their dose-dependent inhibition of the PBD (i.e., (-)-epigallocatechin-3-*O*-gallate, (+)-gallocatechin and (+)-catechin) and preliminary structure–activity relationships were derived. In particular it was experimentally found that additional galloyl moiety gains interactions whilst the stereochemistry of hydroxyl group in (+)-

gallocatechin determines a decreased interaction respect to (-)-epigallocatechin. Similarly, the reduced interaction of (+)-catechin proves the importance of hydroxyl groups on the B ring. Docking studies rationalized the structure-activity relationship by revealing a similar mode of interaction for all catechins. Notably, the presence of galloyl moiety of (-)-epigallocatechin-3-*O*-gallate gained further hydrogen bond and π-π interaction. Conversely the loss of hydrogen bond and the interatomic interferences due to the configuration of 3-hydroxyl group were at the basis of the reduced activity of (+)-gallocatechin and (+)-catechin, respectively. Moreover the results of fluorescence polarization assay indicated that (-)-epigallocatechin is more selective against the PBD of PLK1 than those of PLK2 and PLK3. Also, it was found that (-)-epigallocatechin led to the cell-cycle arrest in the S and G2M phases, and induced growth inhibition of several cancer cell types such as breast, lung and cervical adenocarcinoma cells. Overall the authors presented a precious contribution to better understand the anticancer activities of green tea catechins. Further, the results can be a solid starting point to design novel PLK1 inhibitors against the C-terminal polo-box based on the catechin scaffold.

Instead, within the framework to identify alternative active molecules to reduce the adverse effects of drugs, Islam and co-workers addressed the case study of acetylcholine-esterase (AchE) inhibition by quercetin [48]. Alzheimer's disease is a serious neurodegenerative illness which affects more than 35 million people worldwide [49]. The main cause of Alzheimer's disease is the genesis and deposition of abnormal proteins (i.e., amyloid plaque) which determine oxidative and inflammatory damage and leading to extensive neuronal loss [50]. Notably, the depletion of the neurotransmitter acetylcholine due to reduced number of acetylcholine receptor and increased activity of acetylcholine-esterase are considered as prominent features of Alzheimer's disease. The inhibition of acetylcholine-esterase serves to restore neurotransmission, and inhibitors subsequently play a role in the reduction of amyloid plaque deposition. Moreover, some inhibitors further preserve neuronal loss by exerting antioxidant effects. Nevertheless, the drugs commonly used in Alzheimer's disease treatment show a number of side effects, including hepatotoxicity and the onset of cardiovascular disorders [51]. Interestingly, the flavonoid quercetin showed acetylcholine-esterase inhibitory activity [52]. Moreover, its antioxidant properties are comparable to those of the current inhibitory drugs thus being a strong candidate for drug replacement. Islam proposed an *in silico* comparison between quercetin and the commonly used acetylcholine-esterase inhibitor drugs (namely tacrine, galantamine, rivastigmine and donepezil) [48]. The authors performed an *in silico* quantitative structure-activity relationship (QSAR) study of quercetin to investigate the functional groups that may help to design a better inhibitor of acetylcholine-esterase. Overall, the analysis pointed the high superiority of quercetin respect to the currently used drugs for acetylcholine-esterase inhibition. Furthermore, the methylated form of quercetin at 4-OH position showed a higher drug likeness. Taken together these *in silico* findings set the need for more detailed *in vitro* study for this class of compound, which in the next future might be promising compounds in Alzheimer's disease treatment. Nevertheless further ADMET (Absorption, Distribution, Metabolism, Excretion and Toxicity) investigations are needed for truly accounting quercetin and derivatives for drug replacement.

Again, in the contest of using quantitative structure-activity relationship analysis, Hamzeh-Mivehroud and co-workers [53] addressed the inhibition of aldehyde oxidase enzyme. Such enzyme is involved in the biotransformation of many drugs and xenobiotics and polyphenolic compounds may affect the availability of pharmaceuticals via its inhibition. Consequently, understanding the polyphenols' mode of interaction could be of great interest.

Starting from experimental evaluation of inhibition of several polyphenolic compounds including flavonols, flavanols, flavanones, flavones, and isoflavones, the authors applied QSAR analysis to grab the main structural features underlying the inhibitory properties. Also, they performed docking studies to structurally elucidate the architecture of interaction. Overall, it was found that the size of molecule is an important feature and smaller flavonoids were better inhibitors. Moreover, the authors stated that such an effective QSAR model can be used more broadly to predict the inhibitory activity of flavonoids for application in food–drug and drug–drug interaction studies.

Whereas the majority of *in silico* studies neglect the role of metabolites, Dellafiora and colleagues recently proposed a computational framework to address the potential effect of both human and plant metabolism on polyphenol action toward human estrogen receptors [32] and human topoisomerase I [54]. Indeed, the in-depth analysis of metabolite-dependent effects is paramount for better understanding the role of parent compound in living organism. The estrogen receptors are transcriptional factors that mediate many of the effects of estrogens, but a series of exogenous compounds, including foodborne polyphenols, are able to modulate their activation. For this reason, estrogen receptors are relevant targets in the context of dietary modulation of health. To this end, phytoestrogens are a diverse group of plant-derived chemicals that structurally and functionally mimic estrogens and have been associated with potential benefits for human health, in particular in the context of breast and prostate cancers, cardiovascular diseases, menopausal symptoms, and osteoporosis [55]. Evidence suggests that ellagitannins, a complex subclass of hydrolysable tannins present in berries, walnuts, pomegranate, and oak-aged red wines, may decrease the risk of developing some estrogen-dependent diseases [29, 56]. After consumption, such compounds are hydrolyzed into ellagic acid, which is subsequently metabolized into urolithins by the human gut microbiota [57]. Urolithin A and urolithin B exhibited phytoestrogenic activities on the estrogen-sensitive human breast cancer cells MCF-7, and their ability to bind with both the two isoforms of estrogen receptor has been described [31]. However, urolithins appear in the circulatory system almost exclusively as glucuronide, sulfate and methylated forms as a result of phase II metabolism after absorption in the colon and passage through the liver [58].

Consequently, in addition to aglycones, the vast array of ellagitannin-derived phase II conjugated urolithins, which are the real forms present in the systemic circulation after dietary intake of ellagitannins, should be the target of investigation. A total of 28 ellagitannin-derived metabolites were thus analyzed for their putative estrogenic activity by using an *in silico* procedure involving structure-based virtual screening followed by docking simulations. While all glycoconjugated and sulfated forms were predicted as inactive, a number of methylated metabolites showed appreciable activity toward both estrogen receptor isoforms. Overall, the authors pointed the inactivating role of conjugation and identified hydroxyl group in position 3 as major structural feature for the maintenance of activity. On the other side, topoisomerase is a key enzyme in DNA topology that modulates the state of supercoiling and intertwining. Since topoisomerase enzymes are involved in key steps of cell viability, inhibitors and poisons are widely used as antibacterial and anticancer agents [59]. Both luteolin and quercetin have been reported to inhibit topoisomerase enzymes through poisoning mechanism [60-62], thus supporting their possible chemopreventive activity. After consumption, both compounds undergo to massive phase-II metabolism and thus it would be their glucuronide and sulfate metabolites that might act as potential chemopreventive. However, the role of these circulating metabolites in poisoning topoisomerase enzymes is still understudied. Also, considering that luteolin and quercetin are widely consumed through a normal balanced diet, it is relevant to investigate whether their phase II metabolites are able to act as poisons. Therefore, a total of 11 phase-II metabolites of luetolin and quercetin collected from the literature were analyzed for their poisoning potential by using a knowledge-based strategy involving docking simulation and re-scoring procedure. Strikingly, the most of glucuronidated forms showed high poisoning potential thus strongly suggesting a relevant role of luteolin and quercetin under *in vivo* conditions.

4.3. Predictive Metabolism

The health-protective effects of polyphenols are under study for decades. However, it is becoming increasingly accepted that they may be substrates of cytochrome P450 enzymes and undergo bioactivation to metabolites that likely mediate the most of the biological effects in living organisms. Nonetheless, the metabolism of flavonoids is complex and not completely understood yet and further data on the metabolite-dependent network of biological interactions are also needed. Thus, the profiling of metabolic patterns would be relevant to provide a more informed scenario. Unlikely, the untargeted analysis of metabolomic profiles is still extremely challenging. For this reason, predictive metabolism is a goal of great interest. Two main kinds of approach are currently available for predictive metabolism: ligand-based approaches and structure-based approaches (e.g., ref. [63, 64]). The formers are mainly based on the chemical reactivity of molecules in terms of shape, conformation, and the electronic properties. The latters instead take into account the binding between molecules and CYP450 enzymes thus considering the susceptibility of chemical groups to be properly oriented in respect to the cytochromes' catalytic site.

For example, an intriguing framework addressing the case study of quercetin, rutin, naringenin, and naringin was proposed by Melo-Filho and colleagues [65]. The authors predicted a series of possible phase-I metabolites by using *in silico* predictors and also the ability to interact with the cytochrome CYP2C9 was investigated trough docking simulation. Notably, an integrated use of ligand-based and structure-based predictors was chosen thus obtaining a more comprehensive analysis. The procedure reliability was confirmed by the fact that at least one metabolite for each compounds was also experimentally found. Further, the toxicity prediction of metabolites was performed by searching mutagenic, tumorigenic, irritating, and reproductive effects. Intriguingly, some metabolites of quercetin and rutin presented mutagenic and tumorigenic risks. Overall, the work presented might be a noteworthy first step toward the achievement of a common framework for functional analysis of metabolomes.

4.4. Comprehension of Mechanisms

The advantage of using *in silico* approaches goes beyond the cost-effective screening of huge libraries of compounds and targets. Indeed, it should be kept in mind that in most cases the initiating event underlying the molecules action is not only a consequence of host-guest interaction, but it also involves the stabilization of particular organization states of protein targets. For this reason, the elucidation of host-guest binding architecture can be strongly advisable.

In this framework, Benedik and collaborators [66] addressed the interaction of (-)-epigallocatechin-3-*O*-gallate, (-)-epigallocatechin, kaempferol-3-*O*-glucoside, and kaempferol with Staphylococcal enterotoxin B, which is an exotoxin responsible for food poisoning. Starting from the experimental evidence of such interaction, and in the lack of previous information about sites and architectures of binding, the authors investigated the possible interactions map by using a blind-docking approach. Several putative binding sites were observed and then analyzed to find the most plausible. Interestingly, (-)-epigallocatechin-3-*O*-gallate showed the highest binding affinity. According to *in silico* findings, the authors reported that (-)-epigallocatechin-3-*O*-gallate and similar ligands may bind to the channel at the surface of Staphylococcal enterotoxin B that is responsible for the recognition of the T-cell beta chain fragment and influence the adhesion of Staphylococcal enterotoxin B. On the one side, these results propose a plausible mechanism by which polyphenols may prevent the interaction between enterotoxin and the T cells, thus reducing the production of proinflammatory cytokines and chemokines. On the other side, when the binding architecture is defined, compounds may be modified accordingly to enhance the pattern of interaction.

With a more specific focus, Pandey and collaborators [67] presented instead an *in silico* strategy to explore neuroprotection by quercetin in cerebral ischemia. The neuroprotective role of quercetin has already been established but the mechanism of action is not well understood. Cerebral ischemia is a complex injury involving a great number of events and molecular actors, which leads to cellular extravasation and loss of microvascular integrity. It results in edema, hemorrhagic transformation, activation of resident microglial cells, infiltration of circulating inflammatory cells into the brain and neuronal death [68]. Matrix metalloproteinases and acid sensing ion channels are key proteins for the damage evolution which undergo several regulatory steps including transcriptional and enzymatic inhibition, and indirect effects by regulating the endogenous inhibitors or activators. The authors docked the quercetin within matrix metalloproteinase-2 and -9, and acid sensing ion channel-1a. Since relevant interactions with the catalytic sites were observed, they hypothesized the inhibitions of these proteins as a possible mechanism underlying protective effects.

It is relevant to note that the potentiality of *in silico* studies is not limited to the prediction of protein-ligand interaction. For examples, Košinová and co-workers [69] proposed an intriguing study that addressed the disposition of polyphenols within lipid bilayer. Among the numerous biological activities, polyphenols are strong antioxidant compounds capable of variably inhibiting lipid peroxidation. Positioning of polyphenols in lipid bilayers can provide an insight to the lipid-peroxidation inhibition at the molecular level. The authors indeed determined the location and orientation of quercetin and glucuronidated, methylated, and sulfated metabolites in the lipid bilayer via molecular dynamic simulations. They showed that the degree of penetration and orientation strongly depends by molecules polarity, charge, and hydrogen bonds capacity. In particular, polar interactions with the polar edges of the bilayer may play a relevant role. The partial decrease of lipid peroxidation inhibition observed for 3-*O*'-methylquercetin and quercetin-3-*O*-glucuronide compared to quercetin can be partially explained by their location and orientation in the lipid bilayer that is closer to surface in respect to quercetin. As a consequence, they may be less efficient free-radical scavengers and, further, the regeneration of antioxidants incorporated in the membrane might be reduced. For quercetin-3'-*O*-sulfate, the charged sulfate group pulled the compound out of lipid bilayer, thus explaining its weaker inhibition activity compared to quercetin-3-*O*-glucuronide.

Conclusion

Summarizing, polyphenolic compounds belong to an intriguing sub-class of naturally occurring molecules which may have strong and positive effects on human health and wellbeing. Nevertheless, the comprehension of their mechanisms of action is still unripe, and, due to the wealth of biological targets, their analysis through the canonical experimental investigation may be intrinsically challenging. In this framework, the *in silico* analysis may determine an important step forward toward the in-depth comprehension of the modes of action of polyphenols. It is worthy to note that, in the recent years, a wealth of high quality computational works addressed the case study of polyphenols with various degree of complexity. Actually, computers and digital supports are replacing to such an extent the pencil and notebook – and in some cases the workbench – as the scientists' primary tool [70] that several scientific areas are permeated by computational science. The investigation of bioactive polyphenolic compounds is one of these areas, and it is reasonable to think that in the next future it will continue to take advantages of the computational support.

After all, it should be kept in mind that the investigation of crosstalk between polyphenols and living organisms requires a holistic approach, which inevitably involves the merge and the coordination of multiple branches of knowledge. Nevertheless, the untargeted investigations by using uniquely the experimental practice may be considerably demanding in terms of time and money. Ideally, an upstream framework to drive experimental investigation is strongly advisable, thus avoiding as much as possible to get into red herrings. To this end, the use of computational models can be a straightforward and promising choice.

References

[1] Ezzati, M., Lopez, A. D., Rodgers, A., Vander Hoorn, S., Murray, C. J. L., Selected major risk factors and global and regional burden of disease. *The Lancet,* 2002, 360, 1347-1360.

[2] Harborne, J. B., Baxter, H., Phytochemical Dictionary: A Handbook of Bioactive Compounds from Plants, Taylor & Francis, London, UK 1993.

[3] Manach, C., Scalbert, A., Morand, C., Remesy, C., Jimenez, L., Polyphenols: food sources and bioavailability. *Am. J. Clin. Nutr.*, 2004, 79, 727-747.

[4] Manach, C., Williamson, G., Morand, C., Scalbert, A., Remesy, C., Bioavailability and bioefficacy of polyphenols in humans. I. Review of 97 bioavailability studies. *Am. J. Clin. Nutr.,* 2005, 81, 230S-242S.

[5] Arts, I. C., Hollman, P. C., Polyphenols and disease risk in epidemiologic studies. *Am. J. Clin. Nutr.,* 2005, 81, 317S-325S.

[6] Hertog, M. G., Feskens, E. J., Hollman, P. C., Katan, M. B., Kromhout, D., Dietary antioxidant flavonoids and risk of coronary heart disease: the Zutphen Elderly Study. *Lancet,* 1993, 342, 1007-1011.

[7] Hertog, M. G., Kromhout, D., Aravanis, C., Blackburn, H., et al., Flavonoid intake and long-term risk of coronary heart disease and cancer in the seven countries study. *Arch. Intern. Med.,* 1995, 155, 381 - 386.

[8] Crozier, A., Jaganath, I. B., Clifford, M. N., Dietary phenolics: chemistry, bioavailability and effects on health. *Nat. Prod. Rep.,* 2009, 26, 1001-1043.

[9] Erdman, J. W., Jr., Balentine, D., Arab, L., Beecher, G., et al., Flavonoids and heart health: proceedings of the ILSI North America Flavonoids Workshop, May 31-June 1, 2005, Washington, DC. *J. Nutr.,* 2007, 137, 718S-737S.

[10] Burns, J., Yokota, T., Ashihara, H., Lean, M. E., Crozier, A., Plant foods and herbal sources of resveratrol. *J. Agric. Food Chem.,* 2002, 50, 3337-3340.

[11] Rodriguez-Mateos, A., Vauzour, D., Krueger, C. G., Shanmuganayagam, D., et al., Bioavailability, bioactivity and impact on health of dietary flavonoids and related compounds: an update. *Arch. Toxicol.*, 2014, 88, 1803-1853.

[12] Day, A. J., Canada, F. J., Diaz, J. C., Kroon, P. A., et al., Dietary flavonoid and isoflavone glycosides are hydrolysed by the lactase site of lactase phlorizin hydrolase. *FEBS Lett.,* 2000, 468, 166-170.

[13] Crozier, A., Del Rio, D., Clifford, M. N., Bioavailability of dietary flavonoids and phenolic compounds. *Mol. Aspects Med.,* 2010, 31, 446-467.

[14] Selma, M. V., Espin, J. C., Tomas-Barberan, F. A., Interaction between phenolics and gut microbiota: role in human health. *J. Agric. Food Chem.*, 2009, 57, 6485-6501.

[15] Del Rio, D., Costa, L. G., Lean, M. E., Crozier, A., Polyphenols and health: what compounds are involved? *Nutr. Metab. Cardiovasc. Dis.,* 2010, 20, 1-6.

[16] Ososki, A. L., Kennelly, E. J., Phytoestrogens: a review of the present state of research. *Phytother. Res.*, 2003, 17, 845-869.

[17] Setchell, K. D., Brown, N. M., Lydeking-Olsen, E., The clinical importance of the metabolite equol-a clue to the effectiveness of soy and its isoflavones. *J. Nutr.,* 2002, 132, 3577-3584.

[18] Cosentino, M., Marino, F., Ferrari, M., Rasini, E., et al., Estrogenic activity of 7-hydroxymatairesinol potassium acetate (HMR/lignan) from Norway spruce (Picea abies) knots and of its active metabolite enterolactone in MCF-7 cells. *Pharmacol. Res.,* 2007, 56, 140-147.

[19] Penttinen, P., Jaehrling, J., Damdimopoulos, A. E., Inzunza, J., et al., Diet-derived polyphenol metabolite enterolactone is a tissue-specific estrogen receptor activator. *Endocrinology,* 2007, 148, 4875-4886.

[20] Johnsen, N. F., Olsen, A., Thomsen, B. L., Christensen, J., et al., Plasma enterolactone and risk of colon and rectal cancer in a case-cohort study of Danish men and women. *Cancer Causes Control,* 2010, 21, 153 - 162.

[21] Carreau, C., Flouriot, G., Bennetau-Pelissero, C., Potier, M., Enterodiol and enterolactone, two major diet-derived polyphenol metabolites have different impact on ERalpha transcriptional activation in human breast cancer cells. *J. Steroid Biochem. Mol. Biol.*, 2008, 110, 176-185.

[22] Setchell, K. D., Clerici, C., Equol: pharmacokinetics and biological actions. *J. Nutr.,* 2010, 140, 1363S-1368S.

[23] Kelly, L. A., Seidlova-Wuttke, D., Wuttke, W., O'Leary, J. J., Norris, L. A., Estrogen receptor alpha augments changes in hemostatic gene expression in HepG2 cells treated with estradiol and phytoestrogens. *Phytomedicine,* 2014, 21, 155-158.

[24] Zhang, T., Liang, X., Shi, L., Wang, L., et al., Estrogen receptor and PI3K / Akt signaling pathway involvement in S-(-)equol-induced activation of Nrf2 / ARE in endothelial cells. *PLoS One,* 2013, 8, e79075.

[25] Liu, H., Du, J., Hu, C., Qi, H., et al., Delayed activation of extracellular-signal-regulated kinase 1/2 is involved in genistein- and equol-induced cell proliferation and estrogen-receptor-alpha-mediated transcription in MCF-7 breast cancer cells. *J. Nutr. Biochem.,* 2010, 21, 390-396.

[26] Pugazhendhi, D., Watson, K. A., Mills, S., Botting, N., et al., Effect of sulphation on the oestrogen agonist activity of the phytoestrogens genistein and daidzein in MCF-7 human breast cancer cells. *Journal of Endocrinology,* 2008, 197, 503-515.

[27] Hidalgo, M., Martin-Santamaria, S., Recio, I., Sanchez-Moreno, C., et al., Potential anti-inflammatory, anti-adhesive, anti / estrogenic, and angiotensin-converting enzyme inhibitory activities of anthocyanins and their gut metabolites. *Genes and Nutrition,* 2012, 7, 295-306.

[28] Mena, P., Calani, L., Bruni, R., Del Rio, D., in: Rio, K. T. D. (Ed.), Diet-Microbe Interactions in the Gut, Academic Press, San Diego 2015, pp. 73-101.

[29] Aiyer, H. S., Warri, A. M., Woode, D. R., Hilakivi-Clarke, L., Clarke, R., Influence of berry polyphenols on receptor signaling and cell-death pathways: implications for breast cancer prevention. *J. Agric. Food Chem.*, 2012, 60, 5693-5708.

[30] Papoutsi, Z., Kassi, E., Tsiapara, A., Fokialakis, N., et al., Evaluation of estrogenic / antiestrogenic activity of ellagic acid via the estrogen receptor subtypes ERα and ERβ. *Journal of Agricultural and Food Chemistry,* 2005, 53, 7715-7720.

[31] Larrosa, M., González-Sarrías, A., García-Conesa, M. T., Tomás-Barberán, F. A., Espín, J. C., Urolithins, ellagic acid-derived metabolites produced by human colonic microflora, exhibit estrogenic and antiestrogenic activities. *J. Agric. Food Chem.*, 2006, 54, 1611-1620.

[32] Dellafiora, L., Mena, P., Cozzini, P., Brighenti, F., Del Rio, D., Modelling the possible bioactivity of ellagitannin-derived metabolites. In silico tools to evaluate their potential xenoestrogenic behavior. *Food Funct.,* 2013, 4, 1442-1451.

[33] Ruotolo, R., Calani, L., Brighenti, F., Crozier, A., et al., Glucuronidation does not suppress the estrogenic activity of quercetin in yeast and human breast cancer cell model systems. *Archives of Biochemistry and Biophysics,* 2014, 559, 62-67.

[34] Ruotolo, R., Calani, L., Fietta, E., Brighenti, F., et al., Anti-estrogenic activity of a human resveratrol metabolite. *Nutrition, Metabolism and Cardiovascular Diseases,* 2013, 23, 1086-1092.

[35] Steiner, M. S., Raghow, S., Antiestrogens and selective estrogen receptor modulators reduce prostate cancer risk. *World Journal of Urology,* 2003, 21, 31-36.

[36] Zheng, R., Chen, T. S., Lu, T., A comparative reverse docking strategy to identify potential antineoplastic targets of tea functional components and binding mode. *Int. J. Mol. Sci.,* 2011, 12, 5200-5012.

[37] Adachi, S., Nagao, T., To, S., Joe, A. K., et al., (−)-Epigallocatechin gallate causes internalization of the epidermal growth factor receptor in human colon cancer cells. *Carcinogenesis,* 2008, 29, 1986–1993.

[38] Tachibana, H., Koga, K., Fujimura, Y., Yamada, K., A receptor for green tea polyphenol EGCG. *Nat. Struct. Mol. Biol.,* 2004, 11, 380-381.

[39] Umeda, D., Yano, S., Yamada, K., Tachibana, H., Green tea polyphenol epigallocatechin-3-gallate signaling pathway through 67-kDa laminin receptor. *J. Biol. Chem.*, 2008, 283, 3050–3058.

[40] Kharkar, P. S., Warrier, S., Gaud, R. S., Reverse docking: a powerful tool for drug repositioning and drug rescue. *Future Med. Chem.*, 2014, 6, 333-342.

[41] Gao, Z., Li, H., Zhang, H., Liu, X., et al., PDTD: A web-accessible protein database for drug target identification. *BMC Bioinformatics,* 2008, 9, 104–110.

[42] Chen, H., Yao, K., Nadas, J., Bode, A. M., et al., Prediction of molecular targets of cancer preventing flavonoid compounds using computational methods. *pLoS One,* 2012, 7, e38261.

[43] Phosrithong, N., Ungwitayatorn, J., Molecular docking study on anticancer activity of plant-derived natural products. *Med. Chem. Res.,* 2010, 19, 817-835.

[44] Shan, H. M., Shi, Y., Quan, J., Identification of Green Tea Catechins as Potent Inhibitors of the Polo-Box Domain of Polo-Like kinase 1. *Chem. Med. Chem.*, 2015, 10, 158-163.

[45] Strebhardt, K., Ullrich, A., Targeting polo-like kinase 1 for cancer therapy. *Nat. Rev. Cancer,* 2006, 6, 321-330.

[46] Steegmaier, M., Hoffmann, M., Baum, A., Lénárt, P., et al., BI 2536, a potent and selective inhibitor of polo-like kinase 1, inhibits tumor growth in vivo. *Curr. Biol.,* 2007, 17, 316-322.

[47] Liu, X., Lei, M., Erikson, R. L., Normal cells, but not cancer cells, survive severe Plk1 depletion. *Mol. Cell. Biol.*, 2006, 26, 2093-2108.

[48] Islam, M. R., Zaman, A., Jahan, I., Chakravorty, R., Chakraborty, S., In silico QSAR analysis of quercetin reveals its potential as therapeutic drug for Alzheimer's disease. *Journal of Young Pharmacists,* 2013, 5, 173-179.

[49] Querfurth, H. W., LaFerla, F. M., Alzheimer's disease. *N. Engl. J. Med.,* 2010, 362, 329-344.

[50] Gómez-Isla, T., Hollister, R., West, H., Mui, S., et al., Neuronal loss correlates with but exceeds neurofibrillary tangles in Alzheimer's disease. *Ann. Neurol.*, 1997, 41, 17-24.

[51] Kröger, E., Berkers, M., Carmichael, P. H., Souverein, P., et al., Use of rivastigmine or galantamine and risk of adverse cardiac events: a database study from the Netherlands. *The Journal of Geriatric Pharmacotherapy*, 2012, 10, 373-380.

[52] Jung, M., Park, M., Acetylcholinesterase inhibition by flavonoids from Agrimonia pilosa. *Molecules,* 2007, 12, 2130-2139.

[53] Hamzeh-Mivehroud, M., Rahmani, S., Feizi, M. A., Dastmalchi, S., Rashidi, M. R., In vitro and in silico studies to explore structural features of flavonoids for aldehyde oxidase inhibition. *Arch. Pharm.,* 2014, 347, 738-747.

[54] Dellafiora, L., Mena, P., Del Rio, D., Cozzini, P., Modeling the effect of phase II conjugations on topoisomerase I poisoning: pilot study with luteolin and quercetin. *J. Agric. Food Chem.,* 2014, 62, 5881-5886.

[55] Ososki, A. L., Kennelly, E. J., Phytoestrogens: a review of the present state of research. *Phytother. Res.*, 2003, 17, 845-869.

[56] Larrosa, M., García-Conesa, M. T., Espín, J. C., Tomás-Barberán, F. A., Ellagitannins, ellagic acid and vascular health. *Mol. Aspects. Med.,* 2010, 31, 513-539.

[57] Cerdá, B., Tomás-Barberán, F. A., Espín, J. C., Metabolism of antioxidant and chemopreventive ellagitannins from strawberries, raspberries, walnuts, and oak-aged

wine in humans: identification of biomarkers and individual variability. *J. Agric. Food Chem.*, 2005, 53, 227-235.

[58] González-Barrio, R., Truchado, P., Ito, H., Espín, J. C., Tomás-Barberán, F. A., UV and MS identification of Urolithins and Nasutins, the bioavailable metabolites of ellagitannins and ellagic acid in different mammals. *J. Agric. Food Chem.,* 2011, 59, 1152-1162.

[59] Pommier, Y., Drugging topoisomerases: lessons and challenges. *ACS Chem. Biol.,* 2013, 8, 82-95.

[60] Bandele, O. J., Osheroff, N., Bioflavonoids as poisons of human topoisomerase II alpha and II beta. *Biochemistry,* 2007, 46, 6097-6108.

[61] Boege, F., Straub, T., Kehr, A., Boesenberg, C., et al., Selected novel flavones inhibit the DNA binding or the DNA religation step of eukaryotic topoisomerase I. *J. Bio.l Chem.*, 1996, 271, 2262-2270.

[62] Burgos-Morón, E., Calderón-Montaño, J. M., Orta, M. L., Pastor, N., et al., The coffee constituent chlorogenic acid induces cellular DNA damage and formation of topoisomerase I- and II-DNA complexes in cells. *J. Agric. Food Chem.,* 2012, 60, 7384-7391.

[63] Braga, R. C., Andrade, C. H., QSAR and QM/MM approaches applied to drug metabolism prediction. *Mini Rev. Med. Chem.,* 2012, 12, 573-582.

[64] Stjernschantz, E., Vermeulen, N. P., Oostenbrink, C., Computational prediction of drug binding and rationalisation of selectivity towards cytochromes P450. *Expert Opinion On Drug Metabolism and Toxicology,* 2008, 4, 513-527.

[65] Melo-Filho, C. C., Braga, R. C., Andrade, C. H., Advances in methods for predicting phase I metabolism of polyphenols. *Curr. Drug. Metab.,* 2014, 15, 120-126.

[66] Benedik, E., Skrt, M., Podlipnik, C., Ulrih, N. P., Binding of flavonoids to staphylococcal enterotoxin B. *Food Chem. Toxicol.*, 2014, 74, 1-8.

[67] Pandey, A. K., Verma, S., Bhattacharya, P., Paul, S., et al., An in-silico strategy to explore neuroprotection by quercetin in cerebral ischemia: a novel hypothesis based on inhibition of matrix metalloproteinase (MMPs) and acid sensing ion channel 1a (ASIC1a). *Med. Hypotheses,* 2012, 79, 76-81.

[68] Hamann, G. F., Liebetrau, M., Martens, H., Burggraf, D., et al., Microvascular basal lamina injury after experimental focal cerebral ischemia and reperfusion in the rat. *J. Cereb. Blood Flow Metab.,* 2002, 22, 526-533.

[69] Košinová, P., Berka, K., Wykes, M., Otyepka, M., Trouillas, P., Positioning of antioxidant quercetin and its metabolites in lipid bilayer membranes: implication for their lipid-peroxidation inhibition. *J. Phys. Chem. B.,* 2012, 116, 1309-1318.

[70] Post, D. E., Votta, L. G., Computational Science Demands a New Paradigm. *Physics Today,* 2005, 58, 35-41.

In: From Medicinal Chemistry to Food Science
Editor: Pietro Cozzini
ISBN: 978-1-63483-587-9

Chapter 5

INVERSE DOCKING APPROACHES

***Bernardina Scafuri*[1]*, Angelo Facchiano*[2]**
***and Anna Marabotti*[1,2,3*]**
[1]Dept. Chemistry and Biology, University of Salerno, Fisciano (SA), Italy
[2]Institute of Food Science, CNR, Avellino, Italy
[3]Scientific Institute, IRCCS "E. Medea" – Ass. "La Nostra Famiglia",
Bosisio Parini (LC), Italy

Abstract

This chapter is dedicated to the description of approaches for inverse docking, with an analysis of advantages and disadvantages, potentials and limits of the different tools available to researchers. We will also discuss practical applications of these approaches, and we will present a study specifically made on a food-related subject.

Keywords: inverse docking, reverse docking, inverse virtual screening, proteins

1. Introduction

The terms "inverse" or "reverse" docking identify a procedure that allows to find the macromolecular targets able to interact with one or few small molecules (generally, chemical compounds such as drugs). The adjectives "inverse" or "reverse" emphasize the difference from the traditional docking approach, focused to the study of the interactions that a single (macro) molecule can have with many ligands. In the inverse approach the leading actor is the ligand, and the attention of the researchers is focused on the identification of the more promising receptors for that ligand (Figure 1). At least from a conceptual point of view, however, the final result (i.e., the identification of the partners more likely to interact) is identical in both "inverse" and "forward" approaches, but obviously the procedures and the

[*] Corresponding author: Phone: +39 089 969583 – Fax: +39 089 969603 - E-mail: amarabotti@unisa.it

tools to perform an inverse docking analysis are different, though related to, and in most cases derived from, those used to perform a traditional docking study.

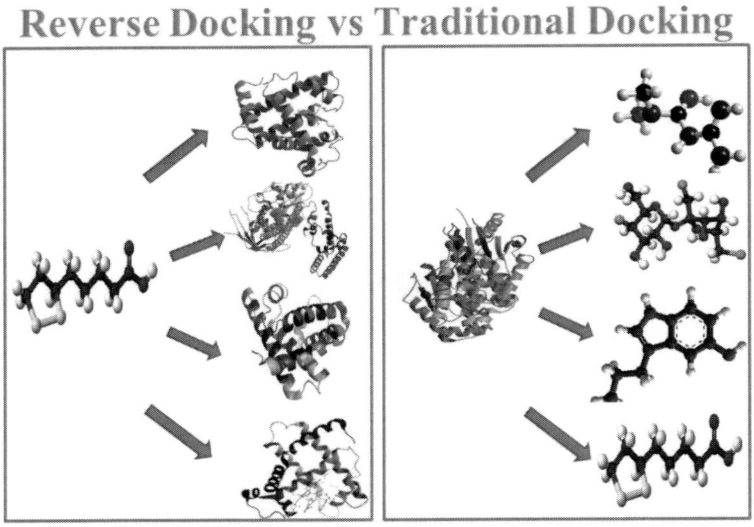

Figure 1. Schematic representation of inverse docking (left) and traditional docking (right).

Although the oldest article present in PubMed introducing the concept of "inverse docking" was published more than 10 years ago [Chen and Zhi, 2001], a search performed in PubMed database for the terms "reverse docking", "inverse docking", "inverse virtual screening" and "reverse virtual screening" (date: December 16th, 2014) retrieved only about 50 unique papers truly related to this subject (intriguingly, the vast majority of them were published by people of Asian affiliation, mainly from China). This means that, despite the potentially wide applications of this procedure, to date inverse docking is not yet entered in the common use of the scientific community (as a comparison, the search for the term "virtual screening" in PubMed, setting the same temporal range, retrieved almost 10,000 articles). This may be due partly to the fact that people prefer to develop and apply strategies to identify ligands for a specific known target protein, considering that in general the studies of protein-ligand complexes have the final aim to develop new drugs for a known specific protein involved in a specific disease or molecular pathway. Additionally, many public and private ligand databases are available, each including millions of active compounds, as combinatorial chemistry [Merrifield, 1963] and other more recent approaches [Abet et al, 2014] have been extensively used to cover all the biologically-relevant chemical space, producing a plethora of chemical substances. Instead, the knowledge of the structures of the proteins, necessary for the inverse docking approach, is much more limited. At the date of December 16, 2014 only about 97,600 protein structures, of which less than 32,000 of human origin, are available in the Protein Data Bank (PDB) [Berman et al, 2003]. Moreover, most of them are redundant: the number of human protein structures with less than 90% sequence similarity is about 11,000. Therefore, the attention of the scientific community during these years has been focused more on developing and applying tools to rapidly screen thousands, if not millions, of compounds towards a single macromolecular target, rather than its opposite.

Another difficulty that probably hindered and still hinders the widespread use of the inverse docking approach in scientific research is related to the size of the systems screened. A chemical compound acting as a ligand for a protein has typically a low molecular weight and a limited conformational variability, with a restricted number of rotatable bonds. For example, ZINC [Irwin et al., 2012], one of the most popular databases of commercially available compounds for structure-based virtual screening, includes molecules with no more than 60 non-H atoms in their structure. Therefore, even a massive screening of these small compounds towards a single macromolecule can be performed relatively quickly and with no need for the use of high-throughout computational infrastructures. On the contrary, a protein is a big molecule with a highly variable shape and size, and with an enhanced conformational freedom. Therefore, the implementation of an inverse docking approach cannot be faced simply by "inverting" the role of the two partners in a common program for virtual screening. There is the need for different approaches to perform the conformational search and to score the best ligand-docking interactions. Likewise, there is the need for computational platforms able to allow the screening of thousands of proteins towards a single chemical compound in a reasonable time frame. Probably, the diffusion of high performance computing solutions, such as supercomputers (http://www.top500.org/), grid computing [Joseph and Fellenstein, 2004] and cloud computing [Armbrust et al., 2009], and their facilitated accessibility for biomedical research will allow inverse docking to be more widely used in the scientific community in the near future [Westermaier et al., 2015].

The aim of this review is to present to the readers an overview of the most used tools to perform an inverse docking research, with their base concepts and limitations, and an overview of the range of their applications and results obtained during these years. A more detailed example of an inverse docking protocol applied to a particular problem in food science will then be discussed.

2. Inverse Docking Protocols

It is interesting to note that to date there is not a leading method in the field of inverse docking, but several tools are available, often freely accessible to users by Web servers, with different features and goals [Westermaier et al., 2015]. Indeed, to develop a structure-based inverse docking method, several challenges must be addressed. First, a database of macromolecules to be tested for their putative interaction with the small chemical must be developed; then, an algorithm to perform the search is needed; finally, a score to rank correctly the results and to find the most probable interaction between the small molecule and the macromolecule is necessary [Kharkar et al., 2014]. In the following paragraphs, each of these aspects will be treated in details together with the pros and cons of different solutions adopted by the different tools developed during last years.

2.1. Macromolecules Database

The universal database of macromolecular structures is the PDB archive [Berman et al., 2003] that presently collects more than 100,000 structures of proteins, DNA, RNA and other kinds of molecules. However, this database is far from ideal to be used "as it is" for an inverse docking strategy. The first problem is redundancy: most of these structures are indeed multiple structures of the same protein, obtained in different conditions (e.g., different pH, temperature, crystallization conditions, solvents...), or with different experimental techniques. Another source of redundancy is the presence of structures of single point variants of the same protein. Notably, most of these single points variant are not natural variants of the proteins, but were obtained artificially, e.g., in order to increase the probability to obtain the crystals of a protein particularly refractory to crystallization. Additionally, other problems intrinsically related to structure quality and experimental techniques to solve protein structures are present in this database. Apart from the low resolution of many structures (more than 20% with $R \geq 2.50$ Å at the date of December 16, 2014), the case of proteins having a structure with missing atoms or residues is very common, and in the most recent structures, frequently it is possible to find residues with multiple conformations. Moreover, a very important question to take into account is the conformational variability of the proteins in the absence or presence of the ligand, and the presence of additional binding cavities on the protein surface, with respect to the "canonical" binding site. Also, the role of cofactors must be considered for the selection of a correct protein structure.

It is clear that the macromolecular database to be used to perform the inverse docking protocol must be created carefully in order to take into account all these factors and increase the chances of success of the procedure. This database must be representative as much as possible of all macromolecular structures of interest to be screened; it must contains structures at high resolution, without gaps, and since the analysis of conformational variability of the protein must be necessarily limited (see next paragraphs), the proteins must be considered in its ligand-binding form, especially in the case of extended conformational changes.

Several strategies have been adopted by different inverse docking tools and research groups to face the problem of selecting macromolecules to be used as reference targets.

In some inverse virtual screening approaches, a very limited (< 200) panel of macromolecular structures extracted from PDB and characterized by their involvement in specific disease processes, was screened against specific natural compounds [Lauro et al., 2011, 2012]. Another database particularly focused on a small pre-selected subgroup of proteins is the one developed for the web server SePreSa, collecting the structural models of 91 human well-known targets for serious adverse drug reactions, including metabolic enzymes, several types of human MHC I proteins mediating drug hypersensitivity, and selected pharmacodynamic proteins matching quality criteria such as completeness of structure and ligand-interacting conformation [Yang et al., 2009a]. In another case, a small in-house database for inverse virtual screening was developed, containing non-protein structures, such as cyclodextrin-based receptors to be tested against selected ligands [Steffen et al., 2007].

Clearly, the approach of selecting a very small database of molecular targets, on one hand speeds up consistently the inverse docking procedure, but on the other hand has the evident limitation of being not exhaustive about finding possible targets for the screened ligands. This approach, however, can be used when the scope is a focused research on a particular subgroup of molecules, or as a preliminary test to understand, for example, if a ligand can be involved (or not) in a specific phenomenon or disease. Moreover, this approach generally takes advantage of the well-known software for traditional docking and indeed often is performed as a traditional docking experiment.

The problem is obviously different when a fully blind search must be performed to find targets for a selected ligand, and therefore when a true inverse docking search must be applied. In this case, during the time many different reference data sets for protein structures have been developed.

The first tool especially developed for inverse docking strategy, INVDOCK [Chen and Zhi, 2001], was based on a protein cavity database developed from the PDB database, to facilitate computer-automated inverse docking searches. It was made of models of individual cavities, represented as a cluster of overlapping spheres following the method of Kuntz and coworkers [Kuntz et al., 1982], for each protein. Each cavity entry was derived from the corresponding PDB entry by a procedure involving the removal of non-pertinent ligands and water molecules, the generation of a protein surface and its coating and filling with clusters of spheres. Cavity entries were divided into classes defined by disease-related species, sequence similarity, and phylogenetic distance to a particular or species class. This database was used, alone or together with INVDOCK tool, to perform several inverse docking experiments [Chen and Ung, 2001; Chen et al., 2003; Ji et al., 2006; Yue et al., 2008; Lu et al., 2010; Feng et al., 2011; Zhao et al., 2012; Ma et al., 2013; Zhang et al., 2013; Pan et al., 2014]. Recently, another database of pockets extracted from all available X-ray structures from human and *Saccharomyces cerevisiae* proteins was developed by Lee and Kim [Lee and Kim, 2012] using the program Fpocket [Schmidtke et al., 2010] in order to create a comprehensive docking profile that improves the sensitivity of inverse docking and helps in finding real binding targets. Other publicly available databases containing information about ligand binding sites are, for example, LigBase [Stuart et al. 2002] and PDBsite [Ivanisenko et al. 2005]. These databases, however, were not specifically developed for inverse docking approaches.

A first approach to set up a database of entire protein structures fitted for inverse docking purposes was performed in 2004 [Paul et al., 2004]. In that work, a 3D protein library called sc-PDB, comprising 2148 entries, was created starting from the available PDB release and analyzing it to retain those entries suitable for molecular docking of small ligands, using in-house developed Perl scripts in order to filter out all those protein structures that did not fit fixed criteria for inclusion. Three sub-databases were then created, collecting the atomic coordinates of the protein, of the binding site of each protein, and of the ligands (Figure 2).

Further development of this database was achieved later [Kellenberger et al., 2006; Meslamani et al., 2011], the database has been made available online (http://bioinfo-pharma.u-strasbg.fr/scPDB), and it was applied to perform some inverse docking studies [Muller et al., 2006; Zahler et al., 2007; Kellenberger et al., 2008]. Recently, this database was also used as a reference starting point to develop the reference database for idTarget web server (http://idtarget.rcas.sinica.edu.tw/) [Wang et al., 2012a], in which a contracted list of targets (2091) was first identified from sc-PDB, and was subsequently expanded in order to include targets that are homologous or contain similar binding sites to the top representatives; this database is also periodically updated to include new protein structures from PDB. Finally, sc-PDB was used also as a benchmark for newly developed inverse docking strategies [Schonburg et al., 2014].

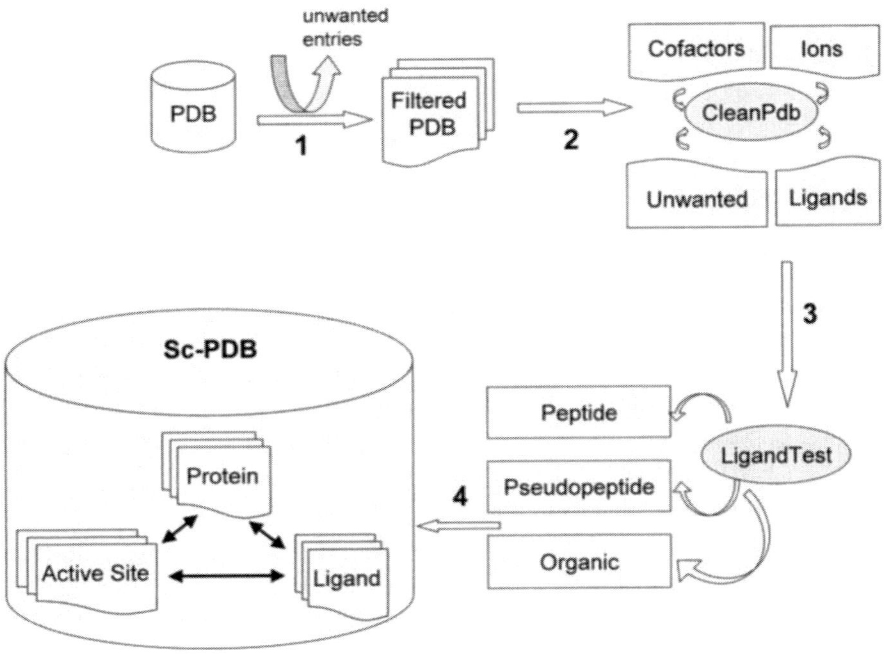

Figure 2. Flow-chart of the sc-PDB protein library set-up. Reprinted with permission from Paul et al., Proteins 2004, 54, 671–680. Copyright 2004 American Chemical Society.

In 2008, another freely accessible database, the Potential Drug Target Database (PTDT) (http://www.dddc.ac.cn/pdtd/) was developed [Gao et al, 2008]. Its aims were not only to give researchers a tool to identify potential binding proteins for an active compound, but also to let them query drug target information. Therefore, PTDT was formed by two sub-databases, the first one storing data about target categories (proteins were divided into 14 categories according to therapeutic areas), biological functions, associated pathways and diseases; the other one storing the coordinates of the selected protein targets in both .pdb and .mol2 formats with AMBER charges [Weiner et al., 1986], together with information about sequence and active site of each protein. PDTD contained >1100 entries, selected in order to retrieve structures complexed with ligands, without mutations and missing residues within 6.5 Å around the bound ligand and of high resolution, with a small degree of redundancy in order to take into account conformational flexibility in selected cases. This database was used, alone or in association with the inverse docking web server TarFisDock (http://www.dddc.ac.cn/tarfisdock/) [Li et al., 2006] to perform several queries [Cai et al., 2006; Jeong et al., 2009; Calvaresi and Zerbetto, 2011; Grinter et al., 2011; Zheng et al., 2011; Eric et al., 2012].

Other programs to perform inverse docking were developed with an integrated customized database of protein targets. For example, a tool developed by researchers from Greenpharma, SELNERGY^{TM}, is based on a database of about 7000 protein targets structures obtained either from crystallographic data or by homology modelling, and annotated according to their source, their protein family and their therapeutic applications [Do et al., 2005]. Recently, a structural human proteome database was created from PDB by removing low resolution (< 3Å) structures and including proteins for which sequences can be mapped to human proteins in Ensembl (Hubbard et al., 2002). The database is formed by 9652 human and human-like high resolution protein structures [Hu et al., 2014].

A completely different database and strategy to find the preferred protein target for a selected ligand was the one developed by Liu and coworkers and implemented in their PharmMapper web server (http://59.78.96.61/pharmmapper/) [Liu et al., 2010]. It cannot be considered strictly an inverse docking strategy, although the aims are the same, as the panel of targets to be screened is not described in molecular terms. The authors developed PharmTargetDB, a large database annotated with all the targets information retrieved from other databases and including approx. 7300 receptor-based pharmacophoric models (spatial arrangements of features describing the binding modes of known ligands), extracted by structural data of macromolecule–ligand complexes in a fully automated way using the program LigandScout [Wolber and Langen, 2005]. Some applications took advantage of this alternative database and of the associated web server [Su et al., 2013; Chen and Ren, 2014; Chen, 2014]. Very recently, another analogous database was developed for iRAISE, a new inverse screening method based on triangle descriptors developed by Schomburg and colleagues [Schomburg et al., 2014]. In this database, protein data are represented in three separated tables containing general information, residue templates and residue coordinates, in order to avoid redundancy of information, to reduce the disk space needed to store data with respect to raw PDB files, and to optimize the time to read them. Ligand-protein matching is abstracted by a triangle descriptor representing pharmacophoric and steric features, consisting on interaction spots at the corners, whereas the triangle side lengths encode the distance between the interaction spots. Rotatable groups are handled by interaction spots with multiple directions or multiple possible interaction spots. The triangle descriptors of the active sites are stored in a compressed bitmap index.

The use of these different kinds of databases for a blind inverse docking strategy strongly affects the results obtained. The creation of curated database with high-quality, non-redundant selected structures can enhance the chances to find a true interaction between the ligand and the target proteins. It is clear that a simplified database responds to the need to overcome the problem of expensive calculations, but obviously limits the reliability of the results, since the active sites taken into account are not truly molecular entities. On the other hand, using the entire protein structure for docking calculations can improve the quality of the results, but considerably increases the computational effort needed to perform the search.

2.2. Search Algorithm

In line of principle, the algorithm to perform the conformational search in an inverse docking procedure should be the same used for "traditional" docking approaches. Indeed, when the panel of targets to be screened is small (e.g., less than one hundred proteins), the algorithm used can be one of those classically employed for traditional virtual screening, such as Autodock Vina [Trott and Olson, 2010], DOCK [Ewing et al., 2001], GlamDOCK [Tietze and Apostolakis, 2007], without substantial modifications. On the other hand, when a truly "blind" research is performed, e.g., when the target receptor must be searched among all possible known protein structures, the algorithm to perform the search must be extremely efficient. In some cases, the algorithms for traditional docking were implemented in a parallel calculation protocol, in order to enhance the computing performances. For example, in the approach of Paul et al. [Paul et al., 2004], one of the best performing algorithm for traditional docking available to date, GOLD [Jones et al., 1997] was applied with fast virtual screening settings implemented in a parallel calculation protocol. The parallelization of a well-known docking algorithm in order to speed up inverse docking calculation was also applied recently when, starting from AutoDockVina, Ellingson and coworkers develop VinaMPI, a massively parallel MPI program used to distribute tasks while multithreading is used to speed-up individual docking tasks [Ellingson et al., 2013]. This program allows for high-throughput inverse docking taking advantage of the largest supercomputers in their entirety. In addition, VinaMPI is an open access code and it is free for users.

In the first example of inverse docking procedure [Chen and Zhi, 2001], a small molecule was flexibly docked into each cavity of the reference database using a shape-based algorithm derived from DOCK, with some modifications to improve the efficiency: a vector-based docking of a ligand to a cavity; a limited conformation optimization on the ligand and side chains belonging to the cavity; energy minimization for all atoms in the binding site and the docking evaluation by molecular mechanics energy functions and comparison with other ligands. The analysis of each ligand-cavity interaction was not exhaustive: when the first successful dock was obtained, the program proceeded to the next protein entry [Chen and Zhi, 2001]. Also another very popular Web server for inverse docking, TarFisDock, was derived from DOCK [Li et al., 2006], whereas the docking protocol of idTarget [Wang et al., 2012a] is based on MEDock [Chang et al., 2010], a global search algorithm that generates initial docking poses of the small ligand with a very high searching efficiency. idTarget adopts a divide-and-conquer approach that permits to analyse the entire protein surface of a possible protein target instead of a part of the surface or the active site only. This is achieved by dividing each protein target into smaller boxes whose size is dynamically determined according to the size of the query ligand. In addition, idTarget provides two options to perform the search: the "scanning mode" in which a conventional docking is performed for each protein structure in the database, and the "fast mode" in which the ligand is first mapped to the binding sites of the homologous proteins through their superposition on precalculated structural alignment information, and then its binding poses are optimized by adaptive local searches to remove close contacts with protein atoms [Wang et al., 2012a]. SELNERGY^{TM}, the tool for inverse docking developed by Greenpharma, is based on FlexX 2.0 engine [Rarey et al., 1996].

Finally, an interesting tool (ILbind) was recently developed by Hu and colleagues, using a consensus-based approach. The authors adapted two ligand-specific predictors of binding pockets to work as inverse ligand binding predictors, and created the consensus by combining selected outputs of these two predictors with an ensemble of 15 support vector machine predictors that use linear kernel functions. The consensus method is shown to perform better with respect to each of the two predictors [Hu et al., 2012].

One of the main problems to be faced by docking approaches is how to deal with the flexibility of both partners. In general, owing to the limited number of its conformational degrees of freedom, the ligand flexibility is usually taken into account, whereas treating the conformational freedom of the protein is still a hard problem to solve. The ideal method to deal with this problem would be to search and optimize the full degrees of freedom of the protein simultaneously to those of the ligand, using Monte Carlo and molecular dynamics simulations. Alternatively, a flexible receptor structure could be represented as an ensemble of multiple static structures. An example of implementation of this last strategy, limited however to small synthetic receptors, is presented in the work by Kamper and colleagues [Kamper et al., 2006]. In that paper, a new docking approach is described that performs a conformational sampling of the host and then sequentially docks the ligand into all receptor conformers using the incremental construction technique of the FlexX software [Rarey et al., 1996]. Obviously, both approaches are computationally very expensive even for traditional docking experiments, and are not applicable in the field of the inverse docking approach with protein targets. Thus, only simplified models to take into account possible protein motions can be applied. In some cases, only the ligand-bound conformation of the protein, when available from PDB, are taken into account [Gao et al., 2008]. In INVDOCK approach, a limited conformation sampling of rotatable bonds of the side chains of the residues in the binding site of the receptor was performed, with each rotatable bond sampled at $\pm 15°$ along with additional energy minimization. However, no attempt was made to explicitly sample the ensemble of protein conformations [Chen and Zhi, 2001].

When a pharmacophoric approach is used to find the best protein target for a ligand of interest, the searching algorithm applied is obviously different from those used in truly molecular docking methods. For example, in the case of PharmMapper [Liu et al., 2010], an algorithm was developed that flexibly aligns the given small molecule onto each pharmacophoric model of proteins in the target list using a strategy of sequential combination of triangle hashing and genetic algorithm optimization. The first stage of the strategy is the generation of multiple conformers of the given ligand and the attribution of seven pharmacophore features according to predefined rules. Then, an approach conceptually similar to “geometric hashing” is applied, creating a hash table for both pharmacophore models and ligand conformers containing triplets of feature points. Finally, each query triangle from the ligand is checked against a pre-computed triangle ensemble of pharmacophore models to verify the correspondence of the features. Users can also specify if they would like to use an additional genetic algorithm to perform post-optimization after the triangle matching stage. A conceptually similar approach was the one developed by Schonburg and colleagues. The basis of their screening technology is a descriptor-based bitmap search, called iRAISE (inverse RApid Index-based Screening Engine). The algorithm acts by matching the triangle descriptors of the query ligand to those of the active sites in the bitmap index. When a match is found, the transformation needed to superpose the triangles is applied to the respective conformations of the query molecule and of the selected protein active site, producing the final pose in the protein’s active site [Schonburg et al., 2014]. Therefore, to date algorithms applied in inverse docking protocols are derived from those developed for traditional docking (or virtual screening). The need to improve their efficiency in finding few receptors among a large database of very big structures has been until now the driving force superseding the need to take into account protein flexibility.

2.3. Scoring Functions

An appropriate scoring function is essential to rank results of an inverse docking procedure and to find the best predicted targets for a small molecule. In the case of tools that perform inverse screening using a pharmacophoric approach, the scoring function is related to the best fitting of the ligand into the pharmacophoric map of the pocket. For example, in PharmMapper [Liu et al., 2010], for each alignment of the small molecule onto each pharmacophore model of proteins in the target list, a "fit value" is calculated with a distance-dependent score related to the relative positions of each pair of feature points from the pharmacophore and ligand, respectively. This fit score can be normalized to the range of 0 to 1 with a binomial function. A penalty score is also calculated to take into account excluded volumes in the pharmacophore, with the same approach used for the fit score, and subtracted to the fit value. Then, the candidate targets are prioritized based on these resulting fit values and the final result is a ranked list of the top hits found by the procedure, sorted by fit score (or normalized fit score) in descending order.

In the "true" inverse docking approaches, instead, once the conformational search is performed, a docking program must apply a scoring function that should represent the protein-ligand interactions as accurately as possible, assessing both steric and chemical complementarities between a ligand and a protein. Ideally, the resulting best scoring solution should be as close as possible to the real conformation of the ligand in the complex (typically deduced by X-ray data), and in general it is assumed that it is the best one in terms of binding energy. Different types of scoring functions are available: i. those derived from classical molecular mechanics force fields, which contain the main interatomic parameters that describe the potential energy of a system of atoms; ii. "knowledge-based" scoring functions, in which the energetic contribution of each pair of individual contacts is obtained from the frequency of contacts derived from a large dataset of structures of protein–ligand complexes, using the inverse Boltzmann's equation; iii. "empirical-based" scoring functions, obtained by experimental data on protein-ligand interactions fitted by different kinds of mathematical functions [Westermaier et al., 2015]. All these kinds of scoring functions have advantages and disadvantages: for example, those related to molecular mechanics force fields can be more generalizable and are of wide use, but they often discard entropy and solvent effects and are computationally more demanding. On the other hand, knowledge-based functions can include solvent effects and are computationally more efficient, but are less accurate and their correctness strongly depends upon the quality of the crystal structures used as a benchmark to define the scoring function itself. Finally, empirical-based functions are strongly conditioned by the training set and the experimental methods by which binding affinities between the protein and the ligand have been determined, and therefore they are less generalizable. Moreover, they represent only the most frequently observed interaction types [Gohlke and Klebe, 2002; Cozzini et al., 2004]. Therefore, an "ideal" scoring function is not yet present in the scenario of docking programs, either traditional or inverse ones. Indeed, several comparisons of performances of docking programs have concluded that the scoring functions is strongly affected by the specific characteristics of both the binding site and the ligand.

The various inverse docking strategies implement different scoring schemes. In INVDOCK, the scoring function is based on a molecular mechanics ligand–protein interaction energy function derived from AMBER force field [Weiner et al., 1986]. As it has been told before, INVDOCK does not perform an exhaustive search, but the first target with a score lower than a fixed threshold is selected. This threshold is an energy value statistically derived from a set of PDB complexes [Chen and Zhi, 2001]. Also in TarFisDock, the scoring function derives from AMBER force field, and is composed of van der Waals and electrostatic interaction terms, where each term is a double sum over ligand atoms and receptor atoms [Li et al., 2006]. In SELNERGY™, the results on ligand/protein pairs found are scored using a method called Multiple Active Site Corrections (MASC) that can be thought of either as a statistical measure of ligand specificity or as a correction for ligand-related bias in the scoring function [Vigers and Rizzi, 2004]. This MASC score is different for each molecule and is specific for the docking program and scoring function used, and once calculated for a molecule, it can be used to correct the docking score for that molecule against any target site of interest.

In idTarget, two different approaches for scoring are combined. A robust scoring function based on the energetic terms and the formulation of AutoDock 4 [Huey et al., 2007] was developed in a preliminary work by the authors [Wang et al., 2011]. This scoring function reports the binding free energy in the experimental scale, allowing the direct comparison among different protein–ligand systems. Additionally, to enhance the confidence level of prediction, an affinity profile analysis is performed for each binding site. The affinity profile of a given pocket is modelled by a Gaussian function, whose width is determined by the range of the predicted binding affinities of the complexes of different ligands in the same binding pocket of the same protein. A Z-score of ligand j to a given protein pocket i is calculated: a large negative Z-score value indicates an important target for the query compound [Wang et al., 2012a].

Probably, the most complex procedure to score results obtained by an inverse docking approach is the one applied by iRAISE [Schonburg et al., 2014]. In this screening method, a "scoring cascade" of five steps is applied to all ligand poses obtained from the descriptor matches for one protein. After this scoring cascade, the score of each pose is used to rank the proteins as targets for the query ligand. The final iRAISE score is then obtained after a further step, by applying a Gaussian score weight to the score of the scoring cascade, to decide if a score is statistically significant for a protein pocket. This procedure allows to discard irrelevant poses and to obtain a score comparable among proteins, avoiding interprotein scoring noise. The effect of the scoring cascade was evaluated towards other scoring schemes, and it was found that the scoring cascade ranks the true targets much better than the simple iRAISE interaction score, but it is only marginally superior, or comparable to the standard scoring schemes of docking programs such as FlexX (Hyde scoring) and GLIDE. The Authors attributed this to the fact that the scoring cascade could be too selective.

When a docking program is used in inverse mode, the implicit assumption is that scoring functions work equally well both in ranking different ligand–protein complexes and in ranking different conformations of the same ligand–protein complex. However, in most cases this is not true, because scoring functions are generally optimized only for conformation ranking. For example, Wang and colleagues performed a test using Glide [Friesner et al., 2004], a widely used docking program that has won many performance comparisons [Li et al., 2010]. Their results showed that the Glide scores had interprotein noises, such as recurrent over- or underestimated scores for certain proteins, leading to false positive or false negative predictions of target proteins. Indeed, in that test, a low success rate (56.9%) was obtained, indicating that the Glide score was not very effective in ranking complexes with different proteins. A deep analysis of the reasons for this failure allowed to suggest a correction for the Glide score (the so-called "Balance-Corrected GlideScores, or BCGlideScores) that reached a higher success rate (72.4%) in ranking different protein targets toward a single ligand, with respect to the original GlideScore [Wang et al., 2012b].

An analysis of the performances of several inverse docking protocols was recently performed, creating a target database containing 1714 entries that covers 1594 drug targets and 18 different biological functions, and integrating many scoring functions into five different inverse docking schemes to evaluate their effectiveness individually. Results confirmed that the most diffused inverse docking scoring functions failed to recognize the most correct targets [Hui-fang et al., 2010]. This result shows that there is still room to improve the scoring schemes of these tools.

Some efforts have been made recently to improve the ranking of different ligand-protein complexes, for example incorporating a topological scoring function based on protein–ligand interaction fingerprints that encode ligand-protein interactions (principally hydrophobic and ionic interactions, aromatic stacking, and H-bonds) into bit strings, allowing to measure the similarity of the binding mode of two ligands to a common ligand binding site as a distance between two bit strings [Kellenberger et al., 2008; Wang et al., 2012b].

3. Examples of Applications of Inverse Docking Procedures to Biological Problems

Applications of inverse docking protocols in literature are mainly focused to medicinal chemistry. The first purpose for the development of this strategy was the prediction, or the molecular explanation, of off-targets effects that are generally associated to toxicity problems for some pharmaceutical compounds. Several examples of inverse docking approaches applied to this aim are present in literature, focused on the prediction of toxicity and side effects of several known and commonly marketed clinical agents [Chen and Ung, 2001; Ji et al., 2006, Yang et al., 2009a,b, 2011; Fan et al., 2012; Pan et al., 2014, Hu et al., 2014]. Strictly related to this scope, inverse docking has been used also for an emerging approach in medicinal chemistry and drug discovery: drug repositioning, in which an old drug is revisited to find new uses outside the scope of the original medical indication [Ashburn and Thor, 2004; Kharkar et al., 2014]. An example of this application is the study on tofisopam, an atypical benzodiazepine that has been found to act as phosphodiesterase 4 inhibitor [Bernard et al., 2008].

The main application of inverse docking, however, is the identification of unknown targets for substances showing a biological activity. The first example of the use of this approach for this purpose is the paper in which for the first time the approach was presented, which predicted the putative protein targets for 4H-tamoxifen and vitamin E [Chen and Zhi, 2001]. Many examples of the application of this approach for these purposes are present in literature. For example, the prediction of kinase targets for three indirubin derivatives [Zahler et al., 2007]; the search for fullerene-binding proteins [Calvaresi and Zerbetto, 2011]; the search for potential targets of PRIMA-1, a small molecule capable of activating mutant p53 protein, restoring its ability to bind to DNA and its tumor suppressor function [Grinter et al., 2011]; the rationalization of cytotoxic effects of novel derivatives of aryl-aminopyridines with anticancer activity, with a proposal of their mechanism of action and suggestions for improvement of studied compounds [Eric et al., 2012]. Another example of successful use of inverse docking approach was the computer-aided optimization of a synthetic receptor for a given guest molecule. In this case, a set of β-cyclodextrin derivatives was evaluated for their ability to bind camptothecin, an anticancer drug. Those receptors predicted to be within the top 10% of the ranking list were selected for synthesis and further characterization. This not only allowed to avoid time consuming trial-and-error searches, but also indicates that inverse virtual screening can support the development of synthetic receptors for a given ligand [Steffen et al., 2007].

Most of the substances tested with inverse docking approach are of natural origin and are part of complex mixtures of active compounds used e.g., in traditional Chinese medicine. For example, several studies have been made to identify the molecular targets of the active compounds of Danshen (*Salvia miltiorrhiza*), one of the most widely used traditional medicine in China, known for its multiple biological effects. An inverse docking study made in 2011 predicted an interaction between salvianolic acid B, a compound of Danshen with protective effect against angina and stroke, and EGFR. The binding between salvianolic acid B and human EGFR was then confirmed using an *in vitro* assay based on surface plasmon resonance (SPR) and other assays to prove the effect of this compound on EGFR internalization and phosphorylation [Feng et al., 2011]. More recently, with the use of inverse docking approach, two other active compounds extracted from Danshen (danshensu and tanshinone IIA), with known antitumoral activity were predicted to bind, respectively, to H-Ras and retinoic acid receptor alpha, thus suggesting that their antitumoral activity may act through the signaling pathways in which these proteins are involved [Chen and Ren, 2014; Chen, 2014].

Ten target compounds isolated from another popular Chinese herbal formula called Shaofu Zhuyu decoction, known for its multiple biological activities, were tested using an inverse docking approach. The resulting 23 potential protein targets were further analysed for their involvement in metabolic pathways to identify the pharmacological mechanism of action of this traditional chinese medicine [Su et al., 2013]. Astragaloside IV, one of the main active compounds of *Astragalus membranaceus* Bunge, another medicinal herb used in traditional Chinese medicine, was investigated for its therapeutic mechanism, and 3 out of 39 putative targets identified were experimentally validated at molecular level [Zhao et al., 2012]. Ganoderic acid D, one of the major component of triterpenes from *Ganoderma lucidum*, a medicinal mushroom that has been used as a home remedy in traditional Chinese medicine for over 2000 years, was characterized for its possible target network using INVDOCK approach, and the binding with 14-3-3 ζ protein was validated using SPR [Yue et al., 2008]. Inverse docking is indicated as a tool for research on Chinese herbal medicines using *in silico* approaches also in a comprehensive review on this subject [Barlow et al, 2012].

Apart from analysis of traditional Chinese medicines, other natural products were tested using an inverse docking approach. For example, berberine, an alkaloid derived from plants and known for its *in vitro* anticancer activity, was tested for the mechanisms of its activity on HeLa cells, using an inverse docking approach further validated by experimental studies [Lu et al., 2010]. The identification of targets of berberin was the subject also of another very recent study [Ma et al., 2013], which suggested four proteins, including calmodulin, able to bind this molecule, and the anticalmodulin property of berberine was demonstrated with an *in vitro* assay. Other natural substances tested with this approach are: nine well-known medicinal plant ingredients (genistein, ginsenoside Rg1, quercetin, acronycine, baicalin, emodin, allicin, catechin and camptothecin) [Chen et al., 2003]; epsilon-viniferin [Do et al., 2005]; meranzin, a major component isolated from *Limnocitrus littoralis* [Do et al., 2007]; gingerol [Jeong et al., 2009]; the molecular components of *Ruta graveolens* [Rollinger et al., 2009]; several natural products extracted from a library of bioactive compounds [Lauro et al., 2011, 2012]; the main functional components of green tea (epigallocatechin gallate, epicatechin gallate, epigallocatechin and epicatechin) [Zheng et al., 2011]; wentilactone B, a tetranorditerpenoid derivative extracted from the marine algae-derived endophytic fungus *Aspergillus wentii* EN-48 [Zhang et al., 2013]; nimbolide, a tetranorterpenoid extracted from the leafs of neem (*Azadirachta indica*), a tree with medicinal properties diffused in the traditional Indian medicine [Chitta et al., 2014].

Finally, an example of reverse docking approach applied not only to identify unknown targets for a known compound, but also as a complementary approach of functional genomics and chemical biology, is shown in the work by Cai and colleagues [Cai et al., 2006]. In this work, a compound isolated from *Ceratostigma willmottianum*, an herb used in traditional medicine, was found to have inhibitory activity against *Helicobacter pylori*, the well-known and major causative factor for gastrointestinal diseases such as gastritis and ulcer, for which no effective therapy exists to eradicate infection. This compounds and an analog with improved MIC value were screened using TarFisDock, and 15 possible protein targets were identified, with only two of them (diaminopimelate decarboxylase and peptide deformylase) having homologous proteins in the genome of the bacterium. Both proteins are attractive targets for the development of novel antibiotics, because of their involvement in pathways essential for bacterial viability and development. Enzymatic assays confirmed that peptide deformylase is actually a target of the compounds tested, and the structures of this protein complexed to its inhibitor were determined by X-ray crystallography. The study of the interactions showed that these compounds bind selectively to type I peptide deformylase, but they do not show inhibitory effect on the homologous enzyme from *E. coli*. This study allowed not only to identify promising lead compounds for the development of new antimicrobial agents, but also to highlight an enzymatic activity that can be used as a promising target for the development of alternative chemical compounds able to block it.

3.1. An Example of Application of a Food-Related Inverse Docking Strategy

It is well known that there is a clear relation between nutrition and health. A correct diet not only gives enough nourishment that helps growth and reproduction, but it promotes health and longevity. Moreover, diet plays a crucial role in preventing chronic diseases such as cardiovascular illness, type II diabetes, metabolic diseases and cancer. These illnesses have a multifactorial genesis and dietary factors can either foster or inhibit inflammation and oxidative stress that are the main mechanisms of the above mentioned diseases.

As a matter of fact, an imbalance between the production of free radicals and the activity of antioxidant species can provoke a massive production of reactive oxygen species, which oxidize the biologic macromolecules (such as lipids proteins and DNA); this process causes alterations in the membranes, inactivity of both enzymes and receptors, genomic damages, etc. [Grassi et al., 2010].

In order to protect the organism from oxidative stress, there are two mechanisms of defense: one endogenous, that involves enzymes, and one exogenous, which is based on the ingestion of antioxidant molecules that are very common in fruits and vegetables. An antioxidant is a substance that, when present at low concentrations compared with those of an oxidizable substrate, significantly delays or prevents oxidation of that substrate [Halliwell, 1995]. The most abundant antioxidants in the diet are polyphenols, secondary metabolites of plants which are particularly helpful against ultraviolet radiation or aggressions by pathogens [Pandey and Rizvi, 2009]. Polyphenol structures are made up by several hydroxyl groups on aromatic rings, and they can be classified into different groups according to the function of the phenol rings that they contain and to the structural elements that bind these rings to each other [Manach et al., 2004].

There is a wide variety of polyphenols in foods, and their absorption and metabolism change according to their type. Their main dietary sources are fruits and plant-derived beverages such as fruit juices, tea, coffee, and red wine. Vegetables, cereals, chocolate, and dry legumes also contribute to the total polyphenol intake. In particular, red wine in its different varieties is extremely rich in polyphenols such as resveratrol, quercetin and derivatives such as rutin, anthocyanis, the monomeric flavanols catechin and epicatechin, (-)-epicatechin-3-O-gallate, chlorogenic acid, dimeric, trimeric and tetrameric procyanidins, myricetin, laricitrin, kaempferol, isorhamnetin and syringetin. Another phenolic compound detected and quantified in grape seed and skin extracts is chlorogenic acid, an ester of caffeic acid and quinic acid, which is considered to be an abundant polyphenol in the human diet [Rockenbach et al., 2011]

Several studies have reported the positive effects that a moderate consumption of red wine has on human health. A well-known example is the so-called "French Paradox", based on the observation that French people have one of the lowest rates of coronary heart disease, despite the high intake of saturated fats in their diet [Ferrières, 2004]; this has been linked to the moderate wine consumption, which has a favourable effect on blood clotting, vascular endothelial cell function (by activation of nitric oxide) and serum lipids by raising the levels of HDL (the so-called "good" cholesterol). Furthermore, epidemiological studies have suggested that moderate wine consumption may reduce the incidence of certain age-related neurological disorders including Alzheimer's disease. Regular dietary intake of foods and/or beverages rich in polyphenols, then, has been associated to 50% reduction in the risk of dementia, a preservation of cognitive performance with ageing, a delay in the onset of Alzheimer's disease and a reduction in the risk of developing Parkinson's disease [Vauzour et al., 2010].

In order to understand possible unknown molecular pathways by which antioxidants exert their positive activity on human body, we applied an inverse docking approach to identify possible protein targets for some antioxidant molecules of red wine. The antioxidants chosen are: resveratrol, quercetin, rutin, (+)catechin, (-)epicatechin, procyanidin B1 and procyanidin B2 (Figure 3).

For this aim, we used the web-server idTarget [Wang et al., 2012a]. The ligands' files were downloaded from PubChem data bank (http://pubchem.ncbi.nlm.nih.gov) in the 3D format .sdf and then converted in .pdb format using UCSF Chimera (http://www.rbvi.ucsf.edu/chimera/), an extensible molecular model system, while the protonic state has been automatically assigned by idTarget. Finally, the scanning mode option has been used for the scanning process. For each ligand, we selected the ten targets with the lowest binding energy retrieved by idTarget. Among these, only human proteins were selected, and non-human proteins were considered only if homologous human ones exist: in fact, the aim was to search human protein targets which antioxidants could bind to. The structures of the selected targets were checked for their correctness. This operation is necessary since protein structures are sometimes incomplete, e.g., they miss amino acids or atoms, to verify that the ligand would not alight in a gap inside the structure that becomes a false ligand pocket, instead of the true binding site. Targets were then analysed for their function and for pathways and pathologies they are involved in. Information was retrieved, respectively, from Uniprot (http://www.uniprot.org) [Magrane and The Uniprot Consortium, 2011] and KEGG (http://www.genome.jp/kegg/) [Kanehisa et al., 2010] databases.

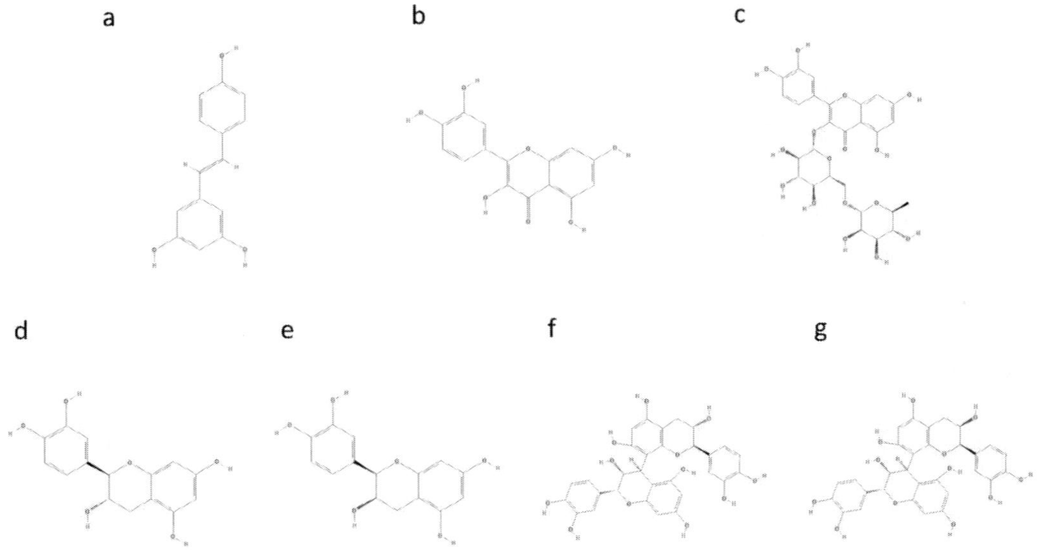

Figure 3. Structures of antioxidants from red wine selected for the inverse docking approach: a) resveratrol, b) quercetin, c) rutin, d) (+)catechin, e) (-)-epicatechin, f) procyanidin B1, g) procyanidin B2. Structures were extracted from PubChem database.

Table 1. Human targets retrieved by idTarget for red wine antioxidant tested. Corresponding structures of the proteins are identified by their PDB code

Antioxidant	Protein target (with EC number)	PDB CODE
Resveratrol	Aldehyde Reductase (EC 1.1.1.21)	4LB4
	Sulfotransferase family cytosolic 1B member 1 (EC 2.8.2)	2Z5F
	Methylthioribose-1-phosphate isomerase (EC 5.3.1.23)	4LDQ
	Dynamin-1 (EC 3.6.5.5)	2X2E
Quercetin	Uridine 5'-monophosphate synthase (EC 4.1.1.23)	2QCG
	Mitogen-activated protein kinase 14 (EC 2.7.11.24)	1WBS
	Aldehyde Reductase (EC 1.1.1.21)	4LB4
Rutin	Lysine-specific histone demethylase 1 (EC 1.14.11.B1)	2IW5
	Amine Oxidase [Flavin-Containing] B (EC 1.4.3.4)	2BK3
	Guanidinoacetate N-methytransferase (EC 2.1.1.2)	3ORH
	Ornithine Aminotransferase (EC 2.6.1.13)	2OAT
(-)Catechin	GTP-ase Hras (EC 3.6.5.2)	3K8Y
	Ras Related protein Rap 1A (EC 2.7.11.1)	1C1Y
	GTP-ase Kras (EC 3.6.5.2)	4OBE
(+)Epicatechin	Matrix metalloproteinase 16 (EC 3.4.24)	1RM8
	GMP reductase 2 (EC 1.7.1.7)	2C6Q
Procyanidin B_1	Renin (EC 3.4.23.15)	2G1Y
	Factor X light chain (EC 3.4.21.6)	3KL6
Procyanidin B_2	3-oxo-5-beta-steroid 4-dehydrogenase (EC 1.3.1.3)	3BUV

The protein targets selected for red wine antioxidants are shown in Table 1. The targets belong mainly to the oxidoreductase enzyme group (6 out of 19) and hydrolase (6 out of 19), and at a minor extent to transferase (5), lyase (1) and isomerase (1). The involvement of enzymes of the oxidoreductase group is in agreement with the expected activity of the ligands investigated, and it is interesting to note that 6 out of 7 ligands have at least one putative target belonging to that enzyme group.

Table 2. List of pathways in which at least two protein targets found for red wine antioxidants are involved

No. of targets	KEGG pathway
6	Metabolic pathways
4	MAPK signaling pathway
4	Neurotrophin signaling pathway
4	Rap1 signaling pathway
3	Alcoholism
3	MicroRNAs in cancer
3	Serotonergic synapse
3	Arginine and proline metabolism
3	Chemokine signaling pathway
3	Endocytosis
3	Fc epsilon RI signaling pathway
3	FoxO signaling pathway
3	GnRH signaling pathway
3	Hepatitis C
3	Long-term potentiation
3	Prolactin signaling pathway
3	Proteoglycans in cancer
3	Ras signaling pathway
3	Renal cell carcinoma
3	Signaling pathways regulating pluripotency of stem cells
3	T cell receptor signaling pathway
3	VEGF signaling pathway
2	Acute myeloid leukemia
2	Axon guidance
2	B cell receptor signaling pathway
2	Bladder cancer
2	Cholinergic synapse
2	Chronic myeloid leukemia
2	Endometrial cancer
2	ErbB signaling pathway
2	Estrogen signaling pathway
2	Gap junction
2	Glioma
2	Glycine, serine and threonine metabolism
2	Hepatitis B
2	HTLV-I infection
2	Insulin signaling pathway
2	Long-term depression
2	Melanogenesis

Table 2. (Continued)

No. of targets	KEGG pathway
2	Melanoma
2	Natural killer cell mediated cytotoxicity
2	Non-small cell lung cancer
2	Oxytocin signaling pathway
2	PI3K-Akt signaling pathway
2	Prostate cancer
2	Regulation of actin cytoskeleton
2	Thyroid cancer
2	Thyroid hormone signaling pathway
2	Tight junction
2	Viral carcinogenesis
2	Bacterial invasion of epithelial cells
2	Dopaminergic synapse
2	Endocrine and other factor-regulated calcium reabsorption
2	Focal adhesion
2	Progesterone-mediated oocyte maturation
2	Synaptic vesicle cycle
2	Platelet activation
2	Leukocyte transendothelial migration

The analysis of the KEGG pathways in which all the targets are involved gives a list of 102 different pathways, and Table 2 reports the 58 pathways found at least two times. In agreement with the expected chemo-preventive effects of antioxidant components of diet, it is interesting to observe that about 1/3 of them are directly related to pathologies, with a prevalence of cancer forms.

Most of the other pathways in the lists are related to specific regulative or metabolic pathways that would require a deeper investigation in order to evaluate the correspondence with known biochemical roles of the ligands under investigation.

Conclusion

In this review, we have discussed different approaches to perform inverse docking, together with their strengths and weaknesses, and their applications to some problems of biological interest, including an example of a food-related research performed in our laboratory. It is clear that finding a protein target for a specific ligand among thousands of structures is still a hard task to reach, but with the increase of available protein structures and of computational resources to perform these simulations efficiently, we think that inverse docking will play a main role in the future to answer to many biological questions.

Acknowledgments

This work has been supported by the Italian Ministry of Health (Targeted Research Funding in Public Health – Young Researchers; grant no. GR-2009-1570296), and it was also made in the framework of Flagship "InterOmics" project (PB.P05) funded and supported by the Italian Ministry of Education, University and Research and Italian National Research Council organizations.

References

Abet V, Mariani A, Truscott FR, Britton S, Rodriguez R. Biased and unbiased strategies to identify biologically active small molecules. *Bioorg. Med. Chem.* 2014, 22, 4474-4489.

Ashburn TT, Thor KB. Drug repositioning: identifying and developing new uses for existing drugs. *Nat. Rev. Drug Discov.* 2004, 3, 673-683.

Armbrust M, Fox A, Griffith R, Joseph AD, Katz R, Konwinski A, Lee G, Patterson D, Rabkin A, Stoica I, Zaharia M. A view of cloud computing. *Commun. ACM* 2010, 53, 50-58.

Barlow DJ, Buriani A, Ehrman T, Bosisio E, Eberini I, Hylands PJ. *In-silico* studies in Chinese herbal medicines' research: Evaluation of *in-silico* methodologies and phytochemical data sources, and a review of research to date. *J. Ethnopharmacol.* 2012, 140, 526-534.

Berman HM, Henrick K, Nakamura H. Announcing the worldwide Protein Data Bank. *Nature Struct. Biol.* 2003, 10, 980.

Bernard P, Dufresne-Favetta C, Favetta P, Do QT, Himbert F, Zubrzycki S, Scior T, Lugnier C. Application of drug repositioning strategy to TOFISOPAM. *Curr. Med. Chem.* 2008, 15, 3196-3203.

Cai J, Han C, Hu T, Zhang J, Wu D, Wang F, Liu Y, Ding J, Chen K, Yue J, Shen X, Jiang H. Peptide deformylase is a potential target for anti-Helicobacter pylori drugs: Reverse docking, enzymatic assay, and X-ray crystallography validation. *Protein Sci.* 2006, 15, 2071–2081.

Calvaresi M, Zerbetto F. Fullerene sorting proteins. *Nanoscale* 2011, 3, 2873-81.

Chang DTH, Lin JH, Hsieh CH, Oyang YJ. On the design of optimization algorithms for prediction of molecular interactions. *Int. J. Artif. Intelligence Tools* 2010, 19, 267–280.

Chen SJ, Ren JL. Identification of a potential anticancer target of danshensu by inverse docking. *Asian Pac. J. Cancer Prev* 2014, 15, 111-116.

Chen SJ. A Potential target of tanshinone IIA for acute promyelocytic leukemia revealed by inverse docking and drug repurposing. *Asian Pac. J. Cancer Prev.* 2014, 15, 4301-4305.

Chen X, Ung CY, Chen YZ. Can an in silico drug-target search method be used to probe potential mechanisms of medicinal plant ingredients? *Nat. Prod. Rep.* 2003, 20, 432–444.

Chen YZ, Ung CY. Prediction of potential toxicity and side effect protein targets of a small molecule by a ligand–protein inverse docking approach. *J. Mol. Graph Model* 2001, 20, 199–218.

Chen YZ, Zhi DG. Ligand–protein inverse docking and its potential use in the computer search of protein targets of a small molecule. *Proteins* 2001, 43, 217–226.

Chitta K, Paulus A, Caulfield TR, Akhtar S, Blake MK, Ailawadhi S, Knight J, Heckman MG, Pinkerton A, Chanan-Khan A. Nimbolide targets BCL2 and induces apoptosis in preclinical models of Waldenströms macroglobulinemia. *Blood Cancer J.*, 2014, 4, e260.

Cozzini P, Fornabaio M, Marabotti A, Abraham DJ, Kellogg GE, Mozzarelli A. Free energy of ligand binding to protein: evaluation of the contribution of water molecules by computational methods. *Curr. Med. Chem.* 2004, 11, 3093-3118.

Do QT, Renimel I, Andre P, Lugnier C, Muller CD, Bernard P. Reverse pharmacognosy: application of selnergy, a new tool for lead discovery. The example of epsilon-viniferin. *Curr. Drug Discov. Technol.* 2005, 2, 161-167.

Do QT, Lamy C, Renimel I, Sauvan N, André P, Himbert F, Morin-Allory L, Bernard P. Reverse pharmacognosy: identifying biological properties for plants by means of their molecule constituents: application to meranzin. *Planta Med.* 2007, 73, 1235-1240.

Ellingson SR, Smith JC, Baudry J. VinaMPI: Facilitating multiple receptor high-throughput virtual docking on high-performance computers. *J. Comput. Chem.* 2013, 34, 2212–2221.

Eric S, Ke S, Barata T, Solmajer T, Stankovic JA, Juranic Z, Savic V, Zloh M. Target fishing and docking studies of the novel derivatives of aryl-aminopyridines with potential anticancer activity. *Bioorg. Med. Chem.* 2012, 20, 5220-5228.

Ewing TJ, Makino S, Skillman GA, Kuntz ID. DOCK 4.0: Search strategies for automated molecular docking of flexible molecule databases. *J. Comput. Aided Mol. Des.* 2001, 15, 411–428.

Fan S, Geng Q, Pan Z, Li X, Tie L, Pan Y, Li X. Clarifying off-target effects for torcetrapib using network pharmacology and reverse docking approach. *BMC Syst. Biol.* 2012, 6, 152.

Feng LX, Jing CJ, Tang KL, Tao L, Cao ZW, Wu WY, Guan SH, Jiang BH, Yang M, Liu X, Guo DA. Clarifying the signal network of salvianolic acid B using proteomic assay and bioinformatic analysis. *Proteomics* 2011, 11, 1473-1485.

Ferrières J. The French paradox: lessons for other countries. *Heart* 2004, 90, 107–111.

Friesner RA, Banks JL, Murphy RB, Halgren TA, Klicic JJ, Mainz DT, Repasky MP, Knoll EH, Shelley M, Perry JK, Shaw DE, Francis P, Shenkin PS. Glide: a new approach for rapid, accurate docking and scoring. 1. Method and assessment of docking accuracy. *J. Med. Chem.* 2004, 47, 1739–1749.

Gao Z, Li H, Zhang H, Liu X, Kang L, Luo X, Zhu W, Chen K, Wang X, Jiang H. PDTD: a web-accessible protein database for drug target identification. *BMC Bioinformatics* 2008, 9104.

Grassi D, Desideri G, Ferri C., Flavonoids: Antioxidants Against Atherosclerosis. *Nutrients* 2010, 2, 889-902.

Gohlke H, Klebe G. Approaches to the description and prediction of the binding affinity of small-molecule ligands to macromolecular receptors. *Angew Chem. Int. Ed. Engl.* 2002, 41, 2644–2676.

Grinter SZ, Liang Y, Huang SY, Hyder SM, Zou X. An inverse docking approach for identifying new potential anti-cancer targets. *J. Mol. Graph Model* 2011, 29, 795-799.

Halliwell B. How to characterize an antioxidant: an update. *Biochem. Soc. Symp.* 1995, 61, 73-101.

Hu G, Gao J, Wang K, Mizianty MJ, Ruan J, Kurgan L. Finding protein targets for small biologically relevant ligands across fold space using inverse ligand binding predictions. *Structure* 2012, 20, 1815-1822.

Hu G, Wang K, Groenendyk J, Barakat K, Mizianty MJ, Ruan J, Michalak M, Kurgan L. Human structural proteome-wide characterization of Cyclosporine A targets. *Bioinformatics* 2014, 30, 3561, 3566.

Huey R, Morris GM, Olson AJ, Goodsell DS. A semiempirical free energy force field with charge-based desolvation. *J. Comput. Chem.* 2007, 28, 1145–1152.

Hui-fang L, Qing S, Jian Z, Wei F. Evaluation of various inverse docking schemes in multiple targets identification. *J. Mol. Graph. Model* 2010, 29, 326-330.

Irwin JJ, Sterling T, Mysinger MM, Bolstad ES, Coleman RG. ZINC: A free tool to discover chemistry for biology. *J. Chem. Info. Model* 2012, 52, 1757-1768.

Ivanisenko VA, Pintus SS, Grigorovich DA, Kolchanov NA: PDBSite: a database of the 3D structure of protein functional sites. *Nucleic Acids Res.* 2005, 33(Database issue):D183-7.

Jeong CH, Bode AM, Pugliese A, Cho YY, Kim HG, Shim JH, Jeon YJ, Li H, Jiang H, Dong Z. [6]-Gingerol suppresses colon cancer growth by targeting leukotriene A4 hydrolase. *Cancer Res.* 2009, 69, 5584-5591.

Ji ZL, Wang Y, Yu L, Han LY, Zheng CJ, Chen YZ. In silico search of putative adverse drug reaction related proteins as a potential tool for facilitating drug adverse effect prediction. *Toxicol. Lett.* 2006, 164, 104–112.

Jones G, Wilett P, Glen RC, Leach AR, Taylor R. Development and validation of a genetic algorithm for flexible docking. *J. Mol. Biol.* 1997, 267, 727–748.

Joseph J, Fellenstein C. Grid Computing. *Prentice Hall*, 2004. 400 pp.

Kamper A, Apostolakis J, Rarey M, Marian CM, Lengauer T. Fully automated flexible docking of ligands into flexible synthetic receptors using forward and inverse docking strategies. *J. Chem. Inf. Model* 2006, 46, 903-911.

Kanehisa M, Goto S, Furumichi M, Tanabe M, Hirakawa M. KEGG for representation and analysis of molecular networks involving diseases and drugs. *Nucleic Acids Res.* 2010, 38(Database issue), D355-D360.

Kellenberger E, Foata N, Rognan D. Ranking targets in structure-based virtual screening of three-dimensional protein libraries: Methods and problems. *J. Chem. Inf. Model* 2008, 48, 1014−1025.

Kellenberger E, Muller P, Schalon C, Bret G, Foata N, Rognan D. sc-PDB: An annotated database of druggable binding sites from the protein databank. *J. Chem. Inf. Model* 2006, 46, 717−727.

Kharkar PS, Warrier S, Gaud RS. Reverse docking: a powerful tool for drug repositioning and drug rescue. *Future Med. Chem.* 2014, 6, 333-342.

Kuntz ID, Blaney JM, Oatley SJ, Langridge R, Ferrin TE. A geometric approach to macromolecule–ligand recognitions. *J. Mol. Biol.* 1982, 161, 269–288.

Lauro G, Masullo M, Piacente S, Riccio R, Bifulco G. Inverse Virtual Screening allows the discovery of the biological activity of natural compounds. *Bioorg. Med. Chem.* 2012, 20, 3596–3602.

Lauro G, Romano A, Riccio R, Bifulco G. Inverse Virtual Screening of antitumor targets: Pilot study on a small database of natural bioactive compounds. *J. Nat. Prod.* 2011, 74, 1401-1407.

Lee M, Kim D. Large-scale reverse docking profiles and their applications. *BMC Bioinformatics* 2012, 13(Suppl 17), S6.

Li H, Gao Z, Kang L, Zhang H, Yang K, Yu K, Luo X, Zhu W, Chen K, Shen J, Wang X, Jiang H: TarFisDock: a web server for identifying drug targets with docking approach. *Nucleic Acids Res.* 2006, 34(Web Server issue), W219-224.

Li X, Li Y, Cheng T, Liu Z, Wang R. Evaluation of the performance of four molecular docking programs on a diverse set of protein-ligand complexes. *J. Comput. Chem.* 2010, 31, 2109–2125.

Liu X, Ouyang S, Yu B, Liu Y, Huang K, Gong J, Zheng S, Li Z, Li H, Jiang H. PharmMapper server: a web server for potential drug target identification using pharmacophore mapping approach. *Nucleic Acids Res.* 2010, 38(Web server issue), W609-W614.

Lu B, Hu M, Liu K, Peng J. Cytotoxicity of berberine on human cervical carcinoma HeLa cells through mitochondria, death receptor and MAPK pathways, and in-silico drug-target prediction. *Toxicol. In Vitro* 2010, 24, 1482-1490.

Ma C, Tang K, Liu Q, Zhu R, Cao Z. Calmodulin as a potential target by which berberine induces cell cycle arrest in human hepatoma Bel7402 cells. *Chem. Biol. Drug Des.* 2013, 81, 775-783.

Magrane M. and the UniProt consortium. UniProt Knowledgebase: a hub of integrated protein data. *Database* 2011, bar009.

Manach C, Scalbert A, Morand C, Rémésy C, Jiménez L. Polyphenols: food sources and bioavailability. *Am. J. Clin. Nutr.* 2004, 79, 727-747.

Merrifield RB. Solid phase peptide synthesis. I. The synthesis of a tetrapeptide. *J Am Chem Soc* 1963, 85, 2149-2154.

Meslamani J, Rognan D, Kellenberger E. sc-PDB: A database for identifying variations and multiplicity of 'druggable' binding sites in proteins. *Bioinformatics* 2011, 27, 1324−1326.

Muller P, Lena G, Boilard E, Bezzine S, Lambeau G, Guichard G, Rognan D. In silico-guided target identification of a scaffold-focused library: 1,3,5-triazepan-2,6-diones as novel phospholipase A2 inhibitors. *J. Med. Chem.* 2006, 49, 6768−6778.

Pan JB, Ji N, Pan W, Hong R, Wang H, Ji ZL. High-throughput identification of off-targets for the mechanistic study of severe adverse drug reactions induced by analgesics. *Toxicol. Appl. Pharmacol.* 2014, 274, 24-34.

Pandey KB, Rizvi SI. Plant polyphenols as dietary antioxidants in human health and disease; *Oxid. Med. Cell Longev.* 2009, 2, 270–278.

Paul N, Kellenberger E, Bret G, Muller P, Rognan D. Recovering the true targets of specific ligands by virtual screening of the Protein Data Bank. *Proteins* 2004, 54, 671–680.

Rarey M, Kramer B, Lengauer T, Klebe G. A fast flexible docking method using an incremental construction algorithm. *J. Mol. Biol.* 1996, 261, 470-489.

Rockenbach II, Gonzaga LV, Rizelio VM, Any Elisa de Souza Schmidt Gonçalves , Genovese MI, Fett R , Phenolic compounds and antioxidant activity of seed and skin extracts of red grape (Vitis vinifera and Vitis labrusca) pomace from Brazilian winemaking. *Food Res. Int.* 2011, 44, 897–901.

Rollinger JM, Schuster D, Danzl B, Schwaiger S, Markt P, Schmidtke M, Gertsch J, Raduner S, Wolber G, Langer T, Stuppner H. In silico target fishing for rationalized ligand discovery exemplified on constituents of Ruta graveolens. *Planta Med.* 2009, 75, 195-204.

Schmidtke P, Le Guilloux V, Maupetit J, Tuffery P: fpocket: online tools for protein ensemble pocket detection and tracking. *Nucleic Acids Res.* 2010, 38, W582-W.589.

Schomburg KT, Bietz S, Briem H, Henzler AM, Urbaczek S, Rarey M. Facing the challenges of structure-based target prediction by inverse virtual screening. *J. Chem. Info. Model* 2014, 54, 1676–1686.

Steffen A, Thiele C, Tietze S, Strassnig C, Kamper A, Lengauer T, Wenz G, Apostolakis J. Improved cyclodextrin-based receptors for camptothecin by inverse virtual screening. *Chemistry* 2007, 13, 6801-6809.

Stuart AC, Ilyin VA, Sali A: LigBase: a database of families of aligned ligand binding sites in known protein sequences and structures. *Bioinformatics* 2002, 18, 200-201.

Su S, Duan J, Cui W, Shang E, Liu P, Bai G, Guo S, Qian D, Tang Y. Network-based biomarkers for cold coagulation blood stasis syndrome and the therapeutic effects of Shaofu Zhuyu decoction in rats. *Evid. Based Complement Alternat Med.* 2013, 2013, 901943.

Tietze S, Apostolakis J. Glamdock: development and validation of a new docking tool on several thousand protein-ligand complexes. *J. Chem. Inf. Model* 2007, 47, 1657–1672.

Trott O, Olson AJ. AutoDock Vina: Improving the speed and accuracy of docking with a new scoring function, efficient optimization, and multithreading. *J. Comp. Chem.* 2010, 31, 455-461.

Vauzour D, Rodriguez-Mateos A, Corona G, Oruna-Concha MJ, Spencer JPE. Polyphenols and human health: Prevention of disease and mechanisms of action. *Nutrients* 2010, 2, 1106–1131.

Vigers GP, Rizzi JP. Multiple active site corrections for docking and virtual screening. *J. Med. Chem.* 2004, 47, 80-89.

Wang JC, Lin JH, Chen CM, Perryman AL, Olson AJ. Robust scoring functions for protein–ligand interactions with quantum chemical charge models. *J. Chem. Inf. Model* 2011, 51, 2528–2537.

Wang JC, Chu PY, Chen CM, Lin JH. idTarget: a web server for identifying protein targets of small chemical molecules with robust scoring functions and a divide-and-conquer docking approach. *Nucleic Acids Res.* 2012a, 40(Web server Issue), W393-W399.

Wang W, Zhou X, He W, Fan Y, Chen Y, Chen X. The interprotein scoring noises in Glide docking scores. *Proteins* 2012b, 80, 169-183.

Weiner SJ, Kollman PA, Nguyen DT, Case DA. An all atom force field for simulations of proteins and nucleic acids. *J. Comput. Chem.* 1986, 7, 230-252.

Westermaier Y, Barril X, Scapozza L. Virtual screening: An in silico tool for interlacing the chemical universe with the proteome. *Methods* 2015, 71, 44-57.

Wolber G, Langer T. LigandScout: 3-D pharmacophores derived from protein-bound ligands and their use as virtual screening filters. *J. Chem. Inf. Model* 2005, 45, 160–169.

Yang L, Luo H, Chen J, Xing Q, He L. SePreSA: a server for the prediction of populations susceptible to serious adverse drug reactions implementing the methodology of a chemical–protein interactome. *Nucleic Acids Res.* 2009a, 37(Web server issue), W406-W412.

Yang L, Chen J, He L. Harvesting candidate genes responsible for serious adverse drug reactions from a chemical-protein interactome. *PLoS Comput. Biol.* 2009b, 5, e1000441.

Yang L, Wang K, Chen J, Jegga AG, Luo H, Shi L, Wan C, Guo X, Qin S, He G, Feng G, He L. Exploring off-targets and off-systems for adverse drug reactions via chemical-protein interactome--clozapine-induced agranulocytosis as a case study. *PLoS Comput. Biol.* 2011, 7, e1002016.

Yue QX, Cao ZW, Guan SH, Liu XH, Tao L, Wu WY, Li YX, Yang PY, Liu X, Guo DA. Proteomics characterization of the cytotoxicity mechanism of ganoderic acid D and computer-automated estimation of the possible drug target network. *Mol. Cell. Proteomics* 2008, 7, 949–961.

Zahler S, Tietze S, Totzke F, Kubbutat M, Meijer L, Vollmar AM, Apostolakis J. Inverse in silico screening for identification of kinase inhibitor targets. *Chem. Biol.* 2007, 14, 1207-1214.

Zhang Z, Miao L, Lv C, Sun H, Wei S, Wang B, Huang C, Jiao B. Wentilactone B induces G2/M phase arrest and apoptosis via the Ras/Raf/MAPK signaling pathway in human hepatoma SMMC-7721 cells. *Cell Death Dis*. 2013, 4, e657.

Zhao J, Yang P, Li F, Tao L, Ding H, Rui Y, Cao Z, Zhang W. Therapeutic effects of astragaloside IV on myocardial injuries: Multi-target identification and network analysis. *PLoS ONE* 2012, 7, e44938.

Zheng R, Chen TS, Lu T. A comparative reverse docking strategy to identify potential antineoplastic targets of tea functional components and binding mode. *Int. J. Mol. Sci.* 2011, 12, 5200-5212.

In: From Medicinal Chemistry to Food Science
Editor: Pietro Cozzini
ISBN: 978-1-63483-587-9

Chapter 6

MOLECULAR DYNAMICS: AN EMERGING TOOL FOR EXPLORING THE MOLECULAR EVENTS IN FOOD RESEARCH

Tiziana Ginex[1], Constantí Seira[2], Carolina Estarellas[2], Axel Bidon-Chanal[2*] and F. Javier Luque[2]

[1]Molecular Modeling Laboratory, Department of Food Science,
University of Parma, Parma, Italy
[2]Department of Chemical Physics, Faculty of Pharmacy,
University of Barcelona, Campus Torribera,
Santa Coloma de Gramenet, Spain

Abstract

The continued progression of computer power and the development of highly efficient computational algorithms have expanded the spatio-temporal scale of molecular dynamics (MD) simulations. Over the last decades, molecular simulations have spread through a number of scientific fields, such as the identification of structure-function relationships in proteins, the structural and dynamical analysis of nucleic acids and their complexes with proteins, the development of novel nanomaterials and the design of bioactive compounds in drug discovery. In spite of the widespread integration across life and material sciences, the role of molecular simulations in food sciences is nevertheless much less appreciated, likely due to the huge complexity inherent to these molecular systems. However, it is reasonable to expect that this scenario will be progressively changed in the next years, as it is already reflected in an increasing number of studies that utilize molecular simulations to gain insight into the molecular factors in a diverse range of topics in food chemistry and nutrition. In this chapter, after a brief description of the current state of molecular dynamics (MD) simulations, we highlight a number of representative cases where MD techniques have been utilized to explore the molecular details of different processes in food science and technology. While MD-based techniques are an integral feature in other scientific disciplines, it can be expected that in the next years these techniques will have an increased impact in food research.

* Corresponding authors: abidonchanalb@ub.edu (AB-C)

Introduction

Conformational flexibility and thermal fluctuations are key factors in our understanding of biomolecular processes, and the relationship between structure and biological function cannot be fully appreciated without taking into account the dynamical behavior of biomolecules [1, 2]. Such a dynamical description is encoded in the conformational transitions exploited by enzymes to carry out their biological function in the native state. This is illustrated by the existence of direct links between micro- to milli-second timescale domain motions and enzymatic activity, and by the role of pico- to nano-second atomic fluctuations in facilitating large-scale motions leading to catalytically competent states in adenylate cyclase [3, 4]. Similarly, the formation of protein-protein and protein-nucleic acid complexes appear to exploit intrinsic deformation patterns of the unbound partners in order to form macromolecular aggregates [5-8]. Allosteric mechanisms also involve the transmission of signals that exploit the existence of dynamical networks in proteins, which depend on protein topology, specific binding events, covalent modifications and cellular environmental conditions [9]. Protein dynamics has also a crucial role in modulating the topological features of inner cavities and tunnels, which in turn has a direct influence on the association and dissociation process of small organic molecules [10]. In turn, the ability to determine the dynamic changes that are triggered upon ligand binding is a key issue in the design of bioactive compounds [11, 12].

The preceding discussion suffices to highlight the need to abandon the static picture that arises from the single structure of biomolecules, and the need to account for a dynamical description, whereby they should be appear as entities able to populate a diversity of conformational states.

The limitations of experimental techniques for providing direct information about biomolecular flexibility and its relationship with the biological function have promoted the development of theoretical approaches, such as molecular dynamics (MD), which is widely used to obtain detailed atomistic information about the structure and dynamics of biomolecules, as well as about the thermodynamics and kinetics of biological processes [13-15]. While MD simulations have had an enormous impact in physics, chemistry and biology, their application to food sciences has not found an equivalent role, and only recently the potential impact of these techniques starts to be recognized [16].

The aim of this contribution is to provide a short review of recent progresses made in the application of atomistic computer simulations to understand the molecular basis of representative processes relevant in food science. Emphasis will not be made on the foundations of these techniques, which can be found in detailed reviews [13-16]. Rather, the major goal is to describe the potential range of applications and the nature of the information that can be extracted by choosing an appropriate set of representative examples recently reported in the literature.

Molecular Dynamics: A Brief Methodological Overview

MD relies on molecular mechanics models, which describe molecular systems as a collection of spherical particles connected by springs that represent the constituent atoms and bonds. Furthermore, the interactions between particles are described by means of simple mathematical functions, which incorporate a number of adjustable parameters. By using this simplified description, the time evolution of a molecular system can be followed by solving Newton's equations of motion, leading to a collection of snapshots that represent the structural fluctuations of the simulated system.

Running a MD simulation requires the adoption of certain assumptions about the level of description of the biomolecular system, the interactions between the classical particles via suitable potential energy functions, and the algorithms utilized for sampling of the conformational landscape.

Description of the simulated system. Atomistic models provide a description of the simulated system at atomic detail, so that the trajectory will permit to follow the time evolution of each single atom of the whole structure subject to the appropriate constraints imposed by the molecular topology. Interactions between "classical" atoms are then described by means of simple analytical expressions, which constitute the force field, where quantum effects are considered in an indirect way via the introduction of suitable parameters. In turn, these parameters are carefully calibrated against high-quality quantum mechanical data and experimental data.

For atoms that are topologically bonded, forming covalent bonds, bond angles and dihedral torsions, biomolecular force fields generally use simple two-body, three-body and four-body expressions, even though the inclusion of cross terms is also feasible. On the other hand, interactions between non-bonded atoms are generally represented by coulombic and Lennard-Jones expressions. This is illustrated by the basic force field implemented in AMBER [17], which includes the analytical expressions given in Eq. 1.

$$E = \sum_{bonds} k_{st}\left(b-b_0\right)^2 + \sum_{angles} k_b\left(\theta-\theta_0\right)^2 + \sum_{torsions} \frac{V_{tor}}{2}\left[1+\cos\left(n\phi-\delta\right)\right]$$
$$+ \sum_{\substack{non-bonded \\ atoms}} \left[\left(\frac{A_{ij}}{r_{ij}^{12}}\right) - \left(\frac{B_{ij}}{r_{ij}^{6}}\right)\right] + \frac{q_i q_j}{\varepsilon r_{ij}} \qquad (1)$$

where b, θ and $\emptyset$ represent the instantaneous bond length, bond angle and dihedral angle between bonded atoms, r_{ij} is the interatomic distance between non-bonded atoms *i* and *j*, while the parameters b_0 and θ_0 stand for the equilibrium bond lengths and angles, V_{tor}, n and δ denote the torsional barrier, periodicity and phase angle of the Fourier term used to describe the torsional energy, q_i is the atomic partial charge of atom *i*, and A_{ij} and C_{ij} stand for the van der Waals parameters that characterize the Lennard-Jones interaction between particles *i* and *j*.

The quality of the results depends directly on the reliability of the parameters introduced in the force field. This has led to the introduction of continued revision of current force fields, leading to the introduction of successive refinements for the treatment of proteins [18, 19], nucleic acids [20-23], glycans [24], lipids [25] and even generalized force field for organic molecules [26]. Even though they are not explicitly described here, it is worth noting that similar refinements have been developed in other biomolecular force fields, such as CHARMM [27] and GROMOS [28, 29].

Current efforts are being made for the explicit inclusion of induction effects using a variety of polarization formalisms [30], which should lead to a better representation of the molecular response to the electric fields created by the anisotropic biological environments. Thus, efforts have led to the implementation of the Drude model in CHARMM [31, 32] and the induced dipole in AMOEBA [33] and AMBER [34, 35]. As an example, we limit ourselves to remark the crucial effect played by the inclusion of induction effects to describe the lipid interface dipole potential, which is responsible for the difference in permeability between negative and positive hydrophobic ions [36].

To set-up the simulation system, one has to specify the nature and composition of the environment, such as the nature of the solvent and the ionic atmosphere. It is now conventional to utilize periodic boundary conditions [13], which permits to replicate the system periodically in all directions, so that the atoms in the central simulation box are imaged in each of the surrounding boxes. This leads to the simulation of a pseudo-infinite system, as this procedure permits each atom to be surrounded by a sizable set of neighboring atoms located either in the central box or in the replicated ones. Furthermore, the physical conditions adequate for the simulation, such as pressure or temperature, must be considered, which will define the ensemble of all microscopic states belonging to a single thermodynamic state, generally corresponding to the canonical (constant temperature and volume) or isobaric-isothermal (constant pressure and temperature) ensembles.

Sampling of the simulated system. The dynamical description of biomolecular systems encompass a large variety of time scales, ranging from extremely fast processes, such as the rotation of solvent-exposed side chains, to extremely slow motions, such as the folding of proteins. The fastest motions correspond to the vibration of bonds, which occur in the time scale of femtoseconds, thus limiting the time step to be used in the integration of Newton's equation of motion in atomistic MD simulations. Accordingly, solving the iterative numerical calculation of the instantaneous forces must be repeated a huge number of times in order to built up a trajectory representative of the accessible microstates for the chosen simulation ensemble.

Freezing the fastest motions by using suitable constraints, such as those implemented in SHAKE [37] and LINCS [38], is a widely adopted strategy to increase the efficiency of computer simulations, as the bond vibrations have generally little interest in conformational analysis of biomolecules. On the other hand, a significant fraction of the computational cost comes from the evaluation of interatomic interactions, specifically due to the long-range nature of Coulombic interactions. As the size of the system is increased in order to incorporate more realistic descriptions of the physiological environment, the number of pairwise interactions to be considered may limit the overall computational efficiency. To overcome this limitation, the spherical truncation approach, by which interactions beyond a given cutoff distance were ignored, was first implemented. However, due to presence of energy conservation and occurrence of unstable simulations led to the adoption of the Ewald summation method [39], which provides a more rigorous framework for the calculation of electrostatic interactions in infinite periodic systems.

In order to enhance the capability of atomistic MD simulations to describe rare events, which may be associated with large-scale structural changes, a number of algorithms have been devised to bias the system toward the target conformational transition [40-45]. Umbrella sampling bias the trajectory along a potential transition coordinate that drives the transition between the corresponding initial and final states through the inclusion of a biasing potential. Targeted MD drives the transition between states by introducing restraints that tend to reduce the root-mean square deviation of the system with regard to a target structure. Accelerated MD adds a boost potential to the potential function of the system, which results in the reduction of the energy barriers that separates different states of the system, with the advantage that there is no need to define a reaction coordinate. Metadynamics inserts memory in the sampling, so that the system is discouraged to visit previous sampled states by filling the free energy wells than have already been visited. Replica-exchange MD employs independent parallel Monte Carlo random walks in several parallel simulations carried out at different temperatures, so that system states have a finite probability to be exchanged depending on temperature and energy differences between selected simulations. The preceding comments suffice to illustrate the diversity of strategies that can be adopted to enhance the conformational sampling, though a detailed explanation escapes the scope of this contribution (for a detailed description of the formalisms the reader is addressed to refs [40-45].

In addition to the developments and refinements in MD algorithms, the advances introduced in computer power have contributed decisively to enhance the sampling capabilities of MD simulation codes. The last versions of MD software packages, such as NAMD [46] and GROMACS [47], have achieved a high performance for exploiting parallelized computing. The development of MD-dedicated massively parallel supercomputers, such as the ANTON computer [48], has also led to an outstanding enhancement in computing power. On the other hand, the high arithmetic performance and intrinsic parallelism of graphical processing units offers an alternative technological edge for enhancing the efficiency of MD simulations, as reflected in the GPU-oriented versions of AMBER [49, 50] and ACEMD [51] codes.

Representative examples of the impact of improvements in force fields, computational algorithms and simulation speed come from the studies that allow direct observation of the dynamics of folding events in atomic detail [52, 53]. These studies are informative about the structure and stability of the folded state, the heterogeinity of folding pathways, the origin of the barriers implicated in protein folding, and the nature of misfolded states. Similarly, MD simulations have shown the spontaneous binding of ligands to their targets without imposing any prior bias toward the binding pocket [54-56], which are valuable not only to identify the binding pathway and energetic barriers that determine binding kinetics, but also to disclose alternative binding sites that might facilitate the design of allosteric drugs.

Simplified treatments of biomolecular systems. Generally the time resolution of experimental techniques cannot go beyond the millisecond-microsecond resolution, while atomistic MD simulations typically cover hundreds of nanoseconds for systems that extend less than 10 nm. There is then urgent need for developing methods capable of bridging the time scale of experimental and computational techniques for realistic simulation systems.

A promising approach to reduce the cost of simulating large macromolecular systems consists in the use of coarse-grained models [57, 58], where 'classical' particles (beads) now represent typically clusters of atoms (e.g., an amino acid is represented by two beads, one for the backbone and the other for the side chain). These models lead to a significant reduction in the degrees of freedom of the system, thus enabling to overcome limitations in size and time scales encountered by atomistic simulations. Models have been developed for different kinds of constituent units in biomolecules within the residue-based coarse-grained approach, including nucleic acids, lipids, proteins and even water [59-64]. Nevertheless, at a coarser level of description, shape-based models, where beads represent protein segments, whose dynamics is described using MD simulations as well, have also been developed [65, 66].

A representative example of the coarse-grained force field is MARTINI [67]. It contains four main types of particles: polar, non-polar, apolar, and charged. Within each type, subtypes are distinguished depending on hydrogen-bond capabilities or the degree of polarity, giving rise to a total of 18 particle types. Non-bonded interactions are described by a Lennard-Jones potential, whereas charged groups also interact via a Coulombic energy function. The non-bonded interactions have been parametrized based on a systematic comparison to experimental thermodynamic data, including free energies of hydration and vaporization, and the partitioning free energies between water and a number of organic phases. Bonded interactions are treated by a standard set of potential energy functions, including harmonic bond and angle potentials, and multimodal dihedral potentials. In this case, parametrization is made using structural data derived from template geometries or from atomistic simulations. Applications of the MARTINI force field cover a wide range of topics, such as lipid polymorphism, lipid membrane properties, membrane protein oligomerization, pore formation in membranes, and the interaction of nanoparticles with membranes [68]. Finally, it is also worth mentioning the package ESPResSO [69], which has been designed as a versatile software for coarse-grained atomistic or bead-spring models of soft matter. The code has been utilized to study membrane remodeling and vesiculation by curvature-inducing model proteins [70].

Molecular Dynamics Simulations: Applications in Food Sciences

Little attention has been paid to the use of MD simulations in the study and characterization of food materials. Indeed, attention has typically relied on the description of foods from the perspective of soft matter physics, mainly due to the large separation of the length scales between mesoscopic structures and the chemical constituents at the molecular scale [71-73]. As a consequence, detailed insight into the processes at the microscopic scale is lost, and the coarse scale model parameters are apparent rather than physical parameters. However, the potential impact of MD simulations in certain areas of food science has been recognized by Pink and Razul, who have reviewed a number of MD-based studies in their recent review [16].

Our aim here is to describe a series of recent studies that illustrate a representative range of potential applications of MD simulations in understanding the molecular basis of a series of processes relevant in food science.

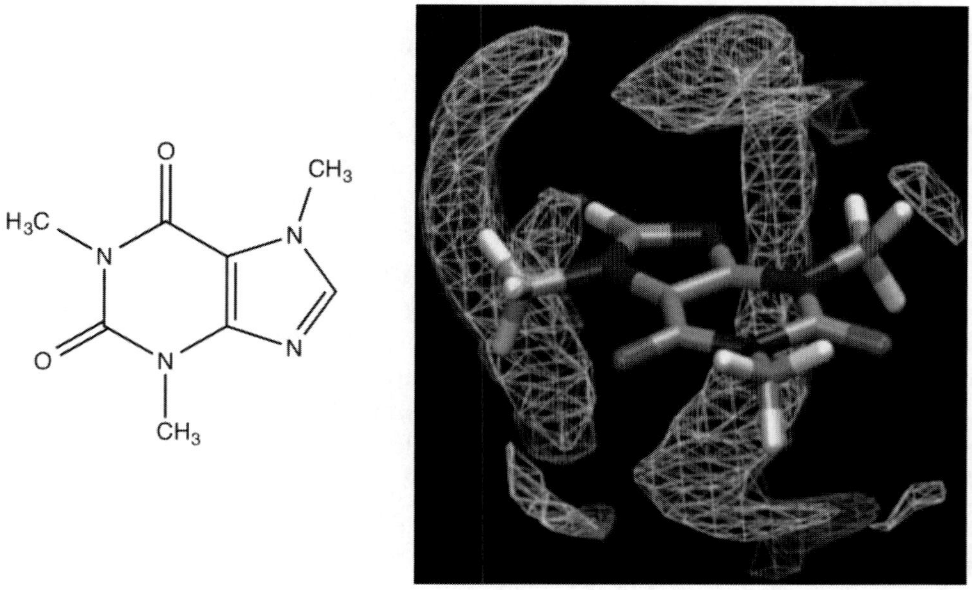

Figure 1. (Left) Chemical structure of caffeine. (Right) isocontour representation of the water distribution around a single caffeine molecule. Isocontour surfaces encloses regions with a water oxygen atom density 1.3 times (yellow) and 1.4 times (red) the bulk density (Reprinted with permission from J. Phys. Chem. B. 2011, 115, 10957-10966. Copyright 2011 American Chemical Society).

Caffeine: Solubility and self-association in water. Coffee is among the most popular beverages in the world. The most important active substance is caffeine (1,3,7-trimethylxanthine; Figure 1), which is also found in chocolate, tea and other kinds of soft drinks. Its huge consumption is related to the role as a stimulant of the central nervous system. In such beverages, caffeine is in solution with many other constituents, such as polysaccharides, lignans, polyphenolics, lipids, melanoidins, and chlorogenic acid [74]. The physicochemical properties of caffeine justify its mild solubility in water, which is estimated to be around 0.11 M at 298 K [75]. Thus, even though it has a significant dipole moment (~3.7 Debyes), the presence of the methyl groups attached to the heterocyclic ring makes the surface of this flat molecule to be only moderately hydrated. Hence, caffeine is sparingly soluble in water at room temperature and the solubility increases with temperature. In fact, caffeine self-associates to form dimers, and to some extent trimers and higher aggregates, and the association is enthalpically driven [75-77]. The apparent self-association constant of the vertical stacking process has been estimated to be 10.5-16.5 M^{-1} at 298 K [78-80].

MD simulations have been recently used to gain insight into the structural details of caffeine hydration and association in aqueous solution [81]. The results show the complex organization of the hydrating water molecules, which appear in three distinct areas around the caffeine solute (Figure 1). Preferred hydration sites are found above and below the molecular plane, close to the carbonylic oxygens. An additional hydration site is localized in the molecular plane facing the lone pair of the pyridine-like nitrogen in the five-membered ring. Finally, a set of weakly bound water molecules is found above and below the molecular plane.

The trajectories also revealed the dynamical exchange of caffeine molecules in self-association. Thus, even though association is guided by stacking of caffeine molecules, steric clashes between methyl groups limits the number of relative arrangements between stacked molecules. The distribution of stacked clusters was found to be in qualitative agreement with the profile expected from the isodesmic model, which assumes that the addition of a monomer to an aggregate is independent of the size of the aggregate.

The effect of cosolutes on the association of caffeine has also been studied. The results reveal distinctive trends in the influence exerted by some inorganic salts on the self-association of caffeine [80]. In particular, the presence of sodium chloride tends to destabilize monomeric caffeine, leading to slightly larger self-association. This trend has been reproduced by MD studies performed for a system comprising 8 caffeine molecules and 4500 water molecules, with a salt (NaCl) concentration ranging from pure water to 0.83 M [82]. The analysis of the caffeine-caffeine center of mass distribution function showed that, on addition of salt, there is an increase in the first peak (located at around 4 Å), but also pointed out the appearance of a second peak (at around 7.2 Å). Furthermore, it was found that higher order clusters of caffeine are formed as the concentration of NaCl increases.

The effect of co-solutes on the caffeine aggregation has also been extended to other substances, such as sucrose, which was found to lower the solubility in water as the sugar concentration increases [75]. The interaction between caffeine and sugars has been examined in a recent study [83], which examined the effect of sugars in simulation systems containing caffeine in concentrated aqueous solutions of sugars (glucose, sucrose). Simulations showed that sugars weakly interact with caffeine by forming face-to-face stacking, with the nonpolar regions of sugars being primarily oriented toward the hydrophobic face of caffeine. Hence, it can be speculated that liberation of water molecules structured over the hydrophobic face of caffeine might facilitate self-aggregation.

While the preceding findings are in agreement with available NMR data (83), they also raise a number of questions, such as the impact of sugars on the interaction of caffeine with other components extracted from coffee, and how the intertwined set of interactions may affect the taste of the beverage. As an illustrative example, it is worth noting the influence of ions on the dissolution and extraction of flavorsome compounds, so that different rates and compositions of coffee extraction can be achieved by varying the nature of the ionic species present in water [84].

Spherification: Role of Calcium Cations in Alginate Aggregation

Alginates are naturally occurring unbranched polysaccharides isolated from brown algal species and from the extracellular matrix of certain bacteria. They are composed of (1→4)-linked residues of β-D-mannuronic acid (M) and its C5-epimer, α-L-guluronic acid (G), and the precise chemical composition depends on the biological source [85]. Alginates are biodegradable, biocompatible and nontoxic. They are commonly used as thickeners in food industry, as well as stabilizing agents and emulsifiers [86]. Hence, there is technological interest in understanding their biochemical and biophysical properties, the nutritional properties of dietary alginates and the potential use as drug carriers [87-91].

The sol-gel transition of sodium alginate is a phase transition induced by the presence of calcium ions, which exhibit a preferential affinity for G units compared to the M ones [92-94]. Thus, the gel forming properties of alginates are derived from their capacity to bind a large number of divalent ions and the gel strength is correlated with the proportion and length of the G residues in the alginate chains. This process is the basis of 'spherification', a technique amply utilized in modernist cuisine pioneered at El Bulli [95]. This technique, which relies on the formation of alginate spheres that contain a physical outer gel membrane with a liquid core, is central to the formation of faux caviar, eggs, gnocchi and ravioli. In the process of spherification a liquid containing sodium alginate is submersed in a bath of calcium or, alternatively, a calcium source is added to the edible liquid and then mixed with a sodium alginate bath. Gelation involves Ca^{2+} cross-linking of alginate chains, which results in the formation of a film surface.

The details of the molecular events implicated in this process are still poorly understood, which in turn explains the increasing interest in gaining insight into the structural and energetic features of the interaction between calcium cation and alginate by means of theoretical studies. An aspect that has attracted much interest is the origin of the differential interaction of calcium cations with G and M units, as the more favorable interaction with G units is believed to arise from the 'egg-box' model proposed around 40 years ago (96). According to this model, by adopting a 2_1 helical conformation, the alginate chain can pack assisted by the presence of the calcium cations between the distinct chains (Figure 2).

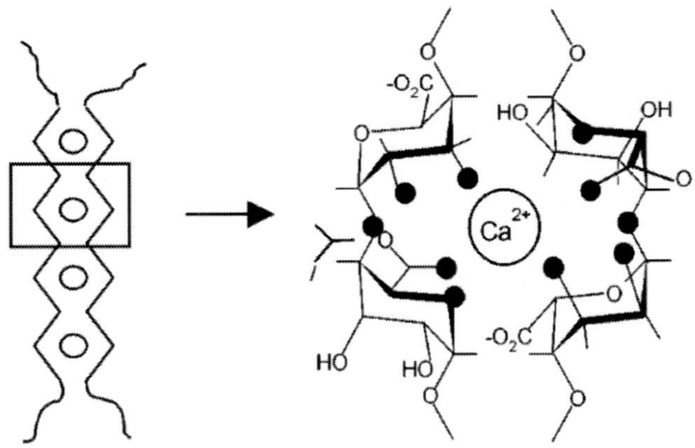

Figure 2. Schematic representation of the calcium coordination in the "egg box" model in calcium alginate junction zones (Reprinted with permission from Biomacromolecules 2001, 2, 1089-1096. Copyright 2011 American Chemical Society).

Several studies have examined the suitability of this model [97-99]. Recently, Plazinskin and Dracht [100] have used hybrid MD-DFT calculations to examine the interaction of calcium cations with a single G molecule or with an oligomeric chain of G units in order to disclose the nature of the structural model and the nature of the interactions involved in cation chelation. They concluded that the preferential interaction of Ca^{2+} include the coordination with negatively charged carboxyl group and water molecules, while excluding a relevant contribution from hydroxyl groups or ring atoms. From an energetic point of view, both mono- and bidentate binding modes are found to be nearly equally probable. Finally, the results point out that only the carboxyl groups of two poly(guluronate) chains contribute to the binding of calcium and to the junction of these two chains.

Recently, by using atomistic MD simulations, Chipot and coworkers have examined the process of how liquid food can be shaped into spheres [101]. Specifically, they analyzed the structure of the calcium alginate membrane and the role played by calcium ions, in comparison with sodium ions, the structural stability of the polysaccharide membrane, and the adaptability of the alginate membrane to encapsulate liquid foods of different nature. The analysis of the trajectories showed that the alginate chains spontaneously form a net-like membrane on the surface of the liquid droplets (Figure 3). Furthermore, they also showed that the joints between adjacent alginate chains possess egg-box motifs, whereby the stabilization can be ascribed to the electrostatic interactions between calcium cations and the carboxylate moieties. Importantly, such stabilization is significantly reduced in the presence of sodium cations, thus pinpointing the crucial role played by the divalent cation in spherification.

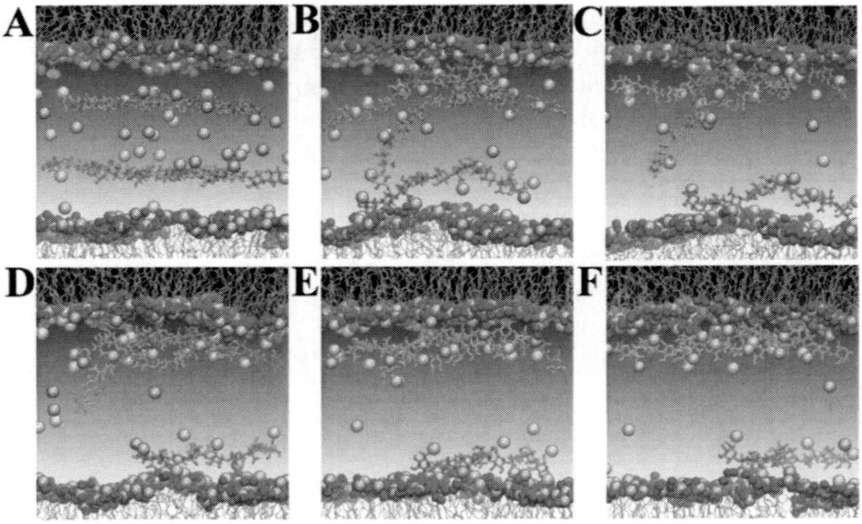

Figure 3. Representation of the spontaneous adsorption of four alginate chains at the oleate-water interface. Panels A-F denote selected snapshots along the 80 ns trajectory. Calcium ions are represented as pink spheres (Reprinted with permission from J. Phys. Chem. B 2014, 118, 11747-11756. Copyright 2011 American Chemical Society).

Green tea catechins: interaction with lipid bilayers. Catechins are polyphenols belonging to the class of flavan-3-ols typically found in cocoa, berries and green tea. Several beneficial effects, partly due to their antioxidant activity, have been reported [102]. Green tea (*Camellia sinensis*) contains high levels of flavan-3-ol monomers, the main components being (-)-epigallocatechin (EGC), (-)-epigallocatechin-3-O-gallate (EGCG) and (-)-epicatechin-3-O-gallate (ECG) [103]. Fermentation of green leaves leads to reduction in flavan-3-ols content as a result of the action of polyphenol oxidase with a consequent accumulation of theaflavins and thearubigins in black tea [104]. Studies on animal models show that EGCG can inhibit carcinogenesis at all stages, viz. initiation, promotion and progression [105].

Understanding the biophysical and biochemical implications of the interaction between catechins and specific cellular targets is of utmost relevance for their pharmacological use. Furthermore, even though catechins have proved to be effective in *in vitro* assays, their poor pharmacokinetic profile makes them unsuitable for direct *in vivo* applications. Thus, they are potentially good "lead" compounds for drug discovery in a nutraceutic scenario.

It has been speculated that the biological activities of catechins could be determined by their interaction with lipid membranes [106, 107]. For instance, differential scanning calorimetry and spectroscopic studies by Catura *et al.* [108] have revealed that ECG and EGCG bind deep in bacterial model membranes. It can be assumed that the interaction of tea catechins with lipid bilayers is influenced by physicochemical factors such as the number of hydroxyl groups on the B-ring, the presence of the galloyl moiety and the stereochemical structure. At this point, EGCG has been found to be the most effective in perturbing the membrane structure of bacteria-like model membranes.

Sirk and co-workers [109] performed atomistic MD simulations to gain insight into the role of hydrogen bonding of catechins with components of cell membranes. The simulation model consisted of a lipid bilayer containing a 1:1 mixture of 1-palmitoyl-2-oleoyl-phosphatidylcholine (POPC) and 1-palmitoyl-2-oleoyl-phosphatidylethanol-amine (POPE), which were considered to build up a membrane model that mimics the HepG2 membrane cell. Structural and dynamic properties for seven catechins (C, EC, EGC, CG, ECG, EGCG and GCG; Figure 4) interacting with the lipid bilayers, such as molecular orientation and hydrogen bonding, were monitored during 50 ns MD simulations.

Each catechin showed distinctive trends with regard to the catechin-bilayer interaction. All of them were able to interact with the lipid bilayer and were rapidly internalized. An important exception to this common behavior was represented by EGCG, since its absorption into the lipid bilayer was preceded by adsorption and fluctuation on the lipid surface. More generally, the adsorption process was influenced by the size and conformational behavior of compounds.

Once into the bilayer, all catechins maintained flexibility: their diffusion though the membrane was characterized by formation and breaking of hydrogen-bond interactions with the lipid headgroups. EGCG was found to be the compound that formed a larger number of interactions. This effect was ascribed to the presence of the gallate unit and its *cis* configuration with regard to the B-ring, which promote multiple hydrogen-bond interactions. This unique behavior among catechins might explain the ability of EGCG to target specific cells [110]. MD simulations also confirmed the higher affinity of EGCG and ECG for lipid bilayers compared to EC and EGC, providing important sub-molecular insights on the capability of catechins to interact with lipid bilayers.

Overall, these studies illustrate the potential impact of MD simulations in gaining insight into the physicochemical basis of the biological effects of polyphenols, which in turn should lead to a better comprehension of their nutraceutic role and the potential implementation in food sciences.

The ubiquity of cholesterol: the feared and crucial molecule. Even though most people relate excess of cholesterol with serious health problems, the importance of this molecule for the correct work of our life is not generally well appreciated. Cholesterol is an essential component of the mammalian cell membranes needed for their structural integrity, fluidity and permeability. Additionally, it is the precursor molecule of vitamin D, bile acids and steroid hormones as cortisol, progesterone, estrogens and testosterone. Due to the significant health implications, cholesterol and its interactions with target molecules and different environments have been intensively studied in the last 30 years [111]. In fact, MD simulations have been largely used to study cholesterol in different biological structures.

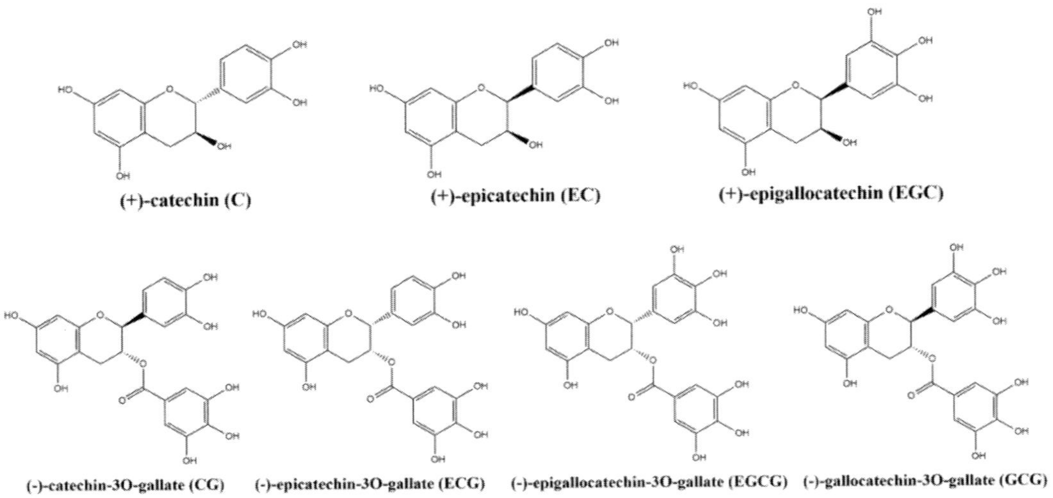

Figure 4. Chemical structures of the seven catechins studied by Sirk and co-workers (109).

The relationship between cholesterol and the properties of the lipid bilayer has been extensively studied by means of MD simulations. Among other topics, attention has been paid to the dependence of the structural and dynamical features of the lipid bilayer on the percentage of cholesterol, considering both interaction patterns [112-115] and the rigidity and membrane order [116-125]. These studies have provided fundamental information about the behavior and properties of the cell membranes. Other relevant subjects are the effect of cholesterol on the permeability and the different hydration degree of the membrane [126, 127], and the influence on the formation and stability of membrane pores [128-130]. The effect of cholesterol analogs on the fluidity, order level and permeability of the membrane have also been examined by MD simulations, including compounds such as dehydroergosterol [131, 132], lanosterol or ergosterol [133-135], among others [136-138].

Cholesterol may also form direct interactions with membrane proteins, particularly with G protein-coupled receptor (GPCRs), as recently reviewed by Oates and Watts [139]. GPCRs form a group of membrane proteins characterized by having seven transmembrane (TM) helical segments connected by three extracellular and three intracellular loops and an alpha-helical C-terminus. GPCRs respond to extracellular stimuli and transmit them inside the cell. Even though the influence of cholesterol on GPCR activity is well known, the nature of the interaction remains to be elucidated, even though the discovery of bound cholesterol in the crystal structure of the β2-adrenergic receptor and the subsequent identification of a consensus cholesterol binding motif supports a direct effect on the receptor [140]. For instance, results from MD simulations of rhodopsin indicate that cholesterol influences the activation elements in the TM1-TM2-TM7-H8 functional network of the receptor, such as the relative angular motion between TM7 and H8, and the relative movement of TM1 and TM7, by affecting the helical kink parameters in TM1, TM2, and TM7 [141]. In particular, the presence of cholesterol near Val1.58, Tyr2.63, and Pro7.38 was found to correlate with the changes in the kink angle values in the TM1, TM2, and TM7, respectively.

Cholesterol also exerts direct effects in other transmembrane proteins, such as ligand-gate ion channels. In the case of the glutamate-gated chloride channel, the X-ray structure of the open state has revealed the existence of a lipophilic agonist, ivermectin, bound to the subunit interface in the transmembrane domain [142]. MD simulations supported a stable binding, though cholesterol exhibits some fluctuations between several orientations [143]. Furthermore, on the relatively short timescale spanned by simulations, cholesterol tends to promote pore opening.

In the case of the nicotinic acetylcholine receptor, MD simulations showed that cholesterol tend to concentrate around TM helix 4, in agreement with photoaffinity labeling studies, typically interacting with the protein via hydrogen bonding and hydrophobic interactions [144]. A putative cholesterol-binding pocket involving TM cysteine residues (Cys445, Cys220) of the β2 subunit was also identified in the open-channel system, while nonannular cholesterol binding sites were found in the closed-channel species (Figure 5). These findings suggest that cholesterol might be involved in the regulation of the channel gating process through direct interactions with critical residues.

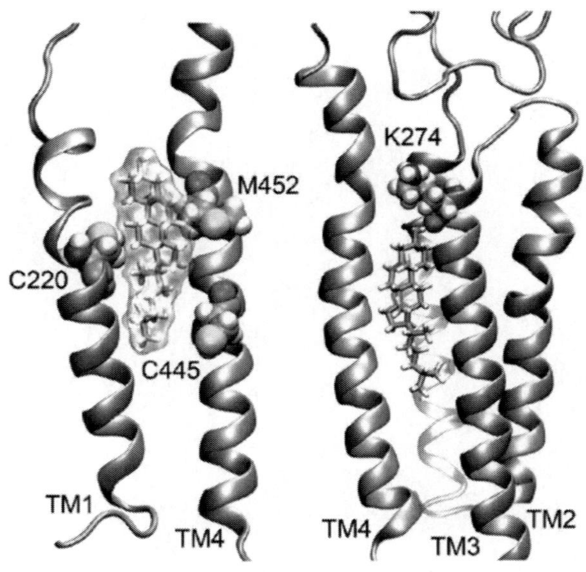

Figure 5. Representation of cholesterol binding sites in the open channel model of the nicotinic acetylcholine receptor. (Left) Cholesterol bound to the three residues containing sulfur atoms (yellow) in the β2 subunit. (Right) Hydrogen bonding of cholesterol to Lys274 in the β2 subunit (Reprinted with permission from J. Phys. Chem. B 2009, 113, 6964-6970. Copyright 2011 American Chemical Society).

Overall, these examples illustrate the potential contribution of MD simulations in clarifying the nature of the interactions formed between cholesterol and membrane proteins, and especially in disclosing the effect of these interactions on the biological function.

Olfactory receptors: the influence of food odor. Olfaction is a complex chemo-biological process that happens when odorants or pheromones bind to Olfactory Receptors (ORs) in the membrane of specialised sensory cells, thereby activating a signalling pathway that is finally transformed into an electrical signal by the opening of ion channels [145, 146]. ORs belong to the superfamily of G protein-coupled receptors, and the binding of odorants activates the consequent signal pathway cascade inside the OR neurons. The relation between odorants and ORs could be defined as promiscuous in the sense that each OR can bind and respond to different odorant molecules and a single odorant can activate multiple ORs that will recognise different structural features of the odorant. As a result, odours are the consequence of the superposition of several signals generated cooperatively by multiple ORs.

Several attempts have been made to establish correspondences between odorants and ORs as well as to classify ligands as agonists or non-agonists and to unravel the determinants of molecular recognition using experimental and computational techniques [147-149]. Singer was the first that used rhodopsin to model the rat's I7 olfactory receptor to analyse the interactions of octanal in its binding site by flexible ligand docking coupled to short (20 ps) restrained MD simulations [150], in the first published work where MD was applied to model an OR. This model was later used to study the interaction of aldehydes with rat OR I7, leading to the identification of an exit pathway from the binding site through MD simulations in vacuum [151]. Identification of the pathway allowed to relate the existence of a clear transit path with OR activation in agreement with experimental data [152].

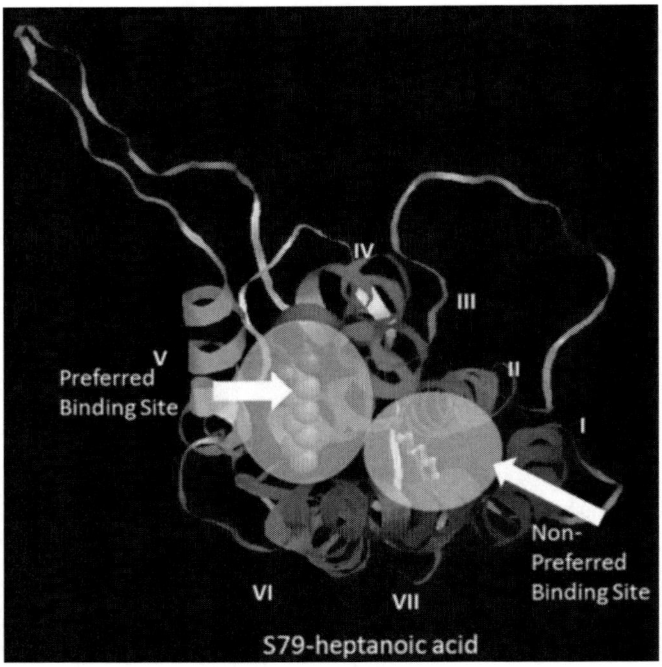

Figure 6. Representation of the structure of OR S70 showing the two putative binding pockets. The circle bound by TMs III, IV, V and VI is the region that favors OR activation, and the circle bound by TMs I, II, III and VII is the nonpreferred binding region (Reprinted with permission from Chem. Senses 2014, 39, 107-123. Copyright 2014 Oxford University Press).

With the recent advances in computing power, Anselmi *et al.* modelled the human OR17-40 using the crystal structures of rhodopsin and the ß2-adrenergic receptor [153]. By using free energy techniques, they showed that although different molecules can bind the OR binding site, only a few of them can bind strongly enough as to elicit an electrophysiological response. A similar strategy has been used to study how the human OR1G1 is able to recognise a broad set of chemically diverse molecules [154], including four known agonists (nonanal, 1-nonanol, 9-decen-1-ol and camphor) and one non-agonist (butanal). This study revealed the importance of the hydrophobic contribution to the binding free energy compared to the electrostatic term, which would explain the adaptability of the ORs to diverse chemical structures. More recently, Lai et *al.* have run MD simulations of two mouse ORs, S97 and S86, forming a complex with different odorants to try to establish a binding pattern common to all of them. Their simulations showed that both ORs have two different binding sites (Figure 6) and that in each case binding to one but not the other has an enhanced activating effect on the receptor [155].

Overall, MD-based simulation techniques are promising tools to decipher the intricate events of OR recognition and activation. However, further efforts need to be made to get a deeper insight into the promiscuity of ORs and the activation mechanisms, which will be valuable to understand how odorants generate odors that we sense as smells.

Conclusion

The refinements introduced in MD algorithms and the sustained increase in computing power has expanded enormously the spatio-temporal scale of processes that are attainable to computer simulations. Thus, MD simulations will be able to disclose the molecular details of relevant biochemical processes using more complete and realistic description of biomolecular systems. Predicting the impact of these technical progresses in food sciences is always difficult, especially keeping in mind that they are in a very incipient state. Nevertheless, it is reasonable to expect that MD simulations will have a similar impact regarding the efforts to understand at the microscopic level the structure and properties of the basic constituents of foods. Up to now, the explicit description of food features at the microscopic scale has been ignored, likely due to the high degree of molecular complexity, thus making it necessary to resort to models that incorporate these features in an average way. Nevertheless, we can be confident that the advances introduced in atomistic simulations will be able to gain deeper insight into the molecular properties and physics of foods and their transformations.

Because of the hierarchically structure of foods, a comprehensive description encompasses features that extend from the molecular scale to the food plant scale. Multiscale models are then necessary to provide a range of sub-models for describing the intricacies of the material behaviour of foods and of the physical phenomena in food engineering at different spatial and time scales. In this context, it can be expected that the advances of MD simulations will also facilitate gaining insight into the fine structure of food materials, which in turn will be valuable to develop more accurate representations for coarser descriptions of foods, thus providing bridges between the microscopic and mesoscopic models. Thus, even though the microscopic simulation of foods can still be considered to be in its infancy, we can be confident that the identification of the hidden relationships between microscopic and mesoscopic scales will provide the required background for food structural engineering.

References

[1] Boehr, D. D.; Nussinov, R.; Wright, P. E. The Role of Dynamic Conformational Ensembles in Biomolecular Recognition. *Nat. Chem. Biol.* 2009, 5, 789-796.

[2] Fuxreiter, M. Fuzziness: Linking Regulation to protein Dynamics. *Mol. BioSyst.* 2012, 8, 168-177.
[3] Henzler-Wildman, K. A.; Thai, V.; Lei, M.; Ott, M.; Wolf-Watz, M.; Fenn, T.; Pozharski, E.; Wilson, M. A.; Petsko, G. A.; Karplus, M.; Hübner, C. G.; Kern, D. A Intrinsic Motions Along an Enzymatic Reaction Trajectory. *Nature* 2007, 450, 838-844.
[4] Henzler-Wildman, K. A.; Lei, M.; Thai, V.; Kerns, S. J.; Karplus, M.; Kern, D. A hierarchy of timescales in protein dynamics is linked to enzyme catalysis. *Nature* 2007, 450, 913-918.
[5] Raimondi, F.; Orozco, M.; Fanelli, F. Deciphering the Deformation Modes Associated with Function Retention and Specialization in Members of the Ras Superfamily. *Structure* 2009, 18, 402-414.
[6] Stein, A.; Rueda, M.; Panjkovich, A.; Orozco, M.; Aloy, P. A Systematic Study of the Energetics Involved in Structural Changes Upon Association and Connectivity in Protein Interaction Networks. *Structure* 2001, 19, 881-889.
[7] Xu, X.; Ge, H.; Gu, C.; Gao, Y. Q.; Wang, S. S.; Thio, B. J. R.; Hynes, J. T.; Xie, X. S.; Cao, J. Modeling Spatial Correlation of DNA Deformation: DNA Allostery in Protein Binding. *J. Phys. Chem.* B 2013, 117, 13378-13387.
[8] Furini, S.; Barbini, P.; Domene, C. DNA-Recognition Process Described by MD Simulations of the Lactose Repressor Protein on a Specific and a Non-Specific DNA Sequence. *Nuc. Acids Res.* 2013, 41, 3963-3972.
[9] del Sol, A.; Tsai, C. J.; Ma, B.; Nussinov, R. The Origin of Allosteirc Functional Modulation: Multiple Pre-Existing Pathways. *Structure* 2009, 17, 1042-1050.
[10] Abbruzzetti, S.; Spyrakis, F.; Bidon-Chanal, A.; Luque, F. J.; Viappiani, C. Ligand Migration Through Hemeprotein Cavities: Insights from Laser Flash Photolysis and Molecular Dynamics Simulations. *Phys. Chem. Chem. Phys.* 2013, 15, 10686-10701.
[11] Cozzini, P.; Kellog, G. E.; Spyrakis, F.; Abraham, D. J.; Costantino, G.; Emerson, A.; Fanelli, F.; Gohlke, H.; Kuhn, L. A.; Morris, G. M.; Orozco, M.; Perthinez, T. A.; Rizzi, M.; Sotriffer, C. A. Target Flexibility: An Emerging Consideration in Drug Discovery and Design. *J. Med. Chem.* 2008, 51, 6237-6255.
[12] Spyrakis, F.; Bidon-Chanal, A.; Barril, X.; Luque, F. J. Protein Flexibility and Ligand Recognition: Challenges for Molecular Modeling. *Curr. Top. Med. Chem.* 2001, 11, 192-210.
[13] Adcock, S. A.; McCammon, J. A. Molecular Dynamics: Survey of Methods for Simulating the Activity of Proteins. *Chem. Rev.* 2006, 106, 1589-1615.
[14] Orozco, M. A Theoretical View of Protein Dynamics. *Chem. Soc. Rev.* 2014, 43, 5051-5066.
[15] Free Energy Calculations. Theory and Applications in Chemistry and Biology. Chipot, C.; Pohorille, A. (Eds.) *Springer Series in Chemical Physics.* Vol. 86. Berlin. 2007.
[16] Pink, D. A.; Razul, M. S. G. Computer Simulation Techniques for Food Science and Engineering: Simulating Atomic Scale and Coarse-Grained Models. *Food Struct.* 2014, 1, 71-90.
[17] Case, D. A.; Cheatham, T.; Darden, T.; Gohlke, H.; Luo, R.; Merz, K. M. Jr.; Onufriev, A.; Simmerling, C.; Wang, B.; Woods, R. The Amber Biomolecular Simulation Programs. *J. Comput. Chem.* 2005, 26, 1668-1688.

[18] Hornak, V.; Abel, R.; Okur, A.; Strockbine, B.; Roitberg, A.; Simmerling, C. Comparison of Multiple Amber Force Fields and Development of Improved Protein Backbone Parameters. *Proteins* 2006, 65, 712-725.

[19] Lindorff-Larsen, K.; Piana, S.; Palmo, K.; Maragakis, P.; Klepeis, J.; Dror, R.O.; Shaw, D. E. Improved Side-Chain Torsion Potentials for the Amber ff99SB Protein Force Field. *Proteins* 2010, 78, 1950-1958.

[20] Perez, A.; Marchan, I.; Svozil, D.; Sponer, J.; Cheatham, T.E.; Laughton, C.A.; Orozco, M. Refinement of the AMBER Force Field for Nucleic Acids: Improving the Description of alpha/gamma Conformers. *Biophys. J.* 2007, 92, 3817-3829.

[21] Zgarbova, M.; Otyepka, M.; Sponer, J.; Mladek, A.; Banas, P.; Cheatham, T.E.; Jurecka, P. Refinement of the Cornell et al. Nucleic Acids Force Field Based on Reference Quantum Chemical Calculations of Glycosidic Torsion Profiles. *J. Chem. Theory Comput.* 2011, 7, 2886-2902.

[22] Yildirim, I.; Stern, H. A.; Kennedy, S.D.; Tubbs, J. D.; Turner, D. H. Reparameterization of RNA chi Torsion Parameters for the AMBER Force Field and Comparison to NMR Spectra for Cytidine and Uridine. *J. Chem. Theory Comput.* 2010, 6, 1520-1531.

[23] Zgarbova, M.; Luque, F. J.; Sponer, J.; Cheatham, T. E., III; Otyepka, M.; Jurecka, P. Toward Improved Description of DNA Backbone: Revisiting epsilon and zeta Torsion Force Field Parameters. *J. Chem. Theory Comput.* 2013, 9, 2339-2354.

[24] Kirschner, K. N.; Yongye, A. B.; Tschampel, S. M.; González-Outeiriño, J.; Daniels, C. R.; Foley, B.L.; Woods, R. J. GLYCAM06: A Generalizable Biomolecular Force Field. Carbohydrates. *J. Comput. Chem.* 2008, 29, 622-655.

[25] Tessier, M. B.; DeMarco, M. L.; Yongye, A. B.; Woods, R. J. Extension of the GLYCAM06 Biomolecular Force Field to Lipids, Lipid Bilayers and Glycolipids. *Mol. Simul.* 2008, 34, 349-363.

[26] Wang, J.; Wolf, R. M.; Caldwell, J. W.; Kollamn, P. A.; Case, D. A. Development and Testing of a General Amber Force Field. *J. Comput. Chem.* 2004, 25, 1157-1174.

[27] Zhu, X.; Lopes, P. E. M.; MacKerell, A. D., Jr. Recent Developments and Applications of the CHARMM Force Fields. *Wires Comput. Mol. Sci.* 2012, 2, 167-185.

[28] Poger, D.; van Gunsteren, W. F.; Mark, A. E. A New Force Field for Simulating Phosphatidylcholine Bilayers. *J. Comput. Chem.* 2010, 31, 1117-1125.

[29] Schmid, N.; Eichenberger, A. P.; Choutko, A.; Riniker, S.; Winger, M.; Mark, A. E.; van Gunsteren, W. F. Definition and Testing of the GROMOS Force-Field Versions 54A7 and 54B7. *Eur. Biophys. J.* 2011, 40, 843-856.

[30] Luque, F. J.; Dehez, F.; Chipot, C.; Orozco, M. Polarization Effects in Molecular Interactions. *WIRES Comput. Mol. Sci.* 2011, 1, 844-854.

[31] Anisimov, V. M.; Lamoureux, G.; Vorobyov, I. V.; Huang, N.; Roux, B.; MacKerell, A. D. Determination of Electrostatic Parameters for a Polarizable Force Field Based on the Classical Drude Oscillator. *J. Chem. Theory Comput.* 2005, 1, 153-168.

[32] Baker, C. M.; Anisimov, V. M.; MacKerell, A. D., Jr. Development of CHARMM Polarizable Force Field for Nucleic Acid Bases Base don the Classical Drude Oscillator Model. *J. Phys. Chem.* B 2011, 115, 580-596.

[33] Ponder, J. W.; Wu, C.; Ren, P.; Pande, V. S.; Chodera, J. D.; Schnieders, M. J.; Haque, I.; Mobley, D. L.; Lambrecht, D. S.; DiStasio, R. A., Jr.; Head-Gordon, M.; Clark, G.

N. I.; Johnson, M. E.; Head-Gordon, T. Current Status of the AMOEBA Polarizable Force Field. *J. Phys. Chem. B 2010, 114, 2549-2564.*

[34] Wang, J.; Cieplak, P.; Li, J.; Hou, T.; Luo, R.; Duan, Y. Development of Polarizable Models for Molecular Mechanical Calculations. 1. Parametrization of Atomic Polarizability. *J. Phys. Chem. B.* 2011, 115, 3091-3099.

[35] Wang, J.; Cieplak, P.; Li, J.; Cai, Q.; Hsieh, M.-J.; Luo, R; Duan, Y. Development of Polarizable Models for Molecular Mechanical Calculations. 4. Van der Waals Parametrization. *J. Phys. Chem. B* 2012, 116, 7088-7101.

[36] Harder, E.; MacKerell, A. D., Jr.; Roux, B. Many-Body Polarization Effects and the Membrane Dipole Potential. *J. Am. Chem. Soc.* 2009, 131, 2760-2761.

[37] Ryckaert, J. P.; Ciccotti, G.; Berendsen, H. J. C. Numerical Integration of the Cartesian Equations of Motion of a System with Constraints: Molecular Dynamics of n-Alkanes. *J. Comput. Phys.* 1977, 23, 327-341.

[38] Hess, B.; Bekker, H.; Berendsen, H. J. C.; Fraaije, J. G. E. M. LINCS: A Linear Constraint Solver for Molecular Simulations. *J. Comput. Chem.* 1997, 18, 1463-1472.

[39] Sagui, C.; Darden, T. A. Molecular Dynamics Simulations of Biomolecules: Long-Range Electrostatics Effects. *Annu. Rev. Biophys. Biomol. Struct.* 1999, 28, 155-179.

[40] Hamelberg, D.; Mongan, J.; McCammon, J. Accelerated Molecular Dynamics: A Promising and Efficient Simulation Method for Biomolecules. *J. Chem. Phys.* 2004, 120, 11919-11929.

[41] Schlick, T. Molecular Dynamics-Based Approaches for Enhancing Sampling of Long-Time, *Large-Scale Conformational Changes in Biomolecules.* F1000 Biol. Rep. 2009, 1, 51.

[42] Abrams, C.; Bussi, G. Enhanced Sampling in Molecular Dynamics Unsing Metadynamics, Replica-Exchange, and Temperature-Acceleration. *Entropy* 2014, 16, 163-199.

[43] Bernardi, R. C.; Melo, M. C. R.; Schulten, K. Enhanced Sampling Techniques in Molecular Dynamics Simulations of Biological Systems. *Biochim. Biophys. Acta. Gen.* In press.

[44] Barducci, A.; Bonomi, M.; Parrinello, M. Metadynamics. *WIRES Comput. Mol. Sci.* 2011, 1, 826-843.

[45] Rocchia, W.; Masetti, M.; Cavalli, A. Enhanced Sampling Methods in Drug Design. In Physico-Chemical and Computational Approaches to Drug Discovery. Luque, F. J.; Barril, X. (Eds.) Royal Society of Chemistry, Drug Discovery Series, No. 23, Cambridge, 2012, pp. 273-301.

[46] Philips, J. C.; Braun, R.; Wang, W.; Gumbart, J.; Tajkhorshid, E.; Villa, E.; Chipot, C.; Skeel, R. D.; Kale, L.; Schulten, K. Scalable Molecular Dynamics with NAMD. *J. Comput. Chem.* 2005, 26, 1781-1802.

[47] Pronk, S.; Páll, S.; Schulz, R.; Larsson, P.; Bjelkmar, P.; Apostolov, R.; Shirts, M. R.; Smith, J. C.; Kasson, P. M.; van der Spoel, D.; Hess, B.; Lindah, E. GROMACS 4.5: A High-Throughput and Highly Parallel Open Source Molecular Simulation Toolkit. *Bioinformatics* 2013, 29, 845-854.

[48] Dror, R. O.; Dirks, R. M.; Grossman, J. P.; Xu, H.; Shaw, D. E. Biomolecular Simulation: A Computational Microscope for Molecular Biology. *Annu. Rev. Biophys.* 2012, 41, 429-452.

[49] Götz, A. W.; Williamson, M. J.; Xu, D.; Poole, D.; Le Grand, S.; Walker, R. C. Routine Microsecond Molecular Dynamics Simulations with AMBER on GPUs. 1. Generalized Born. *J. Chem. Theory Comput.* 2012, 8, 1542-1555.

[50] Le Grand, S.; Götz, A. W.; Walker, R. C. SPFP: Speed without Compromise - A Mixed Precision Model for GPU Accelerated Molecular Dynamics Simulations. *Comput. Phys. Commun.* 2013, 184, 374-380.

[51] Harvey, M. J.; Giupponi, G.; De Fabritiis, G. ACEMD: Accelerating Biomolecular Dynamics in the Microsecond Time Scale. *J. Chem. Theory Comput.* 2009, 5, 1632-1639.

[52] Shaw, D. E.; Maragakis, P.; Lindorff-Larsen, K.; Piana, S.; Dror, R. O.; Eastwood, M. P.; Bank, J. A.; Jumper, J. M.; Salmon, J. K.; Shan, Y.; Wriggers, W. Atomic-level characterization of the structural dynamics of proteins. *Science* 2010, 330, 341-346.

[53] Lindorff-Larsen, K.; Piana, S.; Dror, R. O.; Shaw, D. E. How Fast-Folding Proteins Fold. *Science* 2011, 334, 517-520.

[54] Dror, R. O.; Pan, A. C.; Arlow, D. H.; Borhani, D. W.; Maragakis, P.; Shan, Y.; Xu, H.; Shaw, D. E. Pathway and Mechanism of Drug Binding to G Protein–Coupled Receptors. *Proc. Natl. Acad. Sci. USA* 2011, 108, 13118-13123.

[55] Shan, Y.; Kim, E. T.; Eastwood, M. P.; Dror, R. O.; Seeliger, M. A.; Shaw, D. E. How Does a Drug Molecule Find its Target Binding Site? *J. Am. Chem. Soc.* 2011, 133, 9181-9183.

[56] Buch, I.; Giorgino, T.; De Fabritiis, G. Complete Reconstruction of an Enzyme-Inhibitor Binding Process by Molecular Dynamics Simulations. *Proc. Natl. Acad. Sci. USA* 2011,108, 10184-10189.

[57] Nieve, S. O.; Lopez, C. F.; Srinivas, G.; Klein, M. L. Coarse Grain Models and the Computer Simulation of Soft Materials. *J. Phys.: Condens. Matter* 2004, 16, R481-R512.

[58] Baaden, M.; Marrink, S. J. Coarse-Grain Modelling of Protein-Protein Interactions. *Curr. Opin. Struct. Biol.* 2013, 23, 878-886.

[59] Darré, L.; Machado, M. R.; Pantano, S. Coarse-Grained Models of Water. *WIRES Comput. Mol.* Sci. 2012, 2, 921-930.

[60] Dans, P.; Zeida, A.; Machado, M. R.; Pantano, S. A Coarse Grained Model for Atomic-detailed DNA Simulations with Explicit Electrostatics. *J. Chem. Theory Comput.* 2010, 6, 1711-1715.

[61] Tepper, H. L.; Voth, G. A. A Coarse-Grained Model for Double-Helix Molecules in Solution: Spontaneous Helix Formation and Equilibrium Properties. *J. Chem. Phys.* 2005, 122, 124906.

[62] Tozzini, V. Coarse-Grained Models for Proteins. *Curr. Opin. Struct. Biol.* 2005, 15, 144-150.

[63] Marrink, S. J.; de Vries, A. H.; Mark, A. E. Coarse Grained Model for Semiquantitative Lipid Simulations. *J. Phys. Chem. B* 2004, 108, 750-760.

[64] Shih, A. Y.; Freddolino, P. L.; Arkhipov, A.; Schulten, K. Assembly of Lipoprotein Particles Revealed by Coarse-Grained Molecular Dynamics Simulations. *J. Struct. Biol.* 2007, 157, 579-592.

[65] Arkhipov, A.; Freddolino, P. L.; Schulten, K. Stability and Dynamics of Virus Capsids Described by Coarse-Grained Modeling. *Structure* 2006, 14, 1767-1777.

[66] Arkhipov, A.; Yin, Y.; Schulten, K. Four-Scale Description of Membrane Sculpting by BAR Domains. *Biophys. J.* 2008, 95, 2806-2821.

[67] Marrink, S. J.; Risselada, H. J.; Yefimov, S.; Tieleman, D. P.; de Vries, A. H. The MARTINI Force Field: Coarse Grained Model for Biomolecular Simulations. *J. Phys. Chem. B 2007*, 111, 7812-7824.

[68] Marrink, S. J.; Tieleman, D. P. Perspective on the Martini Model. *Chem. Soc. Rev.* 2013, 42, 6801-6822.

[69] Limbach, H.-J.; Arnold, A.; Mann, B. A.; Holm, C. ESPResSo – An Extensible Simulation Package for Research on Soft Matter Systems. *Comput. Phys. Commun.* 2006, 174, 704-727.

[70] Reynwar, B. J.; Ilya, G.; Harmandaris, V. A.; Müller, M. M.; Kremer, K.; Deserno, M. Aggregation and Vesiculation of Membrane Proteins by Curvature-Mediated Interactions. *Nature* 2007, 447, 461-464.

[71] Limbach, H.-J.; Kremer, K. Multi-Scale Modeling of Polymers: Perspectives for Food Materials. *Trends Food Sci. Tech*. 2006, 17, 215-219.

[72] van der Sman, R. G. M. Soft Matter Approaches to Food Structuring. *Adv. Colloid. Interface Sci*. 2012, 176-177, 18-30.

[73] Ho, Q. T.; Carmeliet, J.; Datta, A. K.; Defraeye, T.; Delele, M. A.; Herremans, E.; Opara, L.; Ramon, H.; Tijskens, E.; van der Sman, R.; van Liedekerke, P.; Verboven, P.; Nicolai, B. M. Multiscale Modeling in Food Engineering. *J. Food Eng*. 2013, 114, 279-291.

[74] Ramalakshmi, K.; Raghavan, B. Caffeine in Coffee: Its Removal. Why and How? *Crit. Rev. Food Sci. Nutr*. 1999, 39, 441-456.

[75] Cesàro, A.; Russo, E.; Crescenzi, V. Thermodynamics of Caffeine Aqueous Solutions. *J. Phys. Chem*. 1976, 80, 335-339.

[76] Falk, M.; Chew, W.; Walter, J. A.; Kwiatkowski, W.; Barclay, K. D.; Klassen, G. A: Molecular Modelling and NMR Studies of the Caffeine Dimer. *Can. J. Chem*. 1998, 76, 48-56.

[77] Lilley, T. H.; Linsdell, H.; Maestre, A. Association of Caffeine in Water and in Aqueous Solutions of Sucrose. *J. Chem. Soc. Faraday Trans.* 1992, 88, 2865-2870.

[78] Fritzsche, H.; Petri, I.; Schütz, H.; Weller, K.; Sedmera, P.; Lang, H. On the Interaction of Caffeine with Nucleic Acids: III. ^{1}NMR Studies of Caffeine-5'-Adenosine Monophosphate and Caffeine-Poly(Riboadenylate) Interactions. *Biophys. Chem*. 1980, 11, 109-119.

[79] Davies, D. B.; Veselkov, D. A.; Djimant, L. N.; Veselkov, A. N. Hetero-association of Caffeine and Aromatic Drugs and Their Competitive Binding with a DNA Oligomer. *Eur. Biophys. J.* 2001, 30, 354-366.

[80] Al-Maaieh, A.; Flanagan, D. R. Salt Effects on Caffeine Solubility, Distribution, and Self-Association. *J. Pharm. Sci. 2002*, 91, 1000-1008.

[81] Tavagnacco, L.; Schnupf, U.; Mason, P. E.; Saboungi, M.-L.; Cesàro, A.; Brady, J. W. Molecular Dynamics Simulations Studies of Caffeine Aggregation in Aqueous Solution. *J. Phys. Chem. B* 2011, 115, 10957-10966.

[82] Sharma, B.; Paul, S. Effects of Dilute Aqueous NaCl Solution on Caffeine Aggregation. *J. Chem. Phys.* 2013, 139, 194504.

[83] Tavagnacco, L.; Engström, O.; Schupf, U.; Saboungi, M.-L.; Himmel, M.; Widmalm, G.; Cesàro, A.; Brady, J. W. Caffeine and Sugars Interact in Aqueous Solutions: A Simulation and NMR Study. *J. Phys. Chem. B* 2012, 116, 11701-11711.

[84] Hendon, C. H.; Colonna-Dashwood, L.; Colonna-Dashwood, M. The Role of Dissolved Cations in Coffee Extraction. *J. Agric. Food Chem.* 2014, 62, 4947-4950.

[85] Draget, K. I.; Moe, S. T.; Skjåk-Braek, G.; Smidsrød, O. Alginates. In: Food Polysaccharides and Their Applications. Stephen, A. M.; Phillips, G. O. (Eds.) 2nd edition, CRC Press, 2006, pp. 289-334.

[86] Brownlee, I. A.; Seal, C. J.; Wilcox, M.; Dettmar, P. W.; Pearson, J. P. Applications of Alginates in Food. In: Alginates: Biology and Application. Rehm, B. H. A. (Ed). Microbiology Monographs, Vol. 13, Springer, 2009, pp 211-228.

[87] Herrero, E. P.; Martín del Valle, E. M.; Peppas, N. A. Protein Imprinting by Means of Alginate-Based Polymer Microcapsules. *Ind. Eng. Chem. Res.* 2010, 49, 9811-9814.

[88] Li, Q.; Liu, C.-G.; Huang, Z.-H.; Xue, F.-F. Preparation and Characterization of Nanoparticles Based on Hydrophobic Alginate Derivative as Carriers for Sustained Release of Vitamin D3. *J. Agric. Food Chem.* 2011, 59, 1962-1967.

[89] Song, S.; Wang, Z.; Qian, Y.; Zhang, L.; Luo, E. The Release Rate of Curcumin from Calcium Alginate Beads Regulated by Food Emulsifiers. *J. Agric. Food Chem.* 2012, 60, 4388-4395.

[90] Zhang, Y.; Jia, X.; Wang, L.; Liu, J.; Ma, G. Preparation of Ca-Alginate Microparticles and Its Application for Phenylketonuria Oral Therapy. *Ind. Eng. Chem. Res.* 2011, 50, 4106-4112.

[91] Jensen, M. G.; Knudsen, J. C.; Viereck, N.; Kristensen, M.; Astrup, A. Functionality of Alginate Based Supplements for Application in Human Appetite Regulation. *Food Chem.* 2012, 132, 823-829.

[92] Lu, L.; Liu, X.; Tong, Z.; Gao, Q. Critical Exponents and Self-Similarity for Sol-Gel Transition in Aqueous Alginate Systems Induced by in situ Release of Calcium Cations. *J. Phys. Chem. B* 2006, 110, 25013-25020.

[93] Fang, Y.; Al-Assaf, S.; Philips, G. O.; Nishinari, K.; Funami, T.; Williams, P. A.; Li, L. Multiple Steps and Critical Behaviors of the Binding of Calcium to Alginate. *J. Phys. Chem.* B 2007, 111, 2456-2462.

[94] Harper, B. A.; Barbut, S.; Lim, L.-T.; Marcone, M. F. Effect of Various Gelling Cations on the Physical Properties of "Wet" Alginate Films. *J. Food Sci.* 2014, 79, E562-E567.

[95] Vega, C.; Castells, P. Spherification. In: Vega, C.; Ubbink, J.; van der Linden, E. (Eds.). The Kitchen as the Laboratory. Columbia University Press. 2012, New York, pp. 25-32.

[96] Grant, G. T.; Morris, E. R.; Ress, D. A.; Smith, P. J. C.; Thom, D. Biological Interactions Between Polysaccharides and Divalent Cation: the Egg-Box Model. *FEBS Lett.* 1973, 32, 195-198.

[97] Braccini, I.; Pérez, S. Molecular Basis of Ca^{2+}-Induced Gelation in Alginates and Pectins: The Egg-Box Model Revisited. *Biomacromolecules* 2001, 2, 1089−1096.

[98] Plazinski, W. Molecular Basis of Calcium Binding by Polyguluronate Chains. Revising the Egg-Box Model. *J. Comput. Chem.* 2011, 32, 2988−2995.

[99] Borgogna, M.; Skjåk-Braek, G.;Paoletti, S.; Donati, I. On the Initial Binding of Alginate by Calcium Ions. The Tilted Egg-Box Hypothesis. *J. Phys. Chem. B* 2013, 117, 7277-7282.

[100] Plazinski, W.; Drach, M. Calcium-□-L-Guluronate Complexes: Ca^{2+} Binding Modes from DFT-MD Simulations. *J. Phys. Chem. B* 2013, 117, 12105-12112.
[101] Fu, H.; Liu, Y.; Adrià, F.; Shao, X.; Cai, W.; Chipot, C. From Material Science to Avant-Garde Cuisine. The Art of Shaping Liquids into Spheres. *J. Phys. Chem. B* 2014, 118, 11747-11756.
[102] Zaveri, N. T. Green Tea and its Polyphenolic Catechins: Medicinal Uses in Cancer and Noncancer Applications. Life Sci. 2006, 78, 2073-2080.
[103] Del Rio, D.; Rodriguez-Mateos, A.; Spencer, J. P.; Tognolini, M.; Borges, G; Crozier, A. Dietary (Poly)phenolics in Human Health: Structures, Bioavailability, and Evidence of Protective Effects Against Chronic Diseases. Antioxid. *Redox Signal.* 2013, 18, 1818-1892.
[104] Del Rio, D.; Stewart, A. J.; Mullen, W.; Burns, J.; Lean, M. E.; Brighenti, F.; Crozier, A. HPLC-MSn Analysis of Phenolic Compounds and Purine Alkaloids in Green and Black Tea. *J. Agric. Food Chem.* 2004, 52, 2807-2815.
[105] Chung, F. L.; Schwartz, J.; Herzog, C. R.; Yang, Y. M. Tea and Cancer Prevention: Studies in Animals and Humans. *J. Nutr.* 2003, 133, 3268S-3274S.
[106] Uekusa, Y.; Kamihira, M.; Nakayama, T. Dynamic Behavior of Tea Catechins Interacting with Lipid Membranes as Determined by NMR Spectroscopy. *J. Agric. Food Chem.* 2007, 55, 9986-9992.
[107] Tamba, Y.; Ohba, S.; Kubota, M.; Yoshioka, H.; Yoshioka, H.; Yamazaki, M. Single GUV Method Reveals Interaction of Tea Catechin (-)-Epigallocatechin Gallate with Lipid Membranes. *Biophys J.* 2007, 92, 3178-3194.
[108] Caturla, N.; Vera-Samper, E.; Villalaín, J.; Mateo, C. R.; Micol, V. The Relationship Between the Antioxidant and the Antibacterial Properties of Galloylated Catechins and the Structure of Phospholipid Model Membranes. *Free Radic. Biol. Med.* 2003, 34, 648-662.
[109] Sirk, T. W.; Brown, E. F.; Sum, A. K.; Friedman, M. Molecular Dynamics Study on the Biophysical Interactions of Seven Green Tea Catechins with Lipid Bilayers of Cell Membranes. J. *Agric. Food Chem.* 2008, 56, 7750-7758.
[110] Nagle, D. G.; Ferreira, D.; Zhou, Y. D. Epigallocatechin-3-gallate (EGCG): Chemical and Biomedical Perspectives. *Phytochemistry* 2006, 67, 1849-1855.
[111] Bennet, W. F. D.; Tieleman, D. P. The Importance of Membrane Defects-Lessons From Simulations. *Acc. Chem. Res.* 2014, 47, 2244-2251.
[112] Martinez-Seara, H.; Rog., T.; Karttunen, M.; Vattulainen, I.; Reigada, R. Cholesterol Induces Specific Spatial and Orientational Order in Cholesterol/Phospholipid Membranes. *PLoS ONE* 2010, 5, e11162.
[113] Bennet, W. F. D; MacCallum, J. L.; Tieleman, D. P. Thermodynamic Analysis of the Effect of Cholesterol on Dipalmitoylphosphatidylcholine Lipid Membranes. *J. Am. Chem. Soc.* 2009, 131, 1972-1978.
[114] Zhu, Q.; Cheng, K. H.; Vaughn, M. W. Molecular Dynamic Studies of the Molecular Structure and Interactions of Cholesterol Superlattices and Random Domains in an Unsaturated Phosphatidylcholine Bilayer Membrane. *J. Phys. Chem. B* 2007, 111, 11021-11031.
[115] Mondal, S.; Mukhopadhyay, C. Molecular Insight of Specific Cholesterol Interactions: A Molecular Dynamics Simulation Study. *Chem. Phys. Lett.* 2007, 439, 166-170.

[116] Alwarawrah, M.; Dai, J. A.; Huang, J. Y. A Molecular View of the Cholesterol Condensing Effect in DOPC Lipid Bilayers. *J. Phys. Chem.* B 2010, 114, 7516-7523.

[117] Hakobyan, D.; Heuer, A. Key Molecular Requirements for Raft Formation in Lipid/Cholesterol Membranes. *PLoS ONE* 2014, 9, e87369.

[118] Khelashvili, G.; Johner, N.; Zhao, G.; Harries, D.; Scott, H. L. Molecular Origins of Bending Rigidity in Lipids with Isolated and Conjugated Double Bonds: The Effect of Cholesterol. *Chem. Phys. Lipids* 2014, 17, 18-26.

[119] Pan, J. J.; Cheng, X. L.; Heberle, F. A.; Mostofian, B.; Kucerka, N.; Drazba, P.; Katsaras, J. Interactions Between Ether Phospholipids and Cholesterol as Determined by Scattering and Molecular Dynamics Simulations. *J. Phys. Chem. B* 2012, 116, 14829-14838.

[120] Hsich, C. J.; Chen, Y. W.; Hwang, D. W. Effects of Cholesterol on Membrane Molecular Dynamics Studied by Fast Field Cycling NMR Relaxometry. *Phys. Chem. Chem. Phys*. 2013, 15, 16634-16640.

[121] Rosetti, C.; Pastorino, C. Comparison of Ternary Bilayer Mixtures with Asymmetric or Symmetric Unsaturated Phosphatidylcholine Lipids by Coarse Grained Molecular Dynamics Simulations. *J. Phys. Chem. B* 2012, 116, 3525-3537.

[122] Plesnar, E.; Subezynski, W. K.; Pasenkiewiez-Gierula, M. Saturation with Cholesterol Increases Vertical Order and Smoothes the Surface of the Phosphatidylcholine Bilayer: A Molecular Simulation Study. *Biochim. Biophys. Acta - Biomembranes* 2012, 1818, 520-529.

[123] Sugar, I. P.; Chong, P. L. G. A Statistical Mechanical Model of Cholesterol/ Phospholipid Mixtures: Linking Condensed Complexes, Superlattices and the Phase Diagram. *J. Am. Chem. Soc*., 2012, 134, 1164-1171.

[124] Choubey, A.; Nomura, K. I.; Kalia, R.; Nakano, A.; Vashishta, P. Cholesterol Flip-Flop Dynamics in a Phospholipid Bilayer: All Atom Molecular Dynamics Simulations. *Biophys. J.* 2012, 102, 241A.

[125] Jo, S.; Rui, H. A.; Lim, J. B.; Klauda, J. B.; Im, W. Cholesterol Flip-Flop: Insights from Free Energy Simulation Studies. *J. Phys. Chem. B* 2010, 114, 13342-13348.

[126] Plesnar, E.; Subezynski, W. K.; Pasenkiewiez-Gierula, M. Comparative Computer Simulation Study of Cholesterol in Hydrated Unary and Binary Lipid Bilayers and in an Anhydrous Crystal. *J. Phys. Chem. B* 2013, 117, 8758-8769.

[127] Saito, H.; Shinoda, W. Cholesterol Effect on Water Permeability Through DPPC and PSM Lipid Bilayers: A Molecular Dynamics Study. *J. Phys. Chem.* B 2011, 115, 15241-15250.

[128] Shigematsu, T.; Koshiyama, K.; Wada, S. Molecular Dynamics Simulations of Pore Formation in Stretched Phospholipid/Cholesterol Bilayers. *Chem. Phys. Lipids* 2014, 183, 43-49.

[129] Xu, W. X.; Wei, G. H.; Su, H. B.; Nordenskiold, L.; Mu, Y. G. Effects of Cholesterol on Pore Formation in Lipid Bilayers Induced by Human Islet Amyloid Polypeptide Fragments: A Coarse-Grained Molecular Dynamics Study. *Phys. Rev*. E 2011, 84, 051922.

[130] Fernandez, M. L.; Marshall, G.; Sagues, F.; Reigada, R. Structural and Kinetic Molecular Dynamics Study of Electroporation in Cholesterol Containing Bilayers. *J. Phys. Chem. B* 2010, 114, 6855-6865.

[131] Pourmousa, M.; Rog, T.; Mikkeli, R.; Vattulainen, I.; Solanko, L. M.; Wustner, D.; List, N. H; Kongsted, J.; Karttunen, M. Dehydroergosterol as an Analogue for Cholesterol: Why It Mimics Cholesterol so Well or Does It? *J. Phys. Chem. B,* 2014, 118, 7345-7357.

[132] Robalo, J. R.; do Canto, A. M. T. M.; Carvalho, A. J. P.; Ramalho, J. P. P.; Loura, L. M. S. Behavior of Fluorescent Cholesterol Analogues Dehydroergosterol and Cholestatrienol in Lipid Bilayers: A Molecular Dynamics Study. *J. Phys. Chem.* B 2013, 117, 5806-5819.

[133] Cournia, Z.; Ullmann, G. M.; Smith, J. C. Differential Effects of Cholesterol, Ergosterol and Lanosterol on a Dipalmitoyl Phosphatidylcholine Membrane: A Molecular Dynamics Simulation Study. *J. Phys. Chem.* B 2007, 111, 1786-1801.

[134] Czub, J.; Baginski, M. Modulation of Amphotericin B Membrane Interaction by Cholesterol and Ergosterols - A Molecular Dynamics Study. *J. Phys. Chem.* B 2006, 110, 16743-16753.

[135] Czub, J.; Baginski, M. Comparative Molecular Dynamics Study of Lipid Membranes Containing Cholesterol and Ergosterol. *Biophys. J.* 2006, 90, 2368-2382.

[136] Robalo, J. R.; Ramalho, J. P. P.; Loura, L. M. S. NBD-labeled Cholesterol Analogues in Phospholipid Bilayers: Insights from Molecular Dynamics. *J. Phys. Chem. B* 2013, 117, 13731-13742.

[137] Rog, T.; Pasenkiewicz-Gierula, M.; Vattulainen, I.; Karttunen, M. Ordering Effects of Cholesterol and its Analogues. *Biochim. Biophys. Acta - Biomembranes* 2009,1788, 97-121.

[138] Rog, T.; Vattulainen, I.; Jansen, M.; Ikonen, E.; Karttunen, M. Comparison of Cholesterol and its Direct Precursors along the Biosynthetic Pathway: Effects of Cholesterol, Desmosterol and 7-Dehydrocholesterol on Saturated and Unsaturated Lipid Bilayers. *J. Chem. Phys*. 2008, 129, 154508.

[139] Oates, J.; Watts, A. Uncovering the Intimate Relationship between Lipids, Cholesterol and GPCR Activation. *Curr. Opin. Struct. Biol.* 2011, 21, 802-807.

[140] Hanson, M. A.; Cherezov, V.; Griffith, M. T.; Roth, C. B.; Jaakola, V. P.; Chien, E. Y.; Velasquez, J.; Kuhn, P.; Stevens, R. C. A Specific Cholesterol Binding Site is Established by the 2.8 Å Structure of the Human beta2-Adrenergic Receptor. *Structure* 2008, 16, 897-905.

[141] Khelashvili, G.; Grossfield, A.; Feller, S. E.; Pitman, M. C.; Weinstein, H. Structural and Dynamic Effects of Cholesterol at Preferred Sites of Interaction with Rhodopsin Identified from Microsecond Lenght Molecular Dynamics Simulations. *Proteins: Struct., Funct., Bioinf.* 2009, 76, 403-417.

[142] Hibbs, R. E.; Gouaux, E. Principles of Activation and Permeation in an Anion-Selective Cys-loop Receptor. *Nature* 2011, 474, 54-60.

[143] Cheng, M. H.; Xu, Y.; Tang, P. Anionic Lipid and Cholesterol Interactions with Alpha4Beta2 nAChR: Insights from MD Simulations. *J. Phys. Chem.* B 2009, 113, 6964-6970.

[144] Henin, J.; Salari, R.; Murlidaran, S.; Brannigan, G. A Predicted Binding Site for Cholesterol on the GABA(A) *Receptor. Biophys. J.* 2014, 106, 1938-1949.

[145] Buck, L.; Axel, R. A Novel Multigene Family May Encode Odorant Receptors: A Molecular Basis for Odor Recognition. Cell 1991, 65, 175-187.

[146] Kaupp, U. B. Olfactory Signalling in Vertebrates and Insects: Differences and Commonalities. *Nat. Rev. Neurosci.* 2010, 11, 188-200.
[147] Crasto, C. J. Computational Biology of Olfactory Receptors. *Curr. Bioinform.* 2009, 4, 8-15.
[148] Lai, P. C.; Crasto, C. J. Beyond Modeling: All-Atom Olfactory Receptor Model Simulations. *Front. Genet.* 2012, 3, 61.
[149] Don, Ch. G.; Riniker, S. Scents and Sense: In Silico Perspectives on Olfactory Receptors. *J. Comput. Chem.* 2014, 35, 2279-2287.
[150] Singer, M. S. Analysis of the Molecular Basis for Octanal Interactions in the Expressed Rat 17 Olfactory Receptor. *Chem. Senses* 2000, 25, 155-165.
[151] Lai, P. C.; Singer, M. S.; Crasto, C. J. Structural Activation Pathways from Dynamic Olfactory Receptor-Odorant Interactions. *Chem. Senses* 2005, 30, 781-792.
[152] Araneda, R.C.; Kini, A.D.; Firestein, S. The Molecular Receptive Range of an Odorant Receptor. *Nat. Neurosci.* 2000, 3, 1248-1255.
[153] Anselmi, C.; Buonocore, A.; Centini, M.; Facino, R. M.; Hatt, H. The Human Olfactory Receptor 17-40: Requisites for Fitting Into the Binding Pocket. *Comp. Biol. Chem.* 2011, 35, 159-168.
[154] Charlier, L.; Topin, J.; Ronin, C.; Kim, S-K.; a Goddard, W.; Efremov, R.; Golebiowski, J. How Broadly Tuned Olfactory Receptors Equally Recognize Their Agonists. Human OR1G1 as a Test Case. *Cell Mol. Life Sci.* 2012, 69, 4205-4213.
[155] Lai, P. C.; Guida, B.; Shi, J.; Crasto, C. J. Preferential Binding of an Odor Within Olfactory Receptors: A Precursor to Receptor Activation. *Chem. Senses* 2014, 39, 107-123.

In: From Medicinal Chemistry to Food Science
Editor: Pietro Cozzini
ISBN: 978-1-63483-587-9

Chapter 7

INCORPORATING PROTEIN FLEXIBILITY IN STRUCTURE-BASED DRUG DESIGN

Francesca Spyrakis[1] and Claudio N. Cavasotto[2*]
[1]Department of Life Sciences,
University of Modena and Reggio Emilia, Modena, Italy
[2]Biomedicine Research Institute of Buenos Aires - CONICET -
Partner Institute of the Max Planck Society (IBioBA-MPSP),
Buenos Aires, Argentina

Abstract

Intrinsic dynamics represents one of the most intriguing properties of proteins and one of the most challenging issues in structure-based drug design. As proteins need to move in order to perform their functions, computational methodologies need to foretell proteins movements in the most accurate and fast way, to give reliable predictions and valuable indications in a drug design perspective.

In this chapter, we survey the currently available techniques to reproduce proteins flexibility with different degrees of approximation and consistency.

Flexible docking strategies, classical and enhanced molecular dynamics simulations, normal modes analysis, Monte Carlo methods and the use of computationally generated conformations in virtual screening are described. Advantages, strengths and weaknesses of each technique are provided, along with different successful applications.

Introduction

It is by now obvious that flexibility is essential to proteins to perform their function and that most of the biochemical and cellular processes are closely coupled to protein fluctuation and intrinsic dynamics. It is probably less known the degree of dynamics required in the different possible scenarios [1], since the range of motions undergone by macromolecules is

* Correspondence to: cnc@cavasotto-lab.net; ccavasotto@ibioba-mpsp-conicet.gov.ar

extremely large in space and time scales, ranging from 10^{-11} to 10^{-8} m, and from femtoseconds to hours, respectively [2]. While faster adjustments are related to ligand binding and enzymatic catalysis events, slower and longer motions are required for modulating allosteric and conformational transitions, for protein folding, and macromolecular association [3]. A minority of proteins, mostly constituted by small and charged residues, undergo only little folding and remain intrinsically disordered [4]. On the contrary, most of the proteins are able to overcome the massive conformational entropy of an unfolded state to generate densely packed crystals through the formation of secondary and tertiary structures [5]. Between these two opposite categories a number of proteins can undergo different levels of motion, according to the number of intra- and inter-molecular interactions and to the capacity of excluding solvent from the inner regions [6, 7]. Moreover, proteins also experience flexibility while associating in quaternary structures, where they can transiently stay either in a more rigid or highly disordered state.

Marsh and Teichmann interestingly reported about the similarities between dynamics and evolutionary processes occurring at significantly different time scales [1]. They observed that: i) local flexibility is highly related to the protein evolutionary rate, ii) protein intrinsic conformational fluctuations remind to the structural variation within protein families, and iii) association and evolution of protein complexes follow similar paths. Flexibility can, thus, be considered as the result of an on-going evolutionary process, since proteins experience point mutations, insertions, deletions and rearrangement of domain organization, which are often translated into structure changes [8]. In particular, the authors compared the sequence-identical ensemble for the human fibroblast growth factor receptor 1 kinase domain (made of different structures of the same domain) with its evolutionary ensemble (made of non-redundant homologous structures), observing that the structural and evolutionary dynamics of the protein follow very similar trends, even if the magnitude of single residue RMSD with respect to the mean positions is different [1].

This correspondence between evolution and dynamics might be fundamental in different computational applications as in homology, since knowing the feasible conformational space should help in constraining the evolutionary space, in structural dynamics predictions, in the absence of sequence-identity ensembles [9], or in protein design where the design process could be considered similar to evolution.

Molecular dynamics (MD) and normal mode (NM) analyses brought a significant contribution in outlining the relationship between dynamics and evolution. Normal mode analysis is, in fact, used to investigate the most stable, conserved and lowest changes in a MD trajectory. These modes can be used for predicting the protein evolution pathway, which will perturb the protein dynamics inducing the most important changes along those same modes.

This intrinsic protein dynamics obviously plays a fundamental role also in protein-ligand recognition and binding and should be taken into account in computational drug design [10-14]. Even if the jiggling's and wiggling's of atoms cannot be completely deciphered and reproduced in computational simulations, it is obvious that the old lock-and-key approach is more anachronistic than ever and has been overcome by models as the induced-fit (based on the induced structural changes occurring upon ligand binding), or the conformational ensemble (based on the presence of an equilibrated and pre-existing ensemble of structures) [15], which not only account for conformational changes, but also for the random fluctuation of receptors and ligands [13, 16, 17].

Ligand binding can be described as a combination of a conformer selection-stage followed by minor adjustments within the binding pocket modulated by the ligand (induced-fit stage) [18], i.e., the structural changes upon ligand binding occur within the ensemble of pre-existing receptor conformers. In this way, some proteins are observed to exhibit conformational changes upon binding to different ligands, which may involve side-chain rotations, wide backbone changes, or a combination of both. From a theoretical point of view, this can be rationalized in terms of linear response theory, whereby structural changes upon ligand binding are represented by the perturbation the ligand induces on the receptor unbound states [19].

Proteins structure. A number of experimental methods are available for determining proteins three-dimensional structures. Among them the most established and accurate method is represented by X-ray crystallography, even if it is highly sensitive to the experimental working conditions, difficult and time consuming. The majority of structures deposited into the Protein Data Bank have been obtained by X-ray diffraction. Even if less accurate and limited to small and medium sized macromolecules, also Nuclear Magnetic Resonance (NMR) is widely used to solve proteins structure. This technique is able to describe the most important proteins motions, giving multiple snapshots of the same entity and important clues about how proteins behave and function [20]. On top of providing information about dynamic protein-ligand interactions, NMR spectroscopy has a particularly relevant role in drug discovery [21]. Other methodologies used to study and get insights about protein conformational changes are represented by infrared and Raman spectroscopies, used to measure the vibrational frequencies of groups of atoms [22], and by inelastic neuron scattering, able to capture the diffusive component of protein internal motions [23]. Also small-angle X-ray scattering has become a valuable tool for the low resolution structural characterization of proteins in solution, thanks to the possibility of performing the analyses onto a wide range of macromolecular sizes and in a variety of experimental conditions [24, 25]. Even if X-ray crystallography still represents the technique of choice, X-ray structures are not completely free of biases and artefacts. Some of them, like those solved at non-biological temperature or co-crystallized by soaking present unnatural peculiarities that could interfere with their potential in drug design studies [26, 27].

Aside to experimental techniques, usually very expensive and time-consuming, a variety of computational methods have been developed to simulate protein behaviour for biophysics, medicinal and medicinal chemistry purposes, always trying to combine accuracy, reliability and computational resources [10, 28-30]. It should be taken into account, however, that some simulated conformations may perform worse than crystallographic structures. Thus, if X-ray structures providing a sufficiently large sampling of the conformational space are available, they should generally represent the best and first choice [31]. Unfortunately, as previously mentioned, experimental information is often limited and sometimes experimentally solved or modelled protein structures may not represent the conformation of interest for drug discovery purposes. Computational simulation has proved a very useful tool to study ligand-modulated conformational changes, and to generate alternative protein structures not available from experimental sources or from simple homology models.

In silico methodologies able to deal with protein flexibility in drug design include soft docking [32, 33], docking with libraries of side-chain rotamers [33], algorithms accounting for induced-fit adjustments [34, 35], Monte Carlo [36] and MD simulations [37]. More recently docking into multiple conformations coming from X-ray [32, 38-44], NMR [45, 46], a combination of both [47, 48] or computationally generated by MD simulations [49-53], normal-mode analysis [54] or homology models [55] has been widely applied and reported. A survey of the aforementioned approaches will be given in the following paragraphs.

Computing protein flexibility. As stated by Blaney and Dixon, docking can be considered as a rational drug design approach seeking to predict the structure and the binding energy of a ligand-receptor complex given only the structures of the free ligand and receptors [56]. The first aim of a docking simulation is to correctly predict and identify the most favourable binding mode of a ligand into its target pocket. The second is to correctly rank the different compounds according with the corresponding experimental binding affinities, while the third to generate likely shortlists of potential inhibitors for experimental characterization [57].

Generally docking algorithms are classified into three main categories according to the considered level of flexibility: i) rigid-body methods, treating both the ligand and the protein as rigid objects and exploring only six degrees of freedom, ii) flexible-ligand docking methods, simulating only ligand flexibility by means of systematic, random or stochastic algorithms [58, 59], iii) flexible-ligand flexible-protein methods, taking into account the motions of both ligands and proteins, and representing the high-end approaches.

Methods modelling both ligand and receptor flexibility can be further classified into two sub-categories. The first implies the treatment of both entities at the same time. In order to limit the computational time a small number of degrees of freedom is usually considered, and the variables driving the conformational changes of the protein along the defined degrees of flexibility are optimized with the ligand translational, rotational and torsional changes. The simplest approaches involve the use of soft potentials, rotameric libraries or local energy minimization to simulate the movements of residues side chains lining the binding pocket [60, 61]. These approaches could be useful when protein motions are limited and when ligands easily fit a fairly static binding site. When ligands select subsets of pre-existing conformations or induce previously unknown states, the inclusion of multiple conformations coming from experimental or computational analyses is, instead, necessary to have reliable predictions.

Soft docking methods only consider a limited level of flexibility allowing a certain structural overlap between the protein and the ligand, by means of soft core potentials, i.e., tolerant scoring functions. The main advantage is represented by the low computational cost and by the possibility of identifying conformations or induced-fit effects normally not experienced my more complex and demanding methodologies. Only subtle changes, mainly confined to side chains and not involving backbone rearrangements can be investigated [12, 62].

A more exhaustive representation of the intrinsic protein flexibility can be obtained by using rotamer libraries, based on experimentally observed and preferred rotameric states for each residue side chain [63, 64]. These methods are applicable to proteins characterized by limited motions at the binding site and for which good structural knowledge about the unbound and bound states are available, possibly for ligands similar to the ones under investigations. For sure artefact complexes could be obtained and, also, no information about the backbone adjustment is provided [65].

Another example of docking tool implementing ligand and protein flexibility at the same time is represented by the SLIDE algorithm [61, 66], developed by Kuhn and co-workers as an evolution of SPECITOPE [67]. The receptor pocket is analysed for the identification of hydrophobic and hydrogen-bonding template points, and all the hydrogen bond donor and acceptor groups of the ligand are combinatorially matched within the binding pocket. Fragmentation is performed to obtain fragments bearing triplets of matching points, positioned by a least square superposition algorithm. The remaining part of the ligand is added and both receptor side chains and ligand rotatable bonds are rotated to remove intermolecular overlaps. In a validation test SPECITOPE screened 140000 different peptide fragments, being able to identify both known inhibitors and potential new ligands [67]. SLIDE was then improved by adding the possibility to look for conserved water molecules in protein binding sites and to predict their energetic contribution to the protein-ligand binding process.

Glide models induced-fit effects by alanine-replacing a given set of residues, which are then reconstructed after ligand placement. Soft docking techniques are used for the preliminary placement of ligands, side chains are then rebuilt and the whole pocket minimized. Finally the ligands are re-docked without using any soft potential. When compared to rigid body docking for a set of 21 relevant targets, Glide performed significantly better in terms of ligands pose [68].

Also RosettaLigand [26, 69] and GalaxyDock2 [70] simultaneously optimize both protein and ligand conformational adjustments using Monte Carlo sampling and space annealing global optimization, respectively. In particular, GalaxyDock2 uses geometry-based docking methods based on the beta-complex [71, 72] that is a structure derived from the Voronoi diagram of receptor atoms, which could enhance both the computational speed and binding mode prediction accuracy.

The second category includes methods using different fixed pre-generated protein conformations (obtained by means of experimental or computational techniques) into which ligands are subsequently docked. Key points are represented by the generation and by the selection of the most representative structures (see the paragraph "How to choose the right conformations").

Molecular dynamics. Despite all the efforts made in the last years the inclusion of protein flexibility in docking still remains one of the most challenging task in drug design, even more than the estimation of the entropy of binding or the prediction of water molecules' role [65, 73, 74]. Detailed quantum-mechanics description of protein motions and of chemical reaction of large molecular system is, in fact, still too demanding in terms of time and computer power. Molecular Dynamics currently represents the best alternative for simulating protein dynamics and, thanks to the constant evolution of computer power and of algorithms performance, will play a more and more leading role in drug design and discovery.

When a ligand enters the binding site of its target in solution has to face an entity in constant motion and not a single frozen entity. X-ray structures are able to represent only some low energy conformations of the target but they completely loose information related to the transitions between one state and the other, they are generally not able to reproduce the higher energy states, and they also completely disregard the possibility of experiencing new ligand-induced rearrangements. Different scenarios might accompany the formation of a protein-ligand complex, according to the intrinsic protein dynamics and to the energy associated to the interaction. There are very few cases in which the ligand almost perfectly fits a static binding pocket as a key into a lock, inducing very small and subtle changes [75]. As previously mentioned, more often ligands select and stabilize subsets of conformations among the possible ones, thus shifting the conformational equilibrium towards those states better stabilizing the interaction with the same ligands [16]. Sometimes ligands can induce rearrangements leading to conformations not pre-existing in ligand absence [76]. MD simulations actually represent the best tool for investigating this conformational space and, in the last years, more and more experimental evidences have supported the capability of MD-generated ensembles to closely replicate the structural dynamics of proteins in solution [77]. Moreover MD methods have demonstrated their capability of identifying druggable allosteric sites, normally not present in rigid static structures [78-80].

Unfortunately, several limitations regarding the investigated range of time and conformational space still affect the possible success of MD-based studies. We have to consider, in fact, that, despite the recent improvements in algorithms and computer power, most MD applications in docking and screening are limited to the post-docking stage, in order to get insights about the stability of the docked conformations, to have additional conformational and energetic insight about ligand binding and, clearly, to refine the poses after docking [81-86].

Since the first MD simulation of a protein [87], the increase in computational power facilitated the growing use of MD in structure-based drug discovery. In a recent study [88], long MD simulations were performed on the undecaprenyl pyrophosphate synthase (UPPS, Figure 1a) to investigate protein flexibility and the impact of protein dynamics on ligand design. MD simulations confirmed that UPPS is a highly flexible protein, with observed conformational changes in agreement with existing crystal structures. Docking of UPPS inhibitors on high- and low-populated conformational states showed that structurally diverse ligands can bind to different sampled conformational states, in agreement with the notion of conformational selection. In this case, docked ligand poses coincide with those observed in ligand: UPPS crystal structures.

It should be noted, however, that enrichments obtained using MD-generated ensembles could be lower than those using crystal structures [31]. Importantly, in some cases the length of MD simulations is not enough to sample biologically relevant processes which happen in the millisecond to second range, since low-energy conformations separated by high energy barriers are seldom crossed, unless the simulation runs for unrealistic time-scales. Unfortunately, some of these conformational changes are important to model ligand: protein interaction. Attempts to run longer MD simulations include the use of *ad hoc* designed hardware [89] and of graphics-processing units (GPUs) [90, 91].

A number of non-conventional MD methodologies able to explore large domain motions and significant active site rearrangements in a computational efficient way have been developed and are listed below [11, 92-94] (see the paragraph "Enhanced sampling techniques"). Even if these strategies introduce some artefacts, they allow proteins to experience conformations that would not have been investigated in the standard MD timescale. In the perspective of increasing the simulation time, great advances have been made with the introduction of GPUs, able to speed molecular dynamics simulation by an order of magnitude [90, 95]. Also, the supercomputer Anton, specifically built to perform high demanding computational simulations, is currently able to run microsecond of simulations per day, allowing the investigation of binding events but also of folding and unfolding processes [96-98].

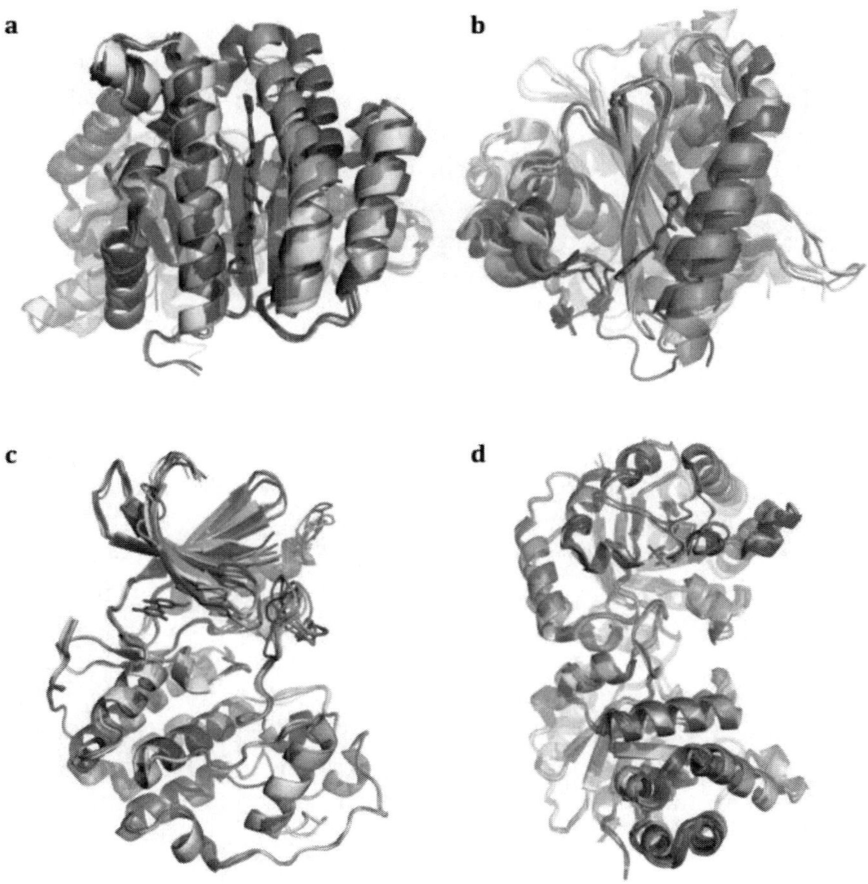

Figure 1. Superimposition of different X-ray structures showing the binding site conformational adjustments of a. *Escherichia coli* undecaprenyl pyrophosphate synthase (UPPS; PDB codes 1v7u, 3qas, 3sgx, 2e98, 4h2j). The binding site is identified by the co-crystallized BPH-1354. b. HIV-1 integrase (PDB codes 1qs4, 3ao5, 3zcm, 4nyf, 4ojr). The binding site is identified by the co-crystallized ligand 1-(5-chloroindol-3-yl)-3-hydroxy-3-(2H-tetrazol-5-yl)-propenone. c. Cyclin-dependent kinase 2 (CDK2; PDB codes 3pxy, 3qqg, 3r7v , 3s2p, 3ti1, 4ez3). The binding site is identified by the co-crystallized ligand JWS648. d. *Trypanosoma Brucei* Triosephosphate isomerase (TIM; PDB codes 1iih, 1tsi, 3tim, 5tim, 6tim). The binding site is identified by the co-crystallized 3-phosphoglycerate.

Another limitation is represented by the force fields, which still require further refinement for providing more reliable energy quantification. Systems where quantum effects are important or, for instance, where transition metals are involved would require more refined force fields. Still, after many years, a well trained polarizable force field has not been developed and MD simulations accounting for electronic polarization are rare [99, 100]. It is, in fact, widely recognized, that docking simulations may fail because of limitations and approximations generally introduced in sampling and scoring [101-104].

In this perspective QM and QM/MM methods have been applied in drug discovery to improve the prediction accuracy for protein-ligand interactions [105]. The advantage is clearly represented by the more reliable treatment of charges and polarization effects, while classic MM algorithms normally assume fixed charges during binding processes. On the contrary they are significantly more time consuming and several simplification schemes have been developed to allow a linear scaling of computing time according with molecular size.

Enhanced sampling techniques. Methodological developments in biomolecular simulation theory have been also undertaken to improve conformational sampling. Replica-exchange molecular dynamics has been used to enhance conformational sampling in biomolecular systems [106, 107].

In accelerated MD (aMD) [108] the Hamiltonian of the system is modified in such a way that the potential energy surface (PES) is smoothed, though maintaining the relative positions of local minima. Since barriers separating low-energy states become lower, those states are more frequently visited by the system. However, since the PES becomes altered, it is not easy to recover the canonical ensemble and, thus, to calculate the thermodynamic properties of the system. In spite of this, several successful applications of this method have been reported [109].

In metadynamics [110, 111] the PES is described in a coarse-grain way using a few collective variables, and a history-dependent potential defined by adding Gaussian contributions to the potential energy along the trajectory of the collective variables, thus forcing the system to avoid already sampled regions. Studying the dissociation process of inhibitor SC-558 in cyclooxigenase-2 with metadynamics, an alternative binding mode was found, which could help rationalizing the slow kinetics of this class of inhibitors within the binding site [112]. Recently, classical and quantum mechanics/molecular mechanics (QM/MM) simulations combined with metadynamics was used to study the formation of the covalent complex between avibactam and beta-lactamase TEM-1 [113], where computational simulations provided details of the molecular mechanism leading to the covalent complex. The use of metadynamics to explore protein flexibility has been recently reviewed [114]. However, the outcome of metadynamics simulations heavily depends on the choice of the collective variables [114, 115], that could not be trivial, especially when there is no knowledge of the final state conformation.

Aside plain MD and enhanced MD, different methodologies more recently emerged, for generating ensembles of protein conformations.

Xu et al. proposed the ligand-model concept to sample the protein conformations most suitable for binding structurally diverse set of ligands [116]. In particular, MD simulations are performed by placing in the binding site a virtual ligand, made by a collection of functional groups and representing a hypothetical ensemble of different chemical species. The presence of the ligand affects and dynamically changes the protein shape and properties during the simulations and provides ligand-induced conformations. The authors demonstrated that their methodology outperformed static docking and also ensemble docking simulations when the conformations are obtained from apo MD simulations, in terms of pose prediction and energy calculations. Unfortunately, only small-scale conformational changes can be reproduced.

Collective degrees of freedom to represent protein flexibility. Collective degrees of freedom can be used to represent protein flexibility and study proteins dynamics [117-119]. Normal modes represent concerted atomic displacements around the position of equilibrium, and are determined by diagonalization of the second-derivative of the energy matrix (Hessian) using an atom mass metric [120]. The corresponding eigenvectors provide the direction of the displacement, while the corresponding eigenvalues represent the amplitude of the energy change upon perturbing the structure along the eigenvectors. Since the actual energy function is not harmonic, normal modes only accurately represent small oscillations around the position of equilibrium. Large-scale domain movements, which are associated with low-energy motions, are usually represented by non-zero low-frequency normal modes, while small though fast movements are represented by high-frequency modes. It should be noted that normal modes calculated for a system where Cα atoms are joined by springs represent very well slow protein dynamics [121]. This technique was further extended by Hinsen [122], and Bahar through the Gaussian network model [123-126]. Normal modes have been shown to correlate well with functional movements in proteins [127] (cf. Ref. [128] for a comparison between protein mobility and normal modes analysis; NMA). Clearly, the key advantage in using NM to study the movement of the system is that the number of relevant degrees of freedom (DOF) is greatly reduced, since only a few low-frequency modes are necessary.

NMA has been applied to study the binding of a Hoechst 33258 analogue to DNA while accounting for protein flexibility [129]. The number of DOF was reduced from ~ 800 to 5. It was observed that starting with the structure of unbound DNA, relaxation of the complex along a subset of low-energy modes significantly improved the fit of the ligand into the binding site. Perturbation along collective modes was performed in the Cartesian coordinate's space, so energy minimization had to be performed to restore covalent geometry. NMA in internal coordinates could also be performed to avoid this problem [130, 131]. The low-mode conformational search procedure (LMOD) [132] complemented with conformational searching (*c*-LMOD) was used to study flexible docking to Purine Nucleoside Phosphorylase (PNP) [133] and to study the binding of inhibitors to HIV integrase (Figure 1b) [134].

It was observed that in protein kinases, mid-scale loop displacements of the glycine-rich and C-term loops cannot be properly represented with the first lowest-energy modes [54] (this was observed in other proteins also [135]). However, calculating NMs using a Cα spring-network model [121] and introducing a measure of relevance of the modes on a given region of interest of the protein, only very few *relevant* low-energy modes (< 10) were necessary to represent backbone binding site flexibility in protein kinases. This approach was used to generate a conformational ensemble of PKA structures by perturbating the structure along the set of relevant normal modes. The receptor conformations thus generated were tested in high-throughput docking and the corresponding protocol in internal coordinates was published [136].

Sperandio et al. also provided their own rules for choosing the suitable receptor conformations, proposing a fast method for generating multiple conformations via Normal Modes Analysis and a dedicated protocol for the selection [137].

More recently, Dietzen et al. extracted the first normal mode from the apo forms of a set of 433 apo/holo pairs and used it for generating ensemble of conformations representing the Cα trace of the ligand-bound protein structure. Despite the good results, the authors observed that the amplitude of the modes significantly changed in the different case studies and underlined that no general rules can be defined about the number and type of normal modes that have to be selected for getting successful docking results [138].

Although MD simulations provide a detailed picture of the evolution of the system through the energy hypersurface, it might be difficult to extract and analyse the collective motions relevant to biomolecular recognition and drug design. A principal component analysis (PCA) can be performed on a large number of MD snapshots to yield collective DOF ranked by their contribution to the motion of the protein [139]. The term essential dynamics is usually used whenever backbone or Cα atoms are used for the PCA [140]. The advantage of PCA-based methods to represent flexibility is that it can go beyond the normal mode approximation. Moreover, the conformational set for PCA can be built using a sufficiently large ensemble of experimental structures. However, mid-scale loop rearrangements might not be well represented with the most significant principal components, in the same way it happens with normal modes.

PCA was used to reduce conformational dimensionality to represent protein flexibility in the HIV-1 protease and aldose reductase [141], where biologically relevant motions were reproduced. Soft modes calculated from PCA after a MD simulation [129] were considered in the docking of FK506 to the unbound structure of FKBP. It should be noted that FK506 fails to dock under the rigid docking approximation.

Monte-Carlo based methods. Using Monte Carlo sampling with minimization [142], the binding mode of the ligand and pocket side chain conformations in the seven-transmembrane proteins bovine rhodopsin and bateriorhodopsin were predicted [143]. The Biased Probability Monte Carlo (BPMC) procedure was used [144], where the selection of random variables in the internal coordinate space [145] is performed according to a predefined continuous probability distribution. The predicted RMSD for the ligand and binding site side-chains was within 0.2 Å and 0.3 Å, respectively. Using a Monte Carlo simulation and the Generalized Born/Surface Area (GB/SA) continuum solvent model, a test set of 14 complexes were evaluated for ligand docking with and without side chain flexibility. The RMSDs of the ligands were comparable in both approaches. However, discrimination of the binding mode according to energy actually deteriorated in the flexible receptor docking procedure. A Monte Carlo-based global-energy minimization method using flexible ligand and receptor was also used to analyse the structural binding determinants of antagonists to the Retinoid X Receptor (RXR) [146].

A "time step" Monte Carlo algorithm to sample side chain conformations using an implicit solvent model was used to study flexibility in ligand binding sites, evaluated on six proteins: DB3 antibody, thermolysin, estrogen receptor, PPAR-γ, PI3 kinase, and CDK2 (Figure 1c). It was found that the binding site conformational ensembles corresponding to the apo proteins correlate well with known induced-fit conformational changes [147]. In another work from the same group, Monte Carlo sampling was used to study protein loop flexibility. Applying the method on the flexible loop in triosephosphate isomerase (Figure 1d) and other loops from different proteins, it was observed that the generated loop ensembles were in good agreement with structural characterizations [148].

The Monte Carlo algorithm above-mentioned [143] using flexible-ligand/flexible-side chains sampling was adopted to generate diverse binding sites in protein kinases [36, 54, 136], which were then used in a small-scale high-throughput docking in the Receptor Ensemble Docking (RED) approach (cf. Refs. [120, 149, 150] for review). A natural expansion of this approach is to incorporate ligand information in the process of homology modelling [151], where existing ligands are explicitly used to shape and optimize the binding site through a docking-based global energy Monte Carlo sampling. This "ligand-steered" homology modelling method was developed and used on a structure-based docking project on Class A GPCR MCH-R1 [152]. Structural models were used for docking, and top-scoring compounds were tested experimentally in a competitive binding assay, where six compounds were identified to be active in the low micromolar range. In a benchmarking study of homology models and docking for Class A GPCRs [153], the ligand-steered homology modelling method was used to cross-model bovine rhodopsin, β_2 adrenergic receptor, and A_{2A} adenosine receptor, then comparing the performance of crystal structures and models on docking. It was shown that the ligand-steered method was able to generate a small set of reliable models, explicitly accounting for protein flexibility in the modelling stage, and that ligand-steered models performed well in high-throughput docking when compared to crystal structures, crude homology models prior to refinement, and to random selection of compounds. The ligand-steered method was also used to explain SAR data of ligand binding to the cannabinoid CB2 receptor [154-156]. According to this idea, GPCR models generated using induced-fit docking [68] were used to assess their performance in high-throughput docking [157].

MD in virtual screening. While docking can introduce a certain level of flexibility into the protein binding site, traditional virtual screening (VS) has for several years relied on single target structures coming from X-ray, NMR or homology modeling studies. As binding sites can experience many valid and druggable conformations, the use of static structures may often lead to discard actual ligands binding to receptor conformations different from the chosen one [158].

The first MD application for drug design was performed more than 30 years ago [87], but the first experimental confirmation of the utility of using multiple computationally generated structures in an ensemble-based approach arrived 23 years later, when Carlson et al. simulated the dynamics of the apo HIV-1 integrase catalytic domain.

Currently two different strategies relying on MD simulations are commonly used, the dynamic pharmacophore model and the receptor complex scheme.

The dynamic pharmacophore model. These methodologies are based on the construction of a dynamic pharmacophore model (DPM), taking into account the information coming from the whole trajectory or from the most representative snapshots of a MD simulation. In 2000 Carlson et al. selected eleven different conformations from the MD-generated trajectory of the apo HIV-1 integrase catalytic domain [159, 160] and used them for defining a dynamic pharmacophore model used to screen the Available Chemical Database. A number of compounds was selected and experimentally tested for measuring their activity. More than 30% of them were able to inhibit the target. The authors also demonstrated that the DPM re-ranking outperformed any single static pharmacophore model. Some years later Meagher and Carlson demonstrated that longer simulations could lead to improved results in CADD as to better discrimination between binders and non-binders [161]. A few years later the DPM strategy was used to simulate the flexibility of MDM2-p53 binding site and to search for new inhibitors. Starting from a library of 35000 compounds, 4 molecules were found to inhibit the target at 50 μM concentration, after being screened against a 6-site pharmacophore model of the active site [162].

Similarly, grid-based approaches pre-compute protein-ligand interaction energy for each structure onto three-dimensional grids which are deformed to take into account the protein conformational changes induced upon ligand binding. Differently from RCS approaches, these methods simulate protein and ligand flexibility at a time and do not require sequential docking to all the protein structures ensemble (EPS) [163, 164].

The relaxed complex scheme. An alternative strategy for using multiple MD-generated structures in drug discovery is the relaxed complex scheme (RCS) [165, 166]. Differently from the construction of an average pharmacophore, RCS methods dock the ligands into several conformations generally extracted from MD simulations [49, 165, 167], harmonic dynamics simulations [168] or normal mode/elastic network analyses [54, 169-173]. These approaches normally combine MD calculations to docking algorithms and scoring functions.

Each ligand will have a number of scores equal to the number of protein conformations used and the ranking will be performed considering, for instance, the average of all the scores.

RCS approaches have demonstrated their capability of identifying new protein inhibitors in several cases. When looking for new FKBP binding protein inhibitors Lin et al. demonstrated how docking performances are extremely sensitive to the receptor conformations [48]. Schames et al. identified a new possible pocket adjacent to the known binding site in HIV-1 integrase. They also found that docking of ligands into this alternative pocket gave better results than docking into the active site, thus opening the possibility for the design of new inhibitors [174]. RCS techniques have been also successfully applied by Durrant et al. in finding inhibitors of *Tripanosoma brucei* uridine diphosphate galactose 4'-epimerase [175] and by Sinko et al. for targeting the antibacterial target undecaprenyl pyrophosphate synthase [88]. New compounds for inhibiting neuraminidase in avian influenza were found by Cheng and co-workers [49], while Amaro et al. used RCS to target RNA-editing ligase 1 and found interesting scaffolds characterized by IC_{50} values equal to 1 μM [176]. Claussen et al. developed the FlexE tool, able to dock flexible ligands into ensembles of protein structures representing the flexibility, the point mutations or the alternative models of a protein. The different structures were superposed and combinatorially joined to create new valid protein models. When tested on a set of 105 crystal structures and 60 ligands the algorithm demonstrated to perform as good as cross docking approaches with a significant lower computing time [177]. Very recently, a novel approach was presented [178], where MD simulations are biased towards existing ligand: protein crystal structures, thus enhancing the sampling of relevant conformations; cross-docking experiments on the structural ensemble thus generated were performed, and various levels of biasing were evaluated on CDK-2 and factor Xa.

Aside to the known and aforementioned limitations typical of MD, RCS-based approaches also suffer from the typical limitations of docking tools and scoring functions [58, 59, 179-181], which usually treat conformational entropy and solvation energy with approximate algorithms, sacrificing accuracy to the calculation speed. More accurate, although slower, algorithms for calculating the energy of binding include thermodynamic integration [182], single-step perturbation [183] and free energy perturbation [184].

Clustering. Even if a number of conformations can be easily generated, obviously not all of them can be used for screening chemical libraries, since the calculation time scales linearly with the number of conformers. Some properly selected structures should be able to cover a large conformational space representing, in principle, the most favourable and stable energetic states. Moreover, it has been demonstrated that the inclusion of too many experimentally- or computationally-determined conformations reduces the screening performances since it increases the probability of selecting false positives [38, 185, 186].

First clustering methodologies were based on the root mean square deviation (RMSD), which should extract the most diverse and dominant conformations with respect to the whole protein or to some specifically selected regions [49]. Interesting results obtained with RMSD-based clustering to both DPM and RCS approaches have been reported by Deng et al., who selected ten different HIV-1 integrase conformations and screened an in house database of about 400 compounds, finding 9 hits active at a concentration equal or less than 100 μM [187], and by Chen et al. who identified 10 molecules active against the avian influenza N1 neuraminidase with K_i lower than 500 μM [49].

Amaro et al. applied on the RNA editing ligase 1 enzyme of *Tripanosoma brucei* an alternative clustering methodology called QR-factorization, and initially proposed by O'Donoghue and Luthey-Schulten, to remove redundancy from multiple structural alignments [188]. The authors observed how this methods led to a drastic reduction of the number of representative conformations which, moreover, would not have been identified through standard screening procedures, and that the QR-reduced set properly described the intrinsic protein flexibility without losing energetic information [176].

How to choose the right conformation. Regardless of the methodology used to run the simulation and of the virtual screening approach, the key point for virtual screening success is represented by the quality of the conformational sampling and by the capability of properly representing the protein conformational space. It is now recognized that RCS approaches can improve virtual screening success, but even better results can be obtained if the right set of receptor conformations is chosen [47, 137]. When multiple conformations are available, in fact, the *a priori* selection of a limited number of forms able to give the best results in terms of screening and identification of new ligand is not trivial. Some strategies have been recently proposed to guide this conformational sampling in order to select the most representative states, not only to increase the calculation speed but, most of all, the performances quality.

One of the first attempts to use multiple conformations in docking was made by Knegtel et al. in 1997. The authors obtained interesting performances when dealing with HIV-1 protease and ras p21, stating that in a real life (non-retrospective) screening study, there is no way to know in advance which the right conformations are [189].

Rueda et al. tried to define the right recipe for selecting experimental conformations for virtual screening by analysing a benchmark of 1086 X-ray conformations for 99 different therapeutically relevant protein targets, and comparing the performances of the single and ensemble-conformation docking [44]. Their conformation selection was then validated through retrospective screening analyses and enrichment calculations. The authors proved that, when possible, a ligand-guided approach represents the best strategy to identify the conformations better able to separate between binders and non-binders. The enrichment factor or the AUC metrics can be exploited to rank *a posteriori* the different conformations to be used in independent dockings [190]. Nevertheless, they also observed that the use of conformational ensembles do not systematically overcome the use of a single well-performing structure and that it is generally not possible to outperform the best single conformation even when the others are handpicked according to their AUCs [38, 191]. Significant improvements by using multiple structures can be reached when poor performing structures or no-experimental structures are available, as in the case of homology models.

When no ligand information is available, an ensemble of randomly chosen 3-5 conformations can provide better results improving ligand recognition with respect to a single static state, as also suggested by Verdonk et al. [192]. In these cases in fact, the score-based uncertainties related to a single receptor state can be reduced and more consistent scores for ligand profiling can be generated. Finally, when different holo structures are available and the computational resources are limited, the authors recommend the use of those structures having the largest co-crystallized ligands.

In a following paper Bottegoni et al. carried out a retrospective screening study against 36 X-ray targets comparing the Single Receptor Conformation (SRC) and the Multiple Receptor Conformation (MRC) VS, focusing on i) the different separation power, ii) the number of actives identified and the chemical diversity of the hits, iii) the contribution of the single receptor conformation to the MRC-VLS overall performance [193]. Their analyses proved that both the MRC-score and MRC-rank methodologies outperformed the SRC and that using multiple receptor conformations with respect to one static structure significantly increases the variety of the chemical scaffolds, which is obviously crucial in a drug design perspective. Further improvements could be achieved introducing more different and ligand-unbiased conformations coming, for instance, from MD simulations. Again, the authors underlined that the conformational selection in absence of any ligand activity information still represents a challenging and major issue, which can greatly affect the VLS results, and that it is still extremely difficult to predict which conformer can provide the best results *a priori* [44].

Similar conclusions were also drawn by Korb et al., who investigated the performances of ensemble docking in virtual screening and pose prediction, as a function of ensemble size, for eight targets using GOLD and three different scoring functions [194]. Again, X-ray structures were used to investigate the proteins conformational space. The authors observed that ensemble docking always outperformed the use of the worst single structure. In some cases ensemble docking gave even better results with respect to the best single protein structure. Since it is normally not possible to foresee *a priori* the performances of single conformations, MPC approaches should at least minimize the risk of bad or worst performances in virtual screening. Good results were also obtained in terms of chemotype diversity. A strong dependence of docking results from scoring functions and sampling accuracy was highlighted, since different performances were observed for the same set of structures when using different scoring functions. The pressing need of developing ensemble selection protocols for choosing the best conformations given a target and a scoring function is again underlined.

Conclusion

From the beginning of the *in silico* era, incredible progresses have been made for including protein structural plasticity in computational simulations. The proper prediction of the conformational space is of paramount importance when looking for ligands targeting pharmaceutically relevant targets, in particular when searching for new scaffolds. Overcoming the ligand induced-fit bias, typically affecting co-crystallized X-ray structures, could lead, in fact, to the investigation and exploitation of the undisclosed, but still real part of the conformational scenario experienced by proteins. This would allow the identification of new binding sites, cryptic or allosteric, as well as the discovery of new molecules different from the already known ones. Alas, deciphering the information encoded into extended simulation is still far from trivial, and reproducing biochemical events occurring in the millisecond or minute scale is still far from reality.

We believe that the progress in algorithms and computational resources will likely allow the description of longer and larger scale events, and that appropriate algorithms for sampling and extracting the right information will help in understanding the rules and role of protein dynamics. The challenge is still open.

References

[1] Marsh, J.A., Teichmann, S.A. 2014. Parallel dynamics and evolution: Protein conformational fluctuations and assembly reflect evolutionary changes in sequence and structure. *Bioessays* 36 (2):209-218.

[2] McCammon, J.A. 1999. Protein dynamics. *Rep. Prog. Phys.* 47 1.

[3] Lukman, S., Verma, C.S., Fuentes, G. 2014. Exploiting protein intrinsic flexibility in drug design. *Adv. Exp. Med. Biol.* 805 245-269.

[4] Forman-Kay, J.D., Mittag, T. 2013. From sequence and forces to structure, function, and evolution of intrinsically disordered proteins. *Structure* 21 (9):1492-1499.

[5] Clothia, C. 1976. The nature of the accessible and buried surfaces in proteins. *J. Mol. Biol.* 105 1-12.

[6] Marsh, J.A. 2013. Buried and accessible surface area control intrinsic protein flexibility. *J. Mol. Biol.* 425 (17):3250-3263.

[7] Marsh, J.A., Teichmann, S.A., Forman-Kay, J.D. 2012. Probing the diverse landscape of protein flexibility and binding. *Curr. Opin. Struct. Biol.* 22 (5):643-650.

[8] Bornberg-Bauer, E., Alba, M.M. 2013. Dynamics and adaptive benefits of modular protein evolution. *Curr. Opin. Struct. Biol.* 23 (3):459-466.

[9] Best, R.B., Lindorff-Larsen, K., DePristo, M.A., Vendruscolo, M. 2006. Relation between native ensembles and experimental structures of proteins. *Proc. Natl. Acad. Sci. U S A* 103 (29):10901-10906.

[10] Cozzini, P., Kellogg, G.E., Spyrakis, F., Abraham, D.J., Costantino, G., Emerson, A., Fanelli, F., Gohlke, H., Kuhn, L.A., Morris, G.M., Orozco, M., Pertinhez, T.A., Rizzi, M., Sotriffer, C.A. 2008. Target flexibility: an emerging consideration in drug discovery and design. *J. Med. Chem.* 51 (20):6237-6255.

[11] Henzler, A., Rarey, M. 2010. In pursuit of fully flexible protein-ligand docking: Modeling the bilateral mechanism of binding. *Mol. Inform.* 29 (3):164-173.

[12] Rao, C., Subramanian, J., Sharma, S.D. 2009. Managing protein flexibility in docking and its applications. *Drug Discov. Today* 14 (7-8):394-400.

[13] Teague, S.J. 2003. Implications of protein flexibility for drug discovery. *Nat Rev Drug Discov* 2 (7):527-541.

[14] Teodoro, M.L., Kavraki, L.E. 2003. Conformational flexibility models for the receptor in structure based drug design. *Curr. Pharm. Des.* 9 (20):1635-1648.

[15] Boehr, D.D., Nussinov, R., Wright, P.E. 2009. The role of dynamic conformational ensembles in biomolecular recognition. *Nat. Chem. Biol.* 5 (11):789-796.

[16] Ma, B., Shatsky, M., Wolfson, H.J., Nussinov, R. 2002. Multiple diverse ligands binding at a single protein site: a matter of pre-existing populations. *Protein Sci.* 11 (2):184-197.

[17] Tsai, C.J., Kumar, S., Ma, B., Nussinov, R. 1999. Folding funnels, binding funnels, and protein function. *Protein Sci.* 8 (6):1181-1190.

[18] Bosshard, H.R. 2001. Molecular recognition by induced fit: how fit is the concept? *News Physiol Sci* 16 171-173.

[19] Ikeguchi, M., Ueno, J., Sato, M., Kidera, A. 2005. Protein structural change upon ligand binding: linear response theory. *Phys. Rev. Lett.* 94 (7):078102.

[20] Baldwin, A.J., Kay, L.E. 2009. NMR spectroscopy brings invisible protein states into focus. *Nat. Chem. Biol.* 5 (11):808-814.

[21] Pellecchia, M., Sem, D.S., Wuthrich, K. 2002. NMR in drug discovery. *Nat. Rev. Drug Discov.* 1 (3):211-219.

[22] Callender, R., Deng, H. 1994. Nonresonance Raman difference spectroscopy: a general probe of protein structure, ligand binding, enzymatic catalysis, and the structures of other biomacromolecules. *Annu. Rev. Biophys. Biomol. Struct.* 23 215-245.

[23] Balog, E., Becker, T., Oettl, M., Lechner, R., Daniel, R., Finney, J., Smith, J.C. 2004. Direct determination of vibrational density of states change on ligand binding to a protein. *Phys. Rev. Lett.* 93 (2):028103.

[24] Mertens, H.D., Svergun, D.I. 2010. Structural characterization of proteins and complexes using small-angle X-ray solution scattering. *J. Struct. Biol.* 172 (1):128-141.

[25] Svergun, D.I. 2010. Small-angle X-ray and neutron scattering as a tool for structural systems biology. *Biol. Chem.* 391 (7):737-743.

[26] Davis, I.W., Baker, D. 2009. RosettaLigand docking with full ligand and receptor flexibility. *J. Mol. Biol.* 385 (2):381-392.

[27] Hassell, A.M., An, G., Bledsoe, R.K., Bynum, J.M., Carter, H.L., 3rd, Deng, S.J., Gampe, R.T., Grisard, T.E., Madauss, K.P., Nolte, R.T., Rocque, W.J., Wang, L., Weaver, K.L., Williams, S.P., Wisely, G.B., Xu, R., Shewchuk, L.M. 2007. Crystallization of protein-ligand complexes. *Acta Crystallogr D Biol Crystallogr* 63 (Pt 1):72-79.

[28] Fuentes, G., Dastidar, S., Madhumalar, A., Verma, C.S. 2011. Role of protein flexibility in the discovery of new drugs. *Drug Dev. Res.* 72 26-35.

[29] Lin, J.H. 2011. Accomodating protein flexibility for structure-based drug design. *Curr. Top Med. Chem.* 11 171-178.

[30] Spyrakis, F., BidonChanal, A., Barril, X., Luque, F.J. 2011. Protein flexibility and ligand recognition: challenges for molecular modeling. *Curr. Top. Med. Chem.* 11 (2):192-210.

[31] Osguthorpe, D.J., Sherman, W., Hagler, A.T. 2012. Exploring protein flexibility: incorporating structural ensembles from crystal structures and simulation into virtual screening protocols. *J. Phys. Chem. B* 116 (23):6952-6959.

[32] Ferrari, A.M., Wei, B.Q., Costantino, L., Shoichet, B.K. 2004. Soft docking and multiple receptor conformations in virtual screening. *J. Med. Chem.* 47 (21):5076-5084.

[33] Kairys, V., Gilson, M. 2002. Enhanced docking with the mining minima optimizer: acceleration and side-chain flexibility. *J. Comput. Chem.* 23 1656-1670.

[34] Nabuurs, S.B., Wagener, M., de Vlieg, J. 2007. A flexible approach to induced fit docking. *J. Med. Chem.* 50 (26):6507-6518.

[35] Sherman, W., Beard, H.S., Farid, R. 2006. Use of an induced fit receptor structure in virtual screening. *Chem. Biol. Drug Des.* 67 (1):83-84.

[36] Cavasotto, C.N., Abagyan, R.A. 2004. Protein flexibility in ligand docking and virtual screening to protein kinases. *J. Mol. Biol.* 337 (1):209-225.

[37] Moitessier, N., Henry, C., Maigret, B., Chapleur, Y. 2004. Combining pharmacophore search, automated docking, and molecular dynamics simulations as a novel strategy for flexible docking. Proof of concept: docking of arginine-glycine-aspartic acid-like compounds into the alphavbeta3 binding site. *J. Med. Chem.* 47 (17):4178-4187.

[38] Barril, X., Morley, S.D. 2005. Unveiling the full potential of flexible receptor docking using multiple crystallographic structures. *J. Med. Chem.* 48 (13):4432-4443.

[39] Craig, I.R., Essex, J.W., Spiegel, K. 2010. Ensemble docking into multiple crystallographically derived protein structures: an evaluation based on the statistical analysis of enrichments. *J. Chem. Inf. Model.* 50 (4):511-524.

[40] Huang, S.Y., Zou, X. 2007. Ensemble docking of multiple protein structures: considering protein structural variations in molecular docking. *Proteins* 66 (2):399-421.

[41] Limongelli, V., Marinelli, L., Cosconati, S., Braun, H.A., Schmidt, B., Novellino, E. 2007. Ensemble-docking approach on BACE-1: pharmacophore perception and guidelines for drug design. *Chem. Med. Chem.* 2 (5):667-678.

[42] Park, S.J., Kufareva, I., Abagyan, R. 2010. Improved docking, screening and selectivity prediction for small molecule nuclear receptor modulators using conformational ensembles. *J. Comput. Aided Mol. Des.* 24 (5):459-471.

[43] Rao, S., Sanschagrin, P.C., Greenwood, J.R., Repasky, M.P., Sherman, W., Farid, R. 2008. Improving database enrichment through ensemble docking. *J. Comput. Aided Mol. Des.* 22 (9):621-627.

[44] Rueda, M., Bottegoni, G., Abagyan, R. 2010. Recipes for the selection of experimental protein conformations for virtual screening. *J. Chem. Inf. Model.* 50 (1):186-193.

[45] Damm, K.L., Carlson, H.A. 2007. Exploring experimental sources of multiple protein conformations in structure-based drug design. *J. Am. Chem. Soc.* 129 (26):8225-8235.

[46] Hritz, J., de Ruiter, A., Oostenbrink, C. 2008. Impact of plasticity and flexibility on docking results for cytochrome P450 2D6: a combined approach of molecular dynamics and ligand docking. *J. Med. Chem.* 51 (23):7469-7477.

[47] Bolstad, E.S., Anderson, A.C. 2008. In pursuit of virtual lead optimization: the role of the receptor structure and ensembles in accurate docking. *Proteins* 73 (3):566-580.

[48] Lin, J.H., Perryman, A.L., Schames, J.R., McCammon, J.A. 2002. Computational drug design accommodating receptor flexibility: the relaxed complex scheme. *J. Am. Chem. Soc.* 124 (20):5632-5633.

[49] Cheng, L.S., Amaro, R.E., Xu, D., Li, W.W., Arzberger, P.W., McCammon, J.A. 2008. Ensemble-based virtual screening reveals potential novel antiviral compounds for avian influenza neuraminidase. *J. Med. Chem.* 51 (13):3878-3894.

[50] Park, I.H., Li, C. 2010. Dynamic ligand-induced-fit simulation via enhanced conformational samplings and ensemble dockings: a survivin example. *J. Phys. Chem. B* 114 (15):5144-5153.

[51] Paulsen, J.L., Anderson, A.C. 2009. Scoring ensembles of docked protein:ligand interactions for virtual lead optimization. *J. Chem. Inf. Model.* 49 (12):2813-2819.

[52] Wong, C.F., Kua, J., Zhang, Y., Straatsma, T.P., McCammon, J.A. 2005. Molecular docking of balanol to dynamics snapshots of protein kinase A. *Proteins* 61 (4):850-858.

[53] Yoon, S., Welsh, W.J. 2004. Identification of a minimal subset of receptor conformations for improved multiple conformation docking and two-step scoring. *J. Chem. Inf. Comput. Sci.* 44 (1):88-96.

[54] Cavasotto, C.N., Kovacs, J.A., Abagyan, R.A. 2005. Representing receptor flexibility in ligand docking through relevant normal modes. *J. Am. Chem. Soc.* 127 (26):9632-9640.

[55] Novoa, E., de Pouplana, L., Barril, X., Orozco, M. 2010. Ensemble docking from homology models. *J. Chem. Theory Comput.* 6 2547-2557.

[56] Blaney, J., Dixon, J. 1993. A good ligand is hard to find: automated docking methods. *Perspect Drug Discov. Des.* 1 301-319.

[57] Bamborough, P., Cohen, F.E. 1996. Modeling protein-ligand complexes. *Curr Opin Struct Biol* 6 (2):236-241.

[58] Spyrakis, F., Cozzini, P., Kellogg, G.E. In *Burger's Medicinal Chemistry, 7th edition*; Wiley-VCH: New York, 2010, pp 601-684.

[59] Spyrakis, F., Kellogg, G.E., Amadasi, A., Cozzini, P. In *Frontiers in Drug Design and Discovery*, 2007; Vol. *3*, pp 317-379.

[60] Jiang, F., Kim, S.H. 1991. "Soft docking": matching of molecular surface cubes. *J. Mol. Biol.* 219 (1):79-102.

[61] Zavodszky, M.I., Kuhn, L.A. 2005. Side-chain flexibility in protein-ligand binding: the minimal rotation hypothesis. *Protein Sci.* 14 (4):1104-1114.

[62] Beier, C., Zacharias, M. 2010. Tackling the challenges posed by target flexibility in drug design. *Expert Opin Drug Discov.* 5 (4):347-359.

[63] Leach, A.R. 1994. Ligand docking to proteins with discrete side-chain flexibility. *J. Mol. Biol.* 235 (1):345-356.

[64] Lovell, S.C., Word, J.M., Richardson, J.S., Richardson, D.C. 2000. The penultimate rotamer library. *Proteins* 40 (3):389-408.

[65] Sousa, S.F., Ribeiro, A.J., Coimbra, J.T., Neves, R.P., Martins, S.A., Moorthy, N.S., Fernandes, P.A., Ramos, M.J. 2013. Protein-ligand docking in the new millennium--a retrospective of 10 years in the field. *Curr Med Chem* 20 (18):2296-2314.

[66] Schnecke, V., Kuhn, L.A. 2000. Virtual screening with solvation and ligand-induced complementarity. *Perspect Drug Discov. Des.* 20 171-190.

[67] Schnecke, V., Swanson, C., Getzoff, E., Tainer, J., Kuhn, L.A. 1998. Screening a peptidyl database for potential ligands to proteins with side-chain flexibility. *Proteins* 33 74-87.

[68] Sherman, W., Day, T., Jacobson, M.P., Friesner, R.A., Farid, R. 2006. Novel procedure for modeling ligand/receptor induced fit effects. *J. Med. Chem.* 49 (2):534-553.

[69] Meiler, J., Baker, D. 2006. RosettaLigand: protein-small molecule docking with full side-chain flexibility. *Proteins* 65 (3):538-548.

[70] Shin, W.H., Kim, J.K., Kim, D.S., Seok, C. 2013. GalaxyDock2: protein-ligand docking using beta-complex and global optimization. *J. Comput. Chem.* 34 (30):2647-2656.

[71] Kim, D.S., Cho, Y., Sugihara, K., Ryu, D., Kim, D. 2010. Three-dimensional beta-shapes and beta-complexes via quasi-triangulation. *Comput . Aided Des.* 42 911-929.

[72] Kim, D.S., Kim, C.M., Won, C.I., Kim, J.K., Ryu, J., Cho, Y., Lee, C., Bhak, J. 2011. BetaDock: shape-priority docking method based on beta-complex. *J. Biomol. Struct. Dyn.* 29 (1):219-242.

[73] Sousa, S.F., Fernandes, P.A., Ramos, M.J. 2006. Protein-ligand docking: current status and future challenges. *Proteins* 65 (1):15-26.

[74] Yuriev, E., Agostino, M., Ramsland, P.A. 2011. Challenges and advances in computational docking: 2009 in review. *J. Mol. Recognit* 24 (2):149-164.

[75] Fisher, E. 1894. Einfluss der configuration auf die wirkung der enzyme. *Ber. Dtsch. Chem. Ges.* 27 2985-2993.

[76] Koshland, D.E. 1958. Application of a Theory of Enzyme Specificity to Protein Synthesis. *Proc Natl Acad Sci U S A* 44 (2):98-104.

[77] Salmon, L., Bouvignies, G., Markwick, P., Lakomek, N., Showalter, S., Li, D.W., Walter, K., Griesinger, C., Bruschweiler, R., Blackledge, M. 2009. Protein conformational flexibility from structure-free analysis of NMR dipolar couplings: quantitative and absolute determination of backbone motion in ubiquitin. *Angew. Chem. Int. Ed. Engl.* 48 (23):4154-4157.

[78] Durrant, J.D., Keranen, H., Wilson, B.A., McCammon, J.A. 2010. Computational identification of uncharacterized cruzain binding sites. *PLoS Negl. Trop Dis.* 4 (5):e676.

[79] Grant, B.J., Lukman, S., Hocker, H.J., Sayyah, J., Brown, J.H., McCammon, J.A., Gorfe, A.A. 2011. Novel allosteric sites on Ras for lead generation. *PLoS One* 6 (10):e25711.

[80] Ivetac, A., McCammon, J.A. 2010. Mapping the druggable allosteric space of G-protein coupled receptors: a fragment-based molecular dynamics approach. *Chem. Biol. Drug Des.* 76 (3):201-217.

[81] Alcaro, S., Artese, A., Ceccherini-Silberstein, F., Ortuso, F., Perno, C.F., Sing, T., Svicher, V. 2009. Molecular dynamics and free energy studies on the wild-type and mutated HIV-1 protease complexed with four approved drugs: mechanism of binding and drug resistance. *J. Chem. Inf. Model.* 49 (7):1751-1761.

[82] Dong, B.L., Liao, Q.H., Wei, J. 2011. Docking and molecular dynamics study on the inhibitory activity of N, N-disubstituted-trifluoro-3-amino-2-propanols-based inhibitors of cholesteryl ester transfer protein. *J Mol Model* 17 (7):1727-1734.

[83] Hao, M., Li, Y., Zhang, S.W., Yang, W. 2011. Investigation on the binding mode of benzothiophene analogues as potent factor IXa (FIXa) inhibitors in thrombosis by CoMFA, docking and molecular dynamic studies. *J. Enzyme Inhib. Med. Chem.* 26 (6):792-804.

[84] Liao, Q.H., Gao, Q.Z., Wei, J., Chou, K.C. 2011. Docking and molecular dynamics study on the inhibitory activity of novel inhibitors on epidermal growth factor receptor (EGFR). *Med. Chem.* 7 (1):24-31.

[85] Maghsoudi, A.H., Khodagholi, F., Hadi-Alijanvand, H., Esfandiarei, M., Sabbaghian, M., Zakeri, Z., Shaerzadeh, F., Abtahi, S., Maghsoudi, N. 2011. Homology modeling, docking, molecular dynamics simulation, and structural analyses of coxsakievirus B3 2A protease: an enzyme involved in the pathogenesis of inflammatory myocarditis. *Int. J. Biol. Macromol.* 49 (4):487-492.

[86] Yang, Z., Nie, Y., Yang, G., Zu, Y., Fu, Y., Zhou, L. 2010. Synergistic effects in the designs of neuraminidase ligands: analysis from docking and molecular dynamics studies. *J. Theor. Biol.* 267 (3):363-374.

[87] McCammon, J.A., Gelin, B.R., Karplus, M. 1977. Dynamics of folded proteins. *Nature* 267 (5612):585-590.

[88] Sinko, W., de Oliveira, C., Williams, S., Van Wynsberghe, A., Durrant, J.D., Cao, R., Oldfield, E., McCammon, J.A. 2011. Applying molecular dynamics simulations to identify rarely sampled ligand-bound conformational states of undecaprenyl pyrophosphate synthase, an antibacterial target. *Chem. Biol. Drug Des.* 77 (6):412-420.

[89] Dror, R.O., Arlow, D.H., Borhani, D.W., Jensen, M.O., Piana, S., Shaw, D.E. 2009. Identification of two distinct inactive conformations of the beta2-adrenergic receptor reconciles structural and biochemical observations. *Proc. Natl. Acad. Sci. U S A* 106 (12):4689-4694.

[90] Friedrichs, M.S., Eastman, P., Vaidyanathan, V., Houston, M., Legrand, S., Beberg, A.L., Ensign, D.L., Bruns, C.M., Pande, V.S. 2009. Accelerating molecular dynamic simulation on graphics processing units. *J. Comput. Chem.* 30 (6):864-872.

[91] Sweet, J.C., Nowling, R.J., Cickovski, T., Sweet, C.R., Pande, V.S., Izaguirre, J.A. 2013. Long Timestep Molecular Dynamics on the Graphical Processing Unit. *J. Chem. Theory Comput.* 9 (8):3267-3281.

[92] Amaro, R.E., Li, W.W. 2010. Emerging methods for ensemble-based virtual screening. *Curr. Top. Med. Chem.* 10 (1):3-13.

[93] Kokh, D., Wade, R., Wenzel, W. 2011. Receptor flexibility in small-molecule docking calculations. *Comput. Mol. Sci.* 1 (2):298-314.

[94] Lill, M.A. 2011. Efficient incorporation of protein flexibility and dynamics into molecular docking simulations. *Biochemistry* 50 (28):6157-6169.

[95] Liu, W., Schmidt, B., Voss, G., Muller-Wittig, W. 2008. Accelerating molecular dynamic simulations using Graphics Proessings Units with CUDA. *Comput Phys Commun* 30 864-872.

[96] Shan, Y., Kim, E.T., Eastwood, M.P., Dror, R.O., Seeliger, M.A., Shaw, D.E. 2011. How does a drug molecule find its target binding site? *J. Am. Chem. Soc.* 133 (24):9181-9183.

[97] Shaw, D.E., Deneroff, M., Dror, R.O., Kuskin, J., Larson, R., Salmon, J., C, Y., al., e. 2008. Anton, a special-purpose machine for molecular dynamics simulation *Commun. ACM* 51 91-97.

[98] Shaw, D.E., Maragakis, P., Lindorff-Larsen, K., Piana, S., Dror, R.O., Eastwood, M.P., Bank, J.A., Jumper, J.M., Salmon, J.K., Shan, Y., Wriggers, W. 2010. Atomic-level characterization of the structural dynamics of proteins. *Science* 330 (6002):341-346.

[99] Cieplak, P., Dupradeau, F.Y., Duan, Y., Wang, J. 2009. Polarization effects in molecular mechanical force fields. *J. Phys. Condens Matter* 21 (33):333102.

[100] Jorgensen, W.L. 2007. Special issue on polarization. *J. Chem. Theory Comput.* 3 1877.

[101] Cheng, T., Li, X., Li, Y., Liu, Z., Wang, R. 2009. Comparative assessment of scoring functions on a diverse test set. *J. Chem. Inf. Model.* 49 (4):1079-1093.

[102] Coupez, B., Lewis, R.A. 2006. Docking and scoring--theoretically easy, practically impossible? *Curr. Med. Chem.* 13 (25):2995-3003.

[103] Kirchmair, J., Markt, P., Distinto, S., Wolber, G., Langer, T. 2008. Evaluation of the performance of 3D virtual screening protocols: RMSD comparisons, enrichment assessments, and decoy selection--what can we learn from earlier mistakes? *J. Comput. Aided Mol. Des.* 22 (3-4):213-228.

[104] Klebe, G. 2006. Virtual ligand screening: strategies, perspectives and limitations. *Drug Discov. Today* 11 (13-14):580-594.

[105] Xu, M., Lill, M.A. 2013. Induced fit docking, and the use of QM/MM methods in docking. *Drug Discov. Today Technol.* 10 (3):e411-418.

[106] Cheng, X., Cui, G., Hornak, V., Simmerling, C. 2005. Modified replica exchange simulation methods for local structure refinement. *J. Phys. Chem. B* 109 (16):8220-8230.

[107] Liu, P., Kim, B., Friesner, R.A., Berne, B.J. 2005. Replica exchange with solute tempering: a method for sampling biological systems in explicit water. *Proc. Natl. Acad. Sci. U S A* 102 (39):13749-13754.

[108] Hamelberg, D., Mongan, J., McCammon, J.A. 2004. Accelerated molecular dynamics: a promising and efficient simulation method for biomolecules. *J. Chem. Phys.* 120 (24):11919-11929.

[109] Markwick, P.R., McCammon, J.A. 2011. Studying functional dynamics in biomolecules using accelerated molecular dynamics. *Phys. Chem. Chem. Phys.* 13 (45):20053-20065.

[110] Laio, A., Gervasio, F. 2008. Metadynamics: a method to simulate rare events and reconstruct the free energy of biophysics, chemistry and material sicence. *Rep. Prog. Phys.* 71 126601.

[111] Laio, A., Parrinello, M. 2002. Escaping free-energy minima. *Proc. Natl. Acad. Sci. U S A* 99 (20):12562-12566.

[112] Limongelli, V., Bonomi, M., Marinelli, L., Gervasio, F.L., Cavalli, A., Novellino, E., Parrinello, M. 2010. Molecular basis of cyclooxygenase enzymes (COXs) selective inhibition. *Proc. Natl. Acad. Sci. U S A* 107 (12):5411-5416.

[113] Sgrignani, J., Grazioso, G., De Amici, M., Colombo, G. 2014. Inactivation of TEM-1 by avibactam (NXL-104): insights from quantum mechanics/molecular mechanics metadynamics simulations. *Biochemistry* 53 (31):5174-5185.

[114] Leone, V., Marinelli, F., Carloni, P., Parrinello, M. 2010. Targeting biomolecular flexibility with metadynamics. *Curr. Opin. Struct. Biol.* 20 (2):148-154.

[115] Barducci, A., Bonomi, M., Parrinello, M. 2011. Metadynamics. *WIRES Comput. Mol. Sci.* 1 826-843.

[116] Xu, M., Lill, M.A. 2011. Significant enhancement of docking sensitivity using implicit ligand sampling. *J. Chem. Inf. Model.* 51 (3):693-706.

[117] Case, D. 1994. Normal mode analysis of protein dynamics. *Curr. Opin. Struct. Biol.* 4 285-290.

[118] Hayward, S., Go, N. 1995. Collective variable description of native protein dynamics. *Annu. Rev. Phys. Chem.* 46 223-250.

[119] Noguti, T., Go, N. 1982. Collective variable description of small-amplitude conformational fluctuations in a globular protein. *Nature* 296 (5859):776-778.

[120] Cavasotto, C.N. 2012. Normal mode-based approaches in receptor ensemble docking. *Methods Mol. Biol.* 819 157-168.

[121] Tirion, M.M. 1996. Large Amplitude Elastic Motions in Proteins from a Single-Parameter, Atomic Analysis. *Phys. Rev. Lett.* 77 (9):1905-1908.

[122] Hinsen, K. 1998. Analysis of domain motions by approximate normal mode calculations. *Proteins* 33 (3):417-429.

[123] Bahar, I., Atilgan, A.R., Erman, B. 1997. Direct evaluation of thermal fluctuations in proteins using a single-parameter harmonic potential. *Fold Des.* 2 (3):173-181.

[124] Bahar, I., Erman, B., Jernigan, R.L., Atilgan, A.R., Covell, D.G. 1999. Collective motions in HIV-1 reverse transcriptase: examination of flexibility and enzyme function. *J. Mol. Biol.* 285 (3):1023-1037.

[125] Haliloglu, T., Bahar, I. 1999. Structure-based analysis of protein dynamics: comparison of theoretical results for hen lysozyme with X-ray diffraction and NMR relaxation data. *Proteins* 37 (4):654-667.

[126] Keskin, O., Jernigan, R.L., Bahar, I. 2000. Proteins with similar architecture exhibit similar large-scale dynamic behavior. *Biophys. J.* 78 (4):2093-2106.

[127] Tama, F., Wriggers, W., Brooks, C.L., 3rd 2002. Exploring global distortions of biological macromolecules and assemblies from low-resolution structural information and elastic network theory. *J. Mol. Biol.* 321 (2):297-305.

[128] Krebs, W.G., Alexandrov, V., Wilson, C.A., Echols, N., Yu, H., Gerstein, M. 2002. Normal mode analysis of macromolecular motions in a database framework: developing mode concentration as a useful classifying statistic. *Proteins* 48 (4):682-695.

[129] Zacharias, M. 2004. Rapid protein-ligand docking using soft modes from molecular dynamics simulations to account for protein deformability: binding of FK506 to FKBP. *Proteins* 54 (4):759-767.

[130] Go, N., Noguti, T., Nishikawa, T. 1983. Dynamics of a small globular protein in terms of low-frequency vibrational modes. *Proc. Natl. Acad. Sci. U S A* 80 (12):3696-3700.

[131] Levitt, M., Sander, C., Stern, P.S. 1985. Protein normal-mode dynamics: trypsin inhibitor, crambin, ribonuclease and lysozyme. *J. Mol. Biol.* 181 (3):423-447.

[132] Kolossvary, I., Guida, W. 1996. Low Mode Search. An efficient, automated computational method for conformational analysis: application to cyclic and acyclic alkanes and cyclic peptide. *J. Am. Chem. Soc.* 118 5011-5019.

[133] Kolossvary, I., Guida, W. 1999. Low-mode Conformational Search Elucidated. Application to C39H80 and Flexible Docking of 9-Deazaguanine Inhibitors to PNP. *J. Comput. Chem.* 20 1671-1684.

[134] Keseru, G.M., Kolossvary, I. 2001. Fully flexible low-mode docking: application to induced fit in HIV integrase. *J. Am. Chem. Soc.* 123 (50):12708-12709.

[135] Petrone, P., Pande, V.S. 2006. Can conformational change be described by only a few normal modes? *Biophys. J.* 90 (5):1583-1593.

[136] Kovacs, J.A., Cavasotto, C.N., Abagyan, R. 2005. Conformational Sampling of Protein Flexibility in Generalized Coordinates: Application to ligand docking. *J. Comp. Theor. Nanosci.* 2 354-361.

[137] Sperandio, O., Mouawad, L., Pinto, E., Villoutreix, B.O., Perahia, D., Miteva, M.A. 2010. How to choose relevant multiple receptor conformations for virtual screening: a test case of Cdk2 and normal mode analysis. *Eur. Biophys. J.* 39 (9):1365-1372.

[138] Dietzen, M., Zotenko, E., Hildebrandt, A., Lengauer, T. 2012. On the applicability of elastic network normal modes in small-molecule docking. *J. Chem. Inf. Model.* 52 (3):844-856.

[139] Berendsen, H.J., Hayward, S. 2000. Collective protein dynamics in relation to function. *Curr. Opin. Struct. Biol.* 10 (2):165-169.

[140] Amadei, A., Linssen, A.B., Berendsen, H.J. 1993. Essential dynamics of proteins. *Proteins* 17 (4):412-425.

[141] Teodoro, M.L., Phillips, G.N., Jr., Kavraki, L.E. 2003. Understanding protein flexibility through dimensionality reduction. *J. Comput. Biol.* 10 (3-4):617-634.

[142] Li, Z., Scheraga, H.A. 1987. Monte Carlo-minimization approach to the multiple-minima problem in protein folding. *Proc. Natl. Acad. Sci. U S A* 84 (19):6611-6615.

[143] Cavasotto, C.N., Orry, A.J., Abagyan, R.A. 2003. Structure-based identification of binding sites, native ligands and potential inhibitors for G-protein coupled receptors. *Proteins* 51 (3):423-433.

[144] Abagyan, R., Totrov, M. 1994. Biased probability Monte Carlo conformational searches and electrostatic calculations for peptides and proteins. *J. Mol. Biol.* 235 (3):983-1002.

[145] Abagyan, R., Totrov, M., Kuznetsov, D. 1994. ICM - a New Method For Protein Modeling and Design - Applications to Docking and Structure Prediction From the Distorted Native Conformation. *J. Comput. Chem.* 15 488-506.

[146] Cavasotto, C.N., Liu, G., James, S.Y., Hobbs, P.D., Peterson, V.J., Bhattacharya, A.A., Kolluri, S.K., Zhang, X.K., Leid, M., Abagyan, R., Liddington, R.C., Dawson, M.I. 2004. Determinants of retinoid X receptor transcriptional antagonism. *J. Med. Chem.* 47 (18):4360-4372.

[147] Nilmeier, J., Jacobson, M. 2008. Multiscale Monte Carlo Sampling of Protein Sidechains: Application to Binding Pocket Flexibility. *J. Chem. Theory Comput.* 4 (5):835-846.

[148] Nilmeier, J., Hua, L., Coutsias, E.A., Jacobson, M.P. 2011. Assessing protein loop flexibility by hierarchical Monte Carlo sampling. *J. Chem. Theory Comput.* 7 (5):1564-1574.

[149] Cavasotto, C.N. In *Virtual Screening. Principles, Challenges and Practical Guidelines.* C, S., Ed.; Wiley-VCH, 2011, pp 245-262.

[150] Cavasotto, C.N., SIngh, N. 2008. Docking and High Throughput Docking: Successes and the Challenge of Protein Flexibility. *Curr Comput Aided Drug Design* 4 221-234.

[151] Cavasotto, C.N. 2011. Homology models in docking and high-throughput docking. *Curr. Top. Med. Chem.* 11 (12):1528-1534.

[152] Cavasotto, C.N., Orry, A.J., Murgolo, N.J., Czarniecki, M.F., Kocsi, S.A., Hawes, B.E., O'Neill, K.A., Hine, H., Burton, M.S., Voigt, J.H., Abagyan, R., Bayne, M.L., Monsma, F.J., Jr. 2008. Discovery of novel chemotypes to a G-protein-coupled receptor through ligand-steered homology modeling and structure-based virtual screening. *J. Med. Chem.* 51 581-588.

[153] Phatak, S.S., Gatica, E.A., Cavasotto, C.N. 2010. Ligand-steered modeling and docking: A benchmarking study in class A G-protein-coupled receptors. *J. Chem. Inf. Model.* 50 (12):2119-2128.

[154] Diaz, P., Phatak, S.S., Xu, J., Astruc-Diaz, F., Cavasotto, C.N., Naguib, M. 2009. 6-Methoxy-N-alkyl isatin acylhydrazone derivatives as a novel series of potent selective cannabinoid receptor 2 inverse agonists: design, synthesis, and binding mode prediction. *J. Med. Chem.* 52 (2):433-444.

[155] Diaz, P., Phatak, S.S., Xu, J., Fronczek, F.R., Astruc-Diaz, F., Thompson, C.M., Cavasotto, C.N., Naguib, M. 2009. 2,3-Dihydro-1-benzofuran derivatives as a series of potent selective cannabinoid receptor 2 agonists: design, synthesis, and binding mode prediction through ligand-steered modeling. *Chem. Med. Chem.* 4 (10):1615-1629.

[156] Petrov, R.R., Knight, L., Chen, S.R., Wager-Miller, J., McDaniel, S.W., Diaz, F., Barth, F., Pan, H.L., Mackie, K., Cavasotto, C.N., Diaz, P. 2013. Mastering tricyclic ring systems for desirable functional cannabinoid activity. *Eur. J. Med. Chem.* 69 881-907.

[157] McRobb, F.M., Capuano, B., Crosby, I.T., Chalmers, D.K., Yuriev, E. 2010. Homology modeling and docking evaluation of aminergic G protein-coupled receptors. *J. Chem. Inf. Model.* 50 (4):626-637.

[158] Durrant, J.D., McCammon, J.A. 2011. Molecular dynamics simulations and drug discovery. *BMC Biol* 9 71.

[159] Carlson, H.A., Masukawa, K.M., Rubins, K., Bushman, F.D., Jorgensen, W.L., Lins, R.D., Briggs, J.M., McCammon, J.A. 2000. Developing a dynamic pharmacophore model for HIV-1 integrase. *J. Med. Chem.* 43 (11):2100-2114.

[160] Lins, R.D., Briggs, J.M., Straatsma, T.P., Carlson, H.A., Greenwald, J., Choe, S., McCammon, J.A. 1999. Molecular dynamics studies on the HIV-1 integrase catalytic domain. *Biophys. J.* 76 (6):2999-3011.

[161] Meagher, K.L., Carlson, H.A. 2004. Incorporating protein flexibility in structure-based drug discovery: using HIV-1 protease as a test case. *J. Am. Chem. Soc.* 126 (41):13276-13281.

[162] Bowman, A.L., Nikolovska-Coleska, Z., Zhong, H., Wang, S., Carlson, H.A. 2007. Small molecule inhibitors of the MDM2-p53 interaction discovered by ensemble-based receptor models. *J. Am. Chem. Soc.* 129 (42):12809-12814.

[163] Kruger, D.M., Bergs, J., Kazemi, S., Gohlke, H. 2011. Target Flexibility in RNA-Ligand Docking Modeled by Elastic Potential Grids. *ACS Med. Chem. Lett.* 2 (7):489-493.

[164] Leis, S., Zacharias, M. 2011. Efficient inclusion of receptor flexibility in grid-based protein-ligand docking. *J. Comput. Chem.* 32 (16):3433-3439.

[165] Amaro, R.E., Baron, R., McCammon, J.A. 2008. An improved relaxed complex scheme for receptor flexibility in computer-aided drug design. *J. Comput. Aided Mol. Des.* 22 (9):693-705.

[166] Lin, J.H., Perryman, A.L., Schames, J.R., McCammon, J.A. 2003. The relaxed complex method: Accommodating receptor flexibility for drug design with an improved scoring scheme. *Biopolymers* 68 (1):47-62.

[167] Frembgen-Kesner, T., Elcock, A.H. 2006. Computational sampling of a cryptic drug binding site in a protein receptor: explicit solvent molecular dynamics and inhibitor docking to p38 MAP kinase. *J. Mol. Biol.* 359 (1):202-214.

[168] Tatsumi, R., Fukunishi, Y., Nakamura, H. 2004. A hybrid method of molecular dynamics and harmonic dynamics for docking of flexible ligand to flexible receptor. *J. Comput. Chem.* 25 (16):1995-2005.

[169] Floquet, N., Marechal, J.D., Badet-Denisot, M.A., Robert, C.H., Dauchez, M., Perahia, D. 2006. Normal mode analysis as a prerequisite for drug design: application to matrix metalloproteinases inhibitors. *FEBS Lett.* 580 (22):5130-5136.

[170] May, A., Zacharias, M. 2008. Protein-ligand docking accounting for receptor side chain and global flexibility in normal modes: evaluation on kinase inhibitor cross docking. *J. Med. Chem.* 51 (12):3499-3506.

[171] Rueda, M., Bottegoni, G., Abagyan, R. 2009. Consistent improvement of cross-docking results using binding site ensembles generated with elastic network normal modes. *J. Chem. Inf. Model.* 49 (3):716-725.

[172] Sander, T., Liljefors, T., Balle, T. 2008. Prediction of the receptor conformation for iGluR2 agonist binding: QM/MM docking to an extensive conformational ensemble generated using normal mode analysis. *J. Mol. Graph. Model.* 26 (8):1259-1268.

[173] Suhre, K., Sanejouand, Y.H. 2004. ElNemo: a normal mode web server for protein movement analysis and the generation of templates for molecular replacement. *Nucleic Acids Res* 32 (Web Server issue):W610-614.

[174] Schames, J.R., Henchman, R.H., Siegel, J.S., Sotriffer, C.A., Ni, H., McCammon, J.A. 2004. Discovery of a novel binding trench in HIV integrase. *J. Med. Chem.* 47 (8):1879-1881.

[175] Durrant, J.D., Urbaniak, M.D., Ferguson, M.A., McCammon, J.A. 2010. Computer-aided identification of Trypanosoma brucei uridine diphosphate galactose 4'-epimerase inhibitors: toward the development of novel therapies for African sleeping sickness. *J. Med. Chem.* 53 (13):5025-5032.

[176] Amaro, R.E., Schnaufer, A., Interthal, H., Hol, W., Stuart, K.D., McCammon, J.A. 2008. Discovery of drug-like inhibitors of an essential RNA-editing ligase in Trypanosoma brucei. *Proc. Natl. Acad. Sci. U S A* 105 (45):17278-17283.

[177] Claussen, H., Buning, C., Rarey, M., Lengauer, T. 2001. FlexE: efficient molecular docking considering protein structure variations. *J. Mol. Biol.* 308 (2):377-395.

[178] Campbell, A.J., Lamb, M.L., Joseph-McCarthy, D. 2014. Ensemble-based docking using biased molecular dynamics. *J. Chem. Inf. Model.* 54 (7):2127-2138.

[179] Kellogg, G.E., Fornabaio, M., Chen, D.L., Abraham, D.J., Spyrakis, F., Cozzini, P., Mozzarelli, A. 2006. Tools for building a comprehensive modeling system for virtual screening under real biological conditions: The Computational Titration algorithm. *J. Mol. Graph. Model.* 24 (6):434-439.

[180] Spyrakis, F., Amadasi, A., Fornabaio, M., Abraham, D.J., Mozzarelli, A., Kellogg, G.E., Cozzini, P. 2007. The consequences of scoring docked ligand conformations using free energy correlations. *Eur. J. Med. Chem.* 42 (7):921-933.

[181] Kitchen, D.B., Decornez, H., Furr, J.R., Bajorath, J. 2004. Docking and scoring in virtual screening for drug discovery: methods and applications. *Nat. Rev. Drug Discov.* 3 (11):935-949.

[182] Adcock, S.A., McCammon, J.A. 2006. Molecular dynamics: survey of methods for simulating the activity of proteins. *Chem. Rev.* 106 (5):1589-1615.

[183] Schwab, F., van Gunsteren, W.F., Zagrovic, B. 2008. Computational study of the mechanism and the relative free energies of binding of anticholesteremic inhibitors to squalene-hopene cyclase. *Biochemistry* 47 (9):2945-2951.

[184] Kim, J.T., Hamilton, A.D., Bailey, C.M., Domaoal, R.A., Wang, L., Anderson, K.S., Jorgensen, W.L. 2006. FEP-guided selection of bicyclic heterocycles in lead optimization for non-nucleoside inhibitors of HIV-1 reverse transcriptase. *J. Am. Chem. Soc.* 128 (48):15372-15373.

[185] [Polgar, T., Keseru, G.M. 2006. Ensemble docking into flexible active sites. Critical evaluation of FlexE against JNK-3 and beta-secretase. *J. Chem. Inf. Model.* 46 (4):1795-1805.

[186] Totrov, M., Abagyan, R. 2008. Flexible ligand docking to multiple receptor conformations: a practical alternative. *Curr. Opin. Struct. Biol.* 18 (2):178-184.

[187] Deng, J., Lee, K.W., Sanchez, T., Cui, M., Neamati, N., Briggs, J.M. 2005. Dynamic receptor-based pharmacophore model development and its application in designing novel HIV-1 integrase inhibitors. *J. Med. Chem.* 48 (5):1496-1505.

[188] O'Donoghue, P., Luthey-Schulten, Z. 2005. Evolutionary profiles derived from the QR factorization of multiple structural alignments gives an economy of information. *J. Mol. Biol.* 346 (3):875-894.

[189] Knegtel, R.M., Kuntz, I.D., Oshiro, C.M. 1997. Molecular docking to ensembles of protein structures. *J. Mol. Biol.* 266 (2):424-440.

[190] Michino, M., Abola, E., Brooks, C.L., 3rd, Dixon, J.S., Moult, J., Stevens, R.C. 2009. Community-wide assessment of GPCR structure modelling and ligand docking: GPCR Dock 2008. *Nat. Rev. Drug Discov.* 8 (6):455-463.

[191] Birch, L., Murray, C.W., Hartshorn, M.J., Tickle, I.J., Verdonk, M.L. 2002. Sensitivity of molecular docking to induced fit effects in influenza virus neuraminidase. *J. Comput. Aided Mol. Des.* 16 (12):855-869.

[192] Verdonk, M.L., Mortenson, P.N., Hall, R.J., Hartshorn, M.J., Murray, C.W. 2008. Protein-ligand docking against non-native protein conformers. *J. Chem. Inf. Model.* 48 (11):2214-2225.

[193] Bottegoni, G., Rocchia, W., Rueda, M., Abagyan, R., Cavalli, A. 2011. Systematic exploitation of multiple receptor conformations for virtual ligand screening. *PLoS One* 6 (5):e18845.

[194] Korb, O., Olsson, T.S., Bowden, S.J., Hall, R.J., Verdonk, M.L., Liebeschuetz, J.W., Cole, J.C. 2012. Potential and limitations of ensemble docking. *J. Chem. Inf. Model.* 52 (5):1262-1274.

In: From Medicinal Chemistry to Food Science
Editor: Pietro Cozzini

ISBN: 978-1-63483-587-9

Chapter 8

How QSAR Changed the World of Drug Design, and Beyond …

***Hardik I. Parikh*[1,2] *and Glen E. Kellogg*[1,2,*]**
[1]Department of Medicinal Chemistry, School of Pharmacy,
Virginia Commonwealth University, Richmond, Virginia US
[2]Institute for Structural Biology and Drug Discovery,
Virginia Commonwealth University, Richmond, Virginia US

Abstract

Quantitative Structure Activity Relationship (QSAR) modeling is grounded in the concept that biological property of a chemical (like enzyme activity, toxicity, etc.) is directly related to its molecular structure, and it is possible to establish a mathematical/quantitative relationship between them. Most QSAR methods were initially developed as a result of medical and pharmaceutical research relating to drug discovery, and more recently have been applied to other fields like food science. In this chapter, we will discuss the evolution of QSAR briefly, highlighting some of the major contributions along the way, how modern computational drug design/discovery approaches are nothing but QSARs, and end with a few applications of QSAR modeling in the field of food chemistry.

The Origins of QSAR

The genesis of quantitative drug design can be traced back to Crum-Brown and Fraser who, in 1868, proposed the dependence of the physiological activity of compounds on their chemical constitution and structure. [1] In 1893, Richet reported the inverse relationship between the narcotic activity of simple chemicals like alcohol, ether and ketones to their water-solubility. [2] The very first systematic physicochemical studies on the distribution of compounds in two immiscible solvents were done by Berthelot and Jungfleisch [3], and

[*] Correspondence to: Glen E. Kellogg, PhD, Professor, Department of Medicinal Chemistry, School of Pharmacy, VCU, Richmond, VA 23298, email: glen.kellogg@vcu.edu

expanded by Nernst. [4] Before the turn of the 20th century, the most notable works of Overton [5] and Meyer [6] showed that the narcotic potency of congeners in tadpoles was correlated to their oil-water and oil-gas partition coefficients. The most significant was Overton's claim that narcosis is due to physical changes effected in the lipid constituents of cells. Meyer showed a quantitative relationship between the narcotic concentrations and partition coefficients, an early and primitive example of the now modern, more sophisticated, quantitative structure activity relationships (QSARs). Ferguson introduced the concept that thermodynamic principles could be related to drug activities by showing the correlation between depressant actions of volatile compounds with their relative saturation in the vehicle in which they were administered. [7] Although the discipline of quantitative rational drug design came into being much later, these studies laid the foundation for it by measuring and rationalizing a compound's physicochemical properties, like partition coefficients and hydrophobicity, which are now known to play central roles in compounds' biological activity.

The early findings were followed by significant contributions in the field of physical organic chemistry, especially in the development of chemical descriptors. Hammett developed the σ constants, a system that quantifies electronic effects of substituents on chemical reactions and ionic equilibration. [8] He showed a correlation between the chemical reactivity of *meta*- and *para*- substituted benzoic acid derivatives to the electronic potential (σ) of the substituent, relative to hydrogen (Eq.1):

$$log\left(\frac{K_x}{K_H}\right) = \sigma, \tag{1}$$

where K_x is the equilibrium constant for the derivative and K_H is the equilibrium constant for the unsubstituted compound. An electron-withdrawing substitution will favor the anion, yielding a positive σ, whereas electron-donating groups will favor the neutral form. The substituent's location on the ring also affects the value of σ, resulting in separate σ_m and σ_p parameters. Importantly, since ionization constants were used to define σ, they can be related to free energies through a linear relationship (Eq. 2):

$$\Delta G = -RTlnK. \tag{2}$$

This work was followed by the studies of Taft and coworkers who developed the steric effect parameters E_s and a polar parameter σ^*. [9, 10] They analyzed the rates of base- and acid-catalyzed hydrolysis of aliphatic esters and separated the polar, steric and resonance effects. These descriptors formed a very strong basis for the effective application of chemical properties of compounds to biological systems and were later seen to be landmarks in the development of Quantitative Structure Activity Relationships (QSARs).

The Hansch-Fujita QSAR Analysis

In simplistic terms, QSARs can be expressed as mathematical models that correlate the biological activity of a chemical compound, as a result of its interaction with a biological structure such as a receptor, enzyme or polynucleotide, to the compound's physicochemical properties. Over fifty years have now passed since the inception of the concept of QSAR by

Corwin Hansch. The very first application of QSAR methodology to a biological system was published in 1962 in *Nature* [11] (later improved in a 1964 *Journal of the American Chemical Society* article [12]) that rationalized the potency variation patterns observed in various mono-substituted phenoxyacetic acids with Hammett substituent constants and partition coefficients. For a number of years, Hansch and his colleagues were developing an SAR (Structure-Activity Relationship) of plant growth regulators, identifying the role of substituent lipophilic character with respect to their activity. As a consequence, they introduced (Eq. 3) the *hydrophobic substituent constant (π)*, which is analogous to Hammett's σ constant:

$$log\left(\frac{P_x}{P_H}\right) = \pi, \quad (3)$$

where P_x is the 1-octanol/water partition coefficient of the unsubstituted compound and P_H is that of the unsubstituted compound. Fujita suggested that if the biological response of a compound was due to the combination of its hydrophobicity and electronic features, the two terms could be combined into a logarithmic relationship (Eq. 4):

$$log\left(\frac{1}{C}\right) = k_1\pi + k_2\sigma + k_3, \quad (4)$$

where C is the molar concentration of compound producing a standard response. Further developments showed that modeling the relationship between physical properties and biological activity was more accurate when using multiple physical properties, and saw the introduction of Taft's steric parameter to Eq. 4, yielding (Eq. 5):

$$log\left(\frac{1}{C}\right) = k_1\pi + k_2\sigma + k_3E_s + k_4. \quad (5)$$

Furthermore, for congeneric series of compounds, where the range of differences in the parameters can be considerable, Hansch postulated that the biological activity (BA) is parabolically related to its parameters. The resulting inclusion of squared terms yielded an equation of the following type (Eq. 6), often known as the Hansch equation [13]:

$$BA = -k_1\pi^2 + k_2\pi + k_3\sigma + k_4E_s + k_5. \quad (6)$$

As the Hansch equation became more known and its applicability to a variety of diverse fields was more appreciated, the community began developing further descriptors that could potentially be correlated to biological activity. In 1964, Charton introduced a new version of the inductive parameter – σ_I. [14] In 1968, Swain and Lupton published a set of field (F) and resonance (R) descriptors. [15] Significant contributions were also made in field of steric parameterization: molar refractivity (MR) by Pauling and Pressman, the v-values by Charton, and Verloop's steric parameters calculated from bond angles and bond distances. [16-18] Kier, who had been applying quantum mechanics principles to drug discovery, and Hall added quite a number of graph theory-based non-empirical descriptors to the pool for describing shape, connectivity and electronic structure. [19-22] By the early 1970s, a highly satisfying arsenal of electronic and steric molecular descriptors was available to scientists for construction of QSARs and the prediction of biological activity for unknown compounds.

With these decades-long community-wide efforts at continuous improvement, QSAR modeling evolved into an indispensible predictive tool for medicinal chemists that enabled them to transform structure and properties of chemicals into knowledge. In fact, by the early 1980s, most of the many (at that time) pharmaceutical research enterprises considered QSAR key drug discovery technology. Although not as ubiquitous, and with many other competing technologies available, the simple Hansch-Fujita QSAR analysis continues to be performed to this day, with applications in various industries from pharmaceutical to agrochemicals, food, flavor and fragrance. Far more importantly, however, is that the basic premises of QSAR – that biological "activity" of a small molecule can be correlated with its structure and that structure can be represented mathematically – are at the heart of most, if not all, molecular modeling applications as they are applied to drug discovery, chemical biology, etc.

Free and Wilson, Topliss and Seiler

Free and Wilson adopted a different approach towards structure-activity correlation. [23] They used multiple regression analysis to correlate biological activity data with the presence or absence of specific substituents at specific locations on a common ring system. The assumption was that effect of substituents on biological activity is additive and independent of the other substituents present. Following (Eq. 7), is an example of the Free and Wilson model:

$$BA = \sum I_{ij} F_{ij} + k, \tag{7}$$

where *i,j* are summed over all possible substituents *i* at positions *j*. F_{ij} is the enhancement factor of substituent type *i* at position *j*, I_{ij} is an "indicator" variable – 0 or 1, indicating the absence or presence of substituent *i* at position *j* on the ring. *k* is the activity of the unsubstituted parent compound. A major advantage of such an approach is that all physicochemical parameters are included in F_{ij}; however, the applicability is limited by the fact that indicator variables are only relevant to the system under study (as opposed to more universal descriptors like π, σ, etc.).

Also in the 1970s, Topliss proposed a systematic approach for analog design to progress as rapidly as possible to the most potent compound in a series, while minimizing the number of compounds synthesized and tested. [24-26] A stepwise decision making approach is undertaken: first, a substituent is added to the parent compound with a significant value of π or σ; then, the analog's biological activity is determined relative to the unsubstituted compound; next, if the analog is more potent, the next compound in the series is designed with increased magnitude of π or σ, or if the analog is less potent, the next compound is designed with increased magnitude of π or σ in the opposite direction. This methodical approach to QSAR analysis allows spanning the maximal descriptor space. The Topliss decision tree approach evolved from the work of Craig, who pointed out the utility of a simple graphical plot of π versus σ (or any two descriptors) to guide the choice of descriptor. [27] Subsequently, the procedures have been widely used to advantage by medicinal chemists as evidenced by numerous literature citations over the past decades.

Seiler, in 1974, improved upon the Hansch and Leo Donor-Acceptor approach and emphasized the importance of aliphatic hydrocarbon-water partition coefficients, especially when correcting QSAR analysis involving passage through the blood-brain barrier (BBB). [28] By means of least-square analysis, he established the I_H parameters for accurate interconversion of lipophilicities from the cyclohexane/water system to the octanol/water system:

$$I_H = \log P\ (oct) - \log P\ (cyclohexane)\ - 0.16. \tag{8}$$

The King of Descriptors: LogP

Despite the plethora of descriptors created over the years, LogP has remained the most desirable and usable. In 1971, Leo et al., published a comprehensive list of partition coefficients of compounds found in biochemical and pharmaceutical literature. [29] The primary reason for the compilation was to build a premise for correlating the hydrophobic bonding in biological systems with partition coefficients. They discuss the additive-constitutive character of partition coefficient, i.e., the intermolecular forces involved in partitioning can be factored into the contributions of its various parts. Also described is the utilization of the Hammett's equation in theoretical calculations of partitioning free energy, forming the basis of C-LogP method of predicting LogP which was developed by Weininger and the Pomona College MedChem Project from the data compiled by Leo. [30-32] This subject, while still in the embryonic stage, allowed for a better understanding of the interaction of small organic molecules with biomacromolecules, and the repercussions of this quantification of hydrophobic interaction energies have been key to drug design projects as well as computational medicinal chemistry.

In the early 1970s, Rekker *et. al.* defined a new methodology for calculating the lipophilicity of a molecule. [33] As opposed to Hansch's substitution principle, Rekker's approach defines a new hydrophobic fragmental constant *f*, which is the lipophilicity contribution of a constituent part of a structure to the total lipophilicity, i.e., (Eq. 9):

$$LogP = \sum_{i=1}^{n} a_i f_i + \sum_{i=1}^{m} b_j F_j, \tag{9}$$

where $LogP$ is the log of the partition coefficient, a is number of fragments, f is the fragmental constant, b_j is the frequency (count in the molecule) of F_j and F_j is the correction factor for the jth fragment. This simplified approach of $LogP$ calculations provided a means for accounting for the effects of the neighboring molecular environment. In the intervening four decades, many other methods for calculating lipophilicity have been described in the literature and reviewed. [34-42] All offer some feature that makes them unique and useful, depending on the problem at hand. For example, some methods are very transparent in the physicochemical effects underlying their predictions, while others are perhaps more accurate in predicting LogP for an arbitrary compound, but are opaque to the user as they are based on well-trained QSAR equations or modeled by artificial neural nets.

Introduction of Computers into Chemical Research

Meanwhile, the use of informatics methods to solve chemical problems began as early as the 1950s and 1960s. [11, 43-45] The early applications involved storing and searching databases for chemical molecules in the form of planar 2D structures, which has now evolved to (and beyond) 3D atom coordinates. The very first user-defined query substructure-searching algorithm was described by Ray and Kirsch, and formed the basis of *graph theory* that represents the atoms and bonds of a chemical molecule by the nodes and edges of a graph. [43] Improvements were made to accelerate the search process – like the set-reduction algorithm of Sussenguth and the refinement algorithm of Ullmann. [46, 47] Further modifications involved indexing the database with *fingerprints*, binary representation of specific fragments within a molecule, which is followed by similarity search between the query substructure and the fingerprints of molecules in the database. Such implementations enabled extremely fast and effective searching of large databases, now containing anything up to 10^7 molecules. Other noteworthy contributions in developing search algorithms include efforts by Lynch et al., *[48, 49]*, Hodes et al., [50, 51], and Morgan [52], the implementations and variations of which continue to be used to the present day.

With advancements in the computer field, chemists began developing algorithms to search databases for chemical reactions. The influential paper by Vleduts [44] described the development of search algorithms for organic reactions by the means of reaction centers, that is, substructures of reactant and product molecules that are involved in the transformation, thus characterizing the reactions. However, much research followed before the first operational reaction database system was deployed in the mid-1980s. [53] This substructure-based reaction indexing idea also led to the implementation of computer-assisted synthesis design (CASD), that is, using computer algorithms to suggest series of reactions to obtain the desired synthetic target, by Corey and Wipke. [54] Another noteworthy achievement during the early days of chemoinformatics applications was the development of algorithms for computer-aided structural elucidation (CASE), wherein analytical (spectral) data is used to suggest the identity of unknown molecules [55, 56].

Returning to QSAR, an important development in the establishment of Hansch-Fujita analysis was the increasing availability of computational software for performing tasks from multiple regressions to statistical validation. The extensions of the milestone paper by Hansch and Fujita [11], which uses a molecule's lipophilicity as a key parameter for prediction of biological activity of new test data / novel molecules, were made possible by the advent of high-speed computing and large storage capacities. One of the inherent weaknesses of classical QSAR is that it is restricted to properties and *two*-dimensional substructural features calculated/predicted for a molecule, even though they are *three*-dimensional objects and their biological activity results from their 3D interaction with biomacromolecules. With more atomic coordinate data becoming available to chemists from crystallography and computational algorithms, an interest arose for moving from 2D to 3D descriptors that give a complete representation of physico-chemical properties, which should result in more meaningful QSAR models.

The 3D Pharmacophore

The concept of *pharmacophore*, the arrangement of structural features in 3D space that are involved in the binding of a ligand to a biological target, was proposed by Kier [19], and inspired the development of tools for 3D geometric descriptors. In turn, the availability of approaches for calculating 3D physicochemical descriptors of molecules was a key factor in the development of 3D-QSAR approaches. The first two 3D QSAR methods to be described were Hypothetical Active Site Lattice (HASL) by Doweyko [57] and Comparative Molecular Field Analysis (CoMFA) by Cramer et al., [58] Both approaches are grid-based, that is, they model the interaction of a ligand with its biological environment by the means of a 3D grid superimposed over the molecule and/or its putative neighborhood in a binding site. In CoMFA, a grid is created with the molecule of interest at its center, and every point of the grid, external to molecule's surface, represents the interactions it makes with a probe atom of specified charge and sterics (as represented by the Lennard-Jones potential [59]). In contrast, the HASL grid-based approach involves representing grid points within the molecule's van der Waal's radii by the electron densities of surrounding atoms. In 1988, molecular quantum similarity indices, which combine quantum theoretical calculations with molecular structure and properties of even large biomolecules, were introduced into QSAR and drug design. [60] This resulted in the development of a new approach for 3D-QSAR of combining the similarity indices with CoMFA, Comparative Molecular Similarity Indices Analysis (CoMSIA), wherein molecular fields are represented by similarity indices among the compounds within the data set [61].

Also key to the development of QSAR, especially for complex systems or higher dimensions, were the growing sophistication of statistical tools for model building. We should highlight Wold's PLS (partial least squares) method [62], which was critical for CoMFA, and the increasing sophistication of genetic algorithm methods, artificial neural net, support vector machines and other model building algorithms. Over its history, the field of QSAR grew from humble, almost seeming to be apocryphal, observations of the late 1800s, into major prominence with the introduction and development of a vast plethora of molecular descriptors, ranging from 0D to 4D, accompanied by a growing robustness and complexity of computational approaches for computing and predicting them [63-67].

Modern Drug Research

Drug research in the present involves two aspects – drug discovery and drug optimization. In the most basic sense, drug discovery involves the identification of small molecules with complementary physicochemical properties to their biological target, often relying on computational modeling techniques. By usage of the term, structural knowledge of the receptor or enzyme of interest is essential for "rational drug design", although there are many medicinal chemistry strategies, e.g., Craig's plot, Topliss tree, or deconstruction-reconstruction-elaboration, that are very *rational.* [24, 26, 27, 68] Once identified, a lead molecule is subjected to (sometimes, many) optimization steps for improving affinity/binding, and thus activity, to drug targets as well as its pharmacokinetic and toxicological properties. This involves an intensive interplay between chemistry and biology, again supported and complemented by computational modeling approaches. When there is sufficient knowledge of 3D structure, structure-based rational drug design has become the norm for identification of novel hits for biological targets of interest. The exponential increase in the number of experimental protein structures being submitted to the Protein Data Bank [69], from a few hundred in the first decade after its inception, to a collection now exceeding 100,000, has further facilitated these efforts.

Two major computational approaches for structure-based drug discovery (SBDD) are – *docking* and *de-novo design*. Molecular docking is a method to computationally simulate the molecular recognition process. The goal of this technique is to identify an ideal orientation and conformation (pose) that a ligand would adopt within a protein's binding site, such that the free energy of the overall system is minimized, which would then correspond to the actual binding pose. In 1982, Kuntz et al., [70] first described their DOCK program for protein-ligand docking, the very first solution to the problem of identifying the optimized orientation of a ligand in a protein binding site, which is often thought of as being similar to identifying a key that might fit into a lock. Briefly, the DOCK program suite constructs an image of the binding site complementary to the properties of residues lining the cavity, which is then matched to the properties of potential ligands. This approach was later coupled with sophisticated molecular mechanics interaction energy scoring functions that account for not only the geometric but also chemical match between the two molecules, to quantify the fit of the ligand within the protein's binding site that can be used to predict its binding affinity [71-73].

In 1985, Goodford introduced the GRID program – a grid-based approach to identify ligand binding-sites on protein surfaces using the physical and chemical properties of probe atoms like water, the methyl group, amine nitrogen, carboxyl oxygen, and hydroxyl, which served to highlight hot spots for interaction of said functional groups with that protein. [74] This approach was of great value to drug design as the contour surfaces delineated in this manner gave a visual representation of protein-ligand interaction energetics. Subsequently, a number of docking approaches were developed that used a grid-based (rather than atom pair-based) scheme to represent the physicochemical properties of the protein binding site, followed by computing the interaction energies to identify the most energetically favorable protein-ligand complex. Examples are Autodock [75] and GLIDE. [76, 77] Owing to the inherent flexibility of many small molecules and protein residues, the traditional "lock-and-key" analogy of protein-ligand interactions can be viewed more realistically with a "hand-in-glove" analogy. [78] More recent docking software algorithms have been designed to account for not only the dynamic behavior of the molecular recognition process, but also for the energetic contributions of solvent molecules in the environment. Examples are programs like GOLD [79], GLIDE [76] and FlexX [80], which being used on a very large scale in industrial lead-discovery projects. With relentless advances in conformational optimization algorithms and scoring functions, docking has evolved into one of the most important and versatile molecular modeling techniques, especially since the real cost of computing has crashed over the past few decades. There are numerous reviews available on docking algorithms and scoring functions, but a chapter in the most recent edition of Burger's Medicinal Chemistry is a good starting place [81].

In contrast to docking techniques that seek to identify existing molecules complementary to the binding site of interest, *de novo* design aims at suggesting novel molecules that are compatible with it. The most common approach of this kind is fragment-based drug design

(FBDD), where the program links/combines together small fragments that may bind only weakly, but make appropriate steric, electrostatic and hydrophobic interactions with their target, to produce a lead molecule with higher affinity. Extensive reviews and books have been published describing various aspects of FBDD [82-85].

Another integral tool of the *in silico* drug discovery arsenal is "virtual" screening, that is, using computational algorithms rather than biological methods to screen libraries of compounds, with the goal of identifying molecules that are more likely than random to bind to the target. [86-88] Virtual screening is a process of evaluating the likeliness of a compound to form favorable interactions with the target biomacromolecule, thereby filtering the enormous chemical space of potential small molecules to a more manageable number of hits to be synthesized/purchased and tested. There are two main techniques for screening molecular databases – ligand-based and structure-based virtual screening. In the ligand-based approach, a pharmacophore model is generated from a single ligand molecule, or better, a set of structurally diverse known ligands that bind to the target. This model describes the number, types, and spatial arrangement of critical features for interaction with the target. All molecules in the database, which can be on the order of millions, are then compared to this model to identify the best candidate that closely resembles the query and hence is more likely to bind and show a similar activity profile. On the other hand, structure-based efforts use similar information from the target's 3D structure to find the most suitable candidates for the binding site. This is generally done by either generating a pharmacophore model complementary to the physicochemical properties of the target site and searching the database for possible matches, or by docking all the molecules within a database into the target site, typically followed by using an energy-based scoring function to estimate the likelihood of each compound to bind at the site. The variety of tools in play for virtual screening are based on a number of algorithms that range from 2D to 3D, thereby including more or less chemoinformatics/QSAR principles. [89-91] While the latter more exhaustive docking/ scoring approach can be computationally expensive compared to filtering by pharmacophore, over three decades of continuous advances has made docking a highly sophisticated approach to virtual screening and continues to be the subject of intense current interest amongst researchers.

At the heart of all virtual screening resides a structural similarity metric that quantifies the degree of resemblance between the molecule of interest in the database and the search query. Many such scoring measures have been developed, but the most common ones by far are those that compare the fragment bit-strings for 2D similarity between the two structures. Adamson and Bush first implemented this approach in 1974, for predicting local anesthetic activity of compounds based on similarity and/or dissimilarity coefficients between pairs of structures. [92-94] However, the first fully operational similarity searching systems were developed after more than a decade of work at Lederle Laboratories [95] and Pfizer UK [96].

How It Fits Together

A computational chemistry parlor game of the early-to-mid 1990s was to identify which marketed drugs came out of CADD (Computer-Aided Drug Discovery) efforts in industry. This was a difficult game then because the pharmaceutical companies were quite reticent to share any details on their discovery processes. Today, the game has changed to identifying which part of a molecule came from what kind of drug discovery process, e.g., was the scaffold discovered by virtual screening suggestions, or by simple high-throughput screening of chemical libraries, were those libraries created by computationally considering diversity, at what point were structure-based design ideas considered, was a crystal structure solved for an early, middle and/or late-stage lead, and maybe, were computational tools invoked as part of the ADME (adsorption-distribution-metabolism-excretion) optimization of the compound as a therapeutic agent? This is all to point out that computational principles and tools have been integrated into drug discovery at nearly all phases of the process and are indispensable to modern pharmaceutical industry.

It should be further emphasized that all of the computational drug discovery technology described above is largely, if not wholly, a result of QSAR. The brilliant principle behind QSAR, that biological activity is a function of molecular structure, and that molecular structure can be quantitated with “descriptors”, is manifested in virtual screening, docking and scoring, etc. Whenever the “fitness” of a compound for some biological purpose is quantitated, that **is** QSAR.

There are a number of similarities between the drug discovery industry and the food industry. First, safety is of utmost concern to both. The consequences of a bad batch of drugs are similar to those of releasing a contaminated foodstuff. But, the problem is magnified in food science because each consumer intakes on the order of one kg of food per day, whereas a minority of people are regular users of medication, and their intake is on the order of 1 g per day. In principle, the manufacture of medicines is continuously monitored for safety and security within factories – from the raw chemical starting materials through packaging – but the production of food starts outside, over vast areas of largely non-protectable land and is gradually consolidated – often through many transportation steps – to “unprocessed” produce and meat or to be processed into (frozen, canned, dried, etc.) meals available for consumers in grocery stores. Second, the key molecular events in food safety and drug efficacy are largely the same: a molecular interaction between a target biomacromolecule and a small molecule leading to a physiological or pharmacological response. In food safety, however, those responses can be largely undesired, whereas in drug efficacy they are the goal. Third, both fields are being “disrupted” by the growing availability of big data, from genomics to structural. Thus, the same principles of computational molecular science can be employed in the contemporary life science industry to generate intelligent hypothesis to gain valuable insights.

Applications of QSAR beyond Drug Discovery/Design

QSAR has evolved into a powerful tool with both realized and potential applications beyond the pharmaceutical industry – from agricultural-related products such as pesticides, to fragrances and flavorings for the food industry. Bioactive peptides, those that exert physiological benefits when consumed, have received great attention in recent years as they can have diverse, often beneficial, activities, like immunomodulatory, antimicrobial, antioxidant, antithrombotic, antihypertensive and anticancer actions. [97-104] Artificial and food-derived bioactive peptides can be found in products like milk, eggs, yogurt, various meats and fishes, and in plant products. We focus in this chapter on a few diverse issues in the food industry where QSAR principles were applied effectively. The first two examples involve peptides, which are naturally found in foodstuffs from virtually all sources. Their taste properties and beneficial antihypertensive roles have been investigated with experiment, modeling and QSAR. The third example we present here involves an application of Quantitative Structure Toxicity Relationships (QSTRs) of small molecule flavoring substances.

Bioactive Peptides and Bitter Taste

One extremely important aspect of food that has received much recent attention is taste, which is affected in large part by the structural properties of the peptides that are present in them. [105-110] Here we review various quantitative modeling techniques that have been extensively used to obtain a better understanding of the relationship between food peptide structures and their activities.

Amino acids contribute significantly to the taste of food products. Early investigations showed that monosodium L-glutamate (MSG) was the key taste-imparting ingredient in Japanese food seasonings like sea tangle. [111] Other traditional Japanese food items such as sake, miso and soy sauce have characteristic tastes that have been attributed to amino acids released from natural proteins during fermentation. [112, 113] Similarly, the amino acids released during ripening characterize the taste of cheeses [114], and peptide fractions with different tastes from fermented or cured meat products can be isolated. [115] Noguchi et al., isolated and identified several acidic oligopeptides from a fish protein hydrolysate that had flavor-enhancing activity [116].

The tastes of amino acids/peptides and their contribution to a food's flavor profile have been the topics of extensive investigation in the nutrition and food industry. Early experiments that determined the taste properties of various protein hydrolysates and systematic investigations of their structural properties with respect to taste have more recently been complemented with quantitative cheminformatics tools. The majority of protein hydrolysates contribute bitter taste to foods. [114, 117] In 1971, Ney proposed the "*Q*-rule", which established a relationship between the average hydrophobicity of amino acids in a peptide, rather than its sequence, and level of bitterness. [118] He proposed a semi-quantitative scale: peptides with *Q*-values less than 1300 cal.res-1 are not bitter, those with *Q*-value greater than 1400 cal.res-1 are bitter, while the bitterness of peptides with values between 1300 and 1400 cannot be predicted. [119-121] Another hydrophobicity-related scale for predicting peptide bitterness was proposed by Bull and Breese based on the effects of constituent amino acids on the surface tension of water (peptides that lower the surface tension of water more are more hydrophobic and in turn more bitter). [122] Ney's theory of hydrophobicity of peptides and bitter taste was further corroborated by, among others, Clegg *et. al.* [123, 124], Schalinatus and Behnke [125-127], and Adler-Nissen [128], who showed that more hydrophobic peptides were better able to interact with taste receptors. Also, the intensity of bitterness increases with the presence of bulky hydrophobic groups at the C terminus. [107-109, 129, 130] In addition to hydrophobicity, the presence of charged amino acids within the sequences plays a role in the binding of the peptides to taste receptors [131-133], and peptide bitterness increases with increasing peptide length. [106] Quantitation of these observations led to a number of studies in the literature utilizing the predictive powers of QSAR modeling to gain insights into the physicochemical properties of peptide structure and their sensory attributes:

Asao et al., investigated the bitter threshold and developed quantitative models that could explain bitterness differences for a literature dataset of amino acids, peptides and their derivatives, using as descriptors hydrophobic parameters (from partition coefficients) and steric parameters. [134] The total length of the peptide (in its extended conformation) was an important predictor – with the intensity of bitterness trending stronger from free amino acids, to di-, tri- and tetra-peptides. Gomez et al., observed a strong (r2 = 0.791) linear correlation between the concentrations of hydrophobic peptides in protein fractions from Hispanico cheese and their bitter flavor characteristics. [135] Gulyaeva et al., calculated the partitioning of amino acids in an aqueous dextran-PEG system: the relative hydrophobicities (up to hexapeptides) were calculated by additivity and used as QSAR descriptors; a quantitative relationship between bitterness threshold and two structural descriptors – the calculated relative hydrophobicity of the peptides and their lipophilicity (logD, octanol-buffer partitioning) – was found. [136] Wolschann et al., obtained QSAR and CoMFA models that explain the aroma of 46 bell-pepper-derived compounds. [137] A strong influence of steric and electronic properties of pyrazine-based compounds to the intensity of bell pepper aromas was reported. Tan and Siebert performed QSAR analysis to reveal a correlation between the flavor thresholds of alcohol, ester, aldehyde, and ketone components in beer and their structure. [138]

In 1989, Jonsson et al., established three descriptive scales – *z1*, *z2* and *z3*, known as the "*z-scales*", to describe the structural variation within a series of related peptides. [139] These can be used to model physical properties of peptides, and therefore enabled more meaningful QSAR analyses. *Z-scales*, along with other descriptors, have been used to establish structure-activity relationships between peptide structures and their taste, e.g.: 1) Kim and Li-Chan used amino acid *z-scales*, total hydrophobicity, residue number and mass to explain the relationship between structure and bitterness of di- to tetradecapeptides; [140] 2) Wu *et. al.* built QSAR models based on amino acid *z-scales* to show that the presence of a hydrophobic amino acid at the C terminus adjacent to a bulky amino acid are the main determinants of bitterness intensity in di- and tri-peptides [141].

As the number of available chemical descriptors increased, electronic descriptors were applied to QSAR models establishing structure-bitterness relationships: 1) Spasov and Hall analyzed a set of dipeptides and showed that bitter taste is explained by the molecular connectivity first-order valence index and the sum of E-state indices for H-bond acceptors; [142] 2) more recently, Soltani et al., reported quantitative structure-bitter taste relationship models for di- to dodeca- peptides, where bitterness increased with increasing peptide size, the mean-square distance index correlated negatively with bitterness, and bitterness increased with E3s index values that describe the size, shape, polarizability, and conformational properties of the peptide; [143] and 3) Wu et al., built statistically significant and validated CoMFA/CoMSIA models to establish SARs between the 3D structures of di- to hexapeptides and bitterness [144].

To summarize, the studies above, and others [140], illustrate that QSAR-derived approaches can reveal the structural basis of taste, which is an intimate relationship between humans and food. The bitter taste phenomenon has also been molecular modeled from the perspective of its receptor structure by a handful of research groups. Floriano et al., used homology approaches to build an all-atom model of the human phenylthiocarbamide (PTC) taste receptor, which is G-protein coupled, from the bovine rhodopsin template. [145] The PTC receptor has both "taster" and "nontaster" forms and these forms and their function were rationalizable in the molecular models. More recently, Dai et al., modeled the human bitter taste receptor hTAS2R1 alone and in complex with ligands. Subjecting these models to molecular dynamics (MD) simulations reveals possible structural bases for receptor activation. [146] Work is extending to other taste receptors. [147, 148] While these latter approaches are much difficult and time consuming than simple 1D, 2D or 3D QSAR, they offer the possibility of detailed understanding of the biological effect, and will be performed much more often in the future as the real cost of computing resource continues to drop.

ACE Inhibitory Peptides

Another well-studied effect of peptides that has a food source connection is related to how Angiotensin-I-converting enzyme (ACE) catalyzes the conversion of angiotensin I to a potent vasoconstrictor, angiotensin II and inactivates the vasodilator bradykinin. Inhibition of ACE has been used as an effective strategy in the treatment of hypertension in humans, as evidenced by drugs like captopril, lisinopril, benazepril, etc., which unfortunately are often associated with side effects. However, studies have shown that enzymatic proteolysis of proteins from food sources release peptides that can exert an ACE inhibitory physiological effect. [149] A number of peptides with diverse molecular properties that possess ACE inhibitory activity have been isolated from diverse natural sources; e.g., snake venom, casein, whey protein, fish and poultry, corn gluten, and soybean protein. [150-158] Identification of nontoxic and efficient alternatives to synthetic antihypertensive drugs from natural food sources is an area of considerable contemporary interest.

The ACE inhibitory activities of peptides are closely related to their amino acid compositions; peptides with hydrophobic residues like leucine, isoleucine and valine at the N terminus showed good antihypertensive activity, while those with proline near the N terminus were not as effective. [104, 159, 160] A study on the composition of 270 ACE inhibitory peptides revealed Tyr, Pro, Trp, Phe and Leu in the C terminus, while Arg, Gly, Val, Ala and Ile were more likely to be found near the N terminus. [159] Again, hydrophobicity is a key factor with respect to the ACE inhibitory activity of peptides. There is a strong correlation between the mass fraction of hydrophobic amino acids within a peptide and its antihypertensive effect. [161] Consequently, the presence of hydrophobic amino acids at N terminus promotes their binding to ACE. [162, 163] Other studies on peptide composition and ACE inhibition showed that positively charged amino acids like lysine and arginine at the C terminus can promote ACE inhibition, suggesting that these charged residues have stronger affinity for the active site [104, 164].

Unsurprisingly, QSAR models for ACE inhibitory peptides have been developed and reported. [144, 162, 165-168] An early QSAR study on ACE inhibitory peptides was performed by Cushman et al., during the development of Captopril. [169] These studies confirmed that the hydrophobicity and size of the C terminus amino acid, as well as the hydrophobicity and electronic properties of the N terminus amino acid are important for ACE inhibition in dipeptides. Using the same dataset, Collantes and Dunn developed new amino acid side chain descriptors – the isotropic surface area and electronic charge index – that were shown to well model the QSAR [170].

Later, QSAR models that correlated the ACE inhibitory activity of peptides derived from food sources to their structural properties were reported. Pripp et al.,'s QSAR models for ACE inhibitory peptides derived from milk proteins showed a relationship between ACE inhibition and the hydrophobicity and positive charge of the C terminal amino acid and size of the adjacent amino acid in peptides as large as 6-mers. [171] Pripp also applied docking and virtual screening to identify bioactive peptide structures by docking a dataset of 58 dipeptides, with experimentally determined ACE inhibition values, into the crystal structure of ACE. [172] Although a meaningful correlation was not observed between docking scores and inhibitory activity, this work indicated that modeling techniques could be employed instead of large-scale experimental screening to identify promising candidate peptides. QSAR analyses also: selected appropriate hydrolysis conditions for production of potent ACE inhibitory peptides from egg proteins, and identified optimal physiochemical properties of ACE inhibitory peptides prepared from defatted wheat germ protein. [173] The model suggested that hydrophobic amino acid at C terminus and abundant hydrophobic character was important for high ACE inhibition. Altogether, QSAR models have been productively applied to the identification and prioritization for development of potential ACE inhibitory peptides from complex matrices that are typically obtained from hydrolyzates of food proteins.

Quantitative Structure-Toxicity Relationship of Flavoring Substances

Much like the activity of drugs is directly a result of their structurally dependent molecular interactions with the *desired* targets, the toxicity of chemical substances is also a result of their *undesirable* interactions with biomacromolecules. Similar to QSAR studies, numerous applications of quantitative structure-toxicity relationship (QSTR) studies to predict and describe the toxicological properties of various groups of chemicals have been published in the literature. Flavor additives, although usually in extremely low concentrations, are omnipresent in current food to alter its taste and/or smell. Since these chemicals are consumed on a daily basis, safety considerations of such, whether natural or synthetic, is of utmost importance.

In 1959, under the Food Additives Amendments to the Federal Food, Drug, and Cosmetic Act (FFDCA), the Flavor and Extract Manufacturers Association of the United States (FEMA) began a program to assess the safety levels of flavor ingredients. With over five decades of research, more than 2600 flavor ingredients have achieved the "generally recognized as safe" (GRAS) status. [174] To reach this consensus, the FEMA Expert Panel applies a number of modern techniques for safety evaluation of flavoring agents, including metabolic studies and structural relationships. The safety evaluation process begins with assigning the chemical to one of known 33 structurally congeneric groups of substances and one of 3 classes of toxic potential. The toxicological and analytical data, both experimental and *in silico*, in the subsequent steps provide a comprehensive basis to evaluate the safety of its intake. [175, 176] In the European Union (EU), it was the Scientific Committee on Food (SCF) that established a program for evaluation of flavoring substances. Another internationally recognized authority that sets the criteria for safety evaluation of flavoring agents is the Joint Food and Agriculture Organization of the United Nations/World Health Organization (FAO/WHO) Expert Committee on Food Additives (JECFA). In order to deal with the large numbers of these substances, JECFA adopted a practical approach of establishing structure-activity relationships and evaluated safety for a series of homologous substances using known data for only a subset of molecules within each group. [177] The EHC70 report by WHO further emphasized the importance of combining exposure levels data and structure-activity relationships to identify priorities for testing. [178] Also, the United States FDA uses structure-activity relationships to establish Concern Levels for food additives and color additives used in foods. The National Toxicology Program (NTP) of the US National Institute of Environmental Health Sciences (NIEHS), NIH, recently adopted QSAR modeling and machine learning for predicting the hepatocarcinogenic potential of alkenylbeneze flavoring agents. [179] A large number of QSAR models for predicting acute and/or chronic cytotoxicity, mutagenicity and/or carcinogenicity data for chemicals have been published in the literature and are widely used by regulatory agencies. [177, 180] While such models are not directly relevant to the regulatory assessment of toxicity, they may provide insights into the mechanisms of cytotoxic action, and may provide useful supporting information. This is clearly a very complex research space, involving not only challenging chemical, toxicological, biological and other science but also regulatory issues from multiple governmental stakeholders.

Conclusion

This chapter has shown, from a largely historical viewpoint, the evolution of quantitative structure-activity relationships (QSARs) as key tools for linking measurable biological responses to chemical structure. High sensitivity assays (*in vivo* and *in vitro*) for experimentally evaluating biological activities of chemicals come with significant costs in terms of time, money, chemicals and animals. There are concomitant environmental and ethical issues. QSAR and other cheminformatics approaches have thus been widely employed in pharmaceutical screening, but they have by no means been restricted to drug discovery, design and development. With the constant evolution of physicochemical descriptors – both theoretical and empirical – within the scientific community in terms of their sophistication and information domain, more accurate *in silico* predictions can now be made regarding the biological activity of various known/unknown classes of chemicals and how they impact human health. Importantly, QSAR modeling has become an integral tool within food science for analysis, data mining, and processing of chemical information. The future certainly suggests that computational approaches to food science will continue to evolve and become even more ubiquitous.

References

[1] Brown, A. C.; Fraser, T. R. On the connection between chemical constitution and physiological action. Part. I.—On the physiological action of the salts of the ammonium

bases, derived from strychnia, brucia, thebaia, codeia, morphia, and nicotia. *Trans. Roy. Soc. Edinburgh* 1868, *25,* 151-203.

[2] Richet, C. On the relationship between the toxicity and the physical properties of substances. *Seances. Soc. Biol.* 1893, *9,* 775-776.

[3] Berthelot, M.; Jungfleisch, E. On the laws that operate for the partition of a substance between two solvents. *Ann. Chim. Phys.* 1872, *26,* 396-407.

[4] Nernst, W. Distribution of a substance between two solvents and between solvent and vapor. *Z. Phys. Chem.* 1891, *8,* 110-139.

[5] Overton, E. Ueber die allgemeinen osmotischen Eigenschafte der Zelie, ihre vermutlichen Ursachen und ihre Bedeuting fur die Physiologie. *Vierteljahresschr. Naturforsch. Ges. Zuerich* 1899, *44,* 87.

[6] Meyer, H. Zur Theorie der Alkoholnarkose, Welche Eigenschaft der Anaesthetika bedingt ihre narkotische Wirkung. *Arch. Exp. Pathol. Pharmakol.* 1899, *42,* 109.

[7] Ferguson, J. The use of chemical potentials as indices of toxicity. *Proc. Royal Soc. London. Ser B* 1939, *127,* 387-404.

[8] Hammett, L. Quantitative relationships involving structure and reactivity. *Phys. Org.Chem.* 1970, 347-390.

[9] Taft, R. W.; Lewis, I. C. Evaluation of Resonance Effects on Reactivity by Application of the Linear Inductive Energy Relationship. V. Concerning a sigmaR Scale of Resonance Effects. *J. Am. Chem. Soc.* 1959, *81,* 5343-5352.

[10] Taft, R. W., Separation of Polar, Steric and Resonance Effects in Reactivity. In *Steric Effects in Organic Chemistry*, Ed. Newman, M. S., John Wiley and Sons, New York: 1973.

[11] Hansch, C.; Maloney, P. P.; Fujita, T.; Muir, R. M. Correlation of biological activity of phenoxyacetic acids with Hammett substituent constants and partition coefficients. *Nature* 1962, *194,* 178-180.

[12] Fujita, T.; Iwasa, J.; Hansch, C. A new substituent constant, π, derived from partition coefficients. *J. Am. Chem. Soc.* 1964, *86,* 5175-5180.

[13] Hansch, C. Quantitative approach to biochemical structure-activity relationships. *Acc. Chem. Res.* 1969, *2,* 232-239.

[14] Charton, M. Definition of "inductive" substituent constants. *J. Org. Chem.* 1964, *29,* 1222-1227.

[15] Swain, C. G.; Lupton, E. C. Field and resonance components of substituent effects. *J. Am. Chem. Soc.* 1968, *90,* 4328-4337.

[16] Pauling, L.; Pressman, D. The serological properties of simple substances. IX. Hapten inhibition of precipitation of antisera homologous to the o-, m-, and p-azophenylarsonic acid groups. *J. Am. Chem. Soc.* 1945, *67,* 1003-1012.

[17] Charton, M. Steric effects. I. Esterification and acid-catalyzed hydrolysis of esters. *J. Am. Chem. Soc.* 1975, *97,* 1552-1556.

[18] Verloop, A.; Hoogenstraaten, W.; Tipker, J., *Development and application of new steric substituent parameters in drug design*. Academic Press, New York: 1976; Vol. 7, p 165-207.

[19] Kier, L., *Molecular orbital theory in drug research*. Academic Press, New York: 1971.

[20] Kier, L. B.; Hall, L. H. An electrotopological-state index for atoms in molecules. *Pharm. Res.* 1990, *7,* 801-807.

[21] Hall, L. H.; Kier, L. B., The Molecular Connectivity Chi Indexes and Kappa Shape Indexes in Structure-Property Modeling. In *Reviews in Computational Chemistry*, Eds. Lipkowitz, K. B.; Boyd, D. B., John Wiley & Sons, Inc.: Hoboken, NJ, USA, 1991; Vol. 2, pp 367-422.
[22] Hall, L. H.; Mohney, B.; Kier, L. B. The electrotopological state: structure information at the atomic level for molecular graphs. *J. Chem. Inf. Comput. Sci.* 1991, *31,* 76-82.
[23] Free, S. M.; Wilson, J. W. A mathematical contribution to structure-activity studies. *J. Med. Chem.* 1964, *7,* 395-399.
[24] Topliss, J. G. Utilization of operational schemes for analog synthesis in drug design. *J. Med. Chem.* 1972, *15,* 1006-1011.
[25] Topliss, J.; Martin, Y., In *Drug Design*, Ed. Ariens, E. J., Academic Press: New York, USA, 1975.
[26] Topliss, J. G. A manual method for applying the Hansch approach to drug design. *J. Med. Chem.* 1977, *20,* 463-469.
[27] Craig, P. N. Interdependence between physical parameters and selection of substituent groups for correlation studies. *J. Med. Chem.* 1971, *14,* 680-684.
[28] Seiler, P. Interconversion of lipophilicities from the hydrocarbon/water systems into the octanol/water system. *Eur. J. Med. Chem.* 1974, *9,* 473-479.
[29] Leo, A.; Hansch, C.; Elkins, D. Partition coefficients and their uses. *Chem. Rev.* 1971, *71,* 525-616.
[30] Leo, A.; Jow, P.; Silipo, C.; Hansch, C. Calculation of hydrophobic constant (log P) from pi and f constants. *J. Med. Chem.* 1975, *18,* 865-868.
[31] Leo, A. J. Hydrophobic parameter: Measurement and calculation. *Methods Enzymol.* 1991, *202,* 544-591.
[32] Leo, A. J. Calculating log Poct from structures. *Chem. Rev.* 1993, *93,* 1281-1306.
[33] Rekker, R. F., *The hydrophobic fragmental constant*. Elsevier, Amsterdam: 1977; Vol. 301.
[34] Chothia, C. Hydrophobic bonding and accessible surface area in proteins. *Nature* 1974, *248,* 338-339.
[35] Hermann, R. B. Theory of hydrophobic bonding. III. Method for the calculation of the hydrophobic interaction based on liquid state perturbation theory and a simple liquid model. *J. Phys. Chem.* 1975, *79,* 163-169.
[36] Sharp, K. A.; Nicholls, A.; Fine, R. F.; Honig, B. Reconciling the magnitude of the microscopic and macroscopic hydrophobic effects. *Science* 1991, *252,* 106-109.
[37] Casari, G.; Sippl, M. J. Structure-derived hydrophobic potential: hydrophobic potential derived from X-ray structures of globular proteins is able to identify native folds. *J. Mol. Biol.* 1992, *224,* 725-732.
[38] Cramer, C. J.; Truhlar, D. G. An SCF solvation model for the hydrophobic effect and absolute free energies of aqueous solvation. *Science* 1992, *256,* 213-217.
[39] Oobatake, M.; Ooi, T. Hydration and heat stability effects on protein unfolding. *Prog. Biophys. Mol. Biol.* 1993, *59,* 237-284.
[40] Hummer, G. Hydrophobic force field as a molecular alternative to surface-area models. *J. Am. Chem. Soc.* 1999, *121,* 6299-6305.
[41] Kellogg, G. E.; Abraham, D. J. Hydrophobicity: is LogP o/w more than the sum of its parts? *Eur. J. Med. Chem.* 2000, *35,* 651-661.

[42] Sarkar, A.; Kellogg, G. E. Hydrophobicity–shake flasks, protein folding and drug discovery. *Curr. Topics Med. Chem.* 2010, *10,* 67-83.

[43] Ray, L. C.; Kirsch, R. A. Finding chemical records by digital computers. *Science* 1957, *126,* 814-819.

[44] Vleduts, G. E. Concerning one system of classification and codification of organic reactions. *Inf. Stor. Ret.* 1963, *1,* 117-146.

[45] Lederberg, J. Topological mapping of organic molecules. *Proc. Nat. Acad. Sci. U. S. A.* 1965, *53,* 134-139.

[46] Sussenguth, E. H. A graph-theoretic algorithm for matching chemical structures. *J. Chem. Doc.* 1965, *5,* 36-43.

[47] Ullmann, J. R. An algorithm for subgraph isomorphism. *J. Assoc. Comput. Machinery* 1976, *23,* 31-42.

[48] Crowe, J. E.; Lynch, M. F.; Town, W. G. Analysis of structural characteristics of chemical compounds in a large computer-based file. Part I. Non-cyclic fragments. *J. Chem. Soc. C* 1970, 990-996.

[49] Adamson, G. W.; Cowell, J.; Lynch, M. F.; McLure, A. H.; Town, W. G.; Yapp, A. M. Strategic considerations in the design of a screening system for substructure searches of chemical structure files. *J. Chem. Doc.* 1973, *13,* 153-157.

[50] Feldman, A.; Hodes, L. An efficient design for chemical structure searching. I. The screens. *J. Chem. Inf. Comput. Sci.* 1975, *15,* 147-152.

[51] Hodes, L. Selection of descriptors according to discrimination and redundancy. Application to chemical structure searching. *J. Chem. Inf. Comput. Sci.* 1976, *16,* 88-93.

[52] Morgan, H. L. The generation of a unique machine description for chemical structures - A technique developed at chemical abstracts service. *J. Chem. Doc.* 1965, *5,* 107-113.

[53] Willett, P., *Modern approaches to chemical reaction searching*. Gower Publishing Company: 1986.

[54] Corey, E.; Wipke, W. T. Computer-assisted design of complex organic syntheses. *Science* 1969, *166,* 178-192.

[55] Bremser, W.; Klier, M.; Meyer, E. Mutual assignment of subspectra and substructures—A way to structure elucidation by 13C NMR spectroscopy. *Org. Magnet. Res.* 1975, *7,* 97-106.

[56] Lindsay, R. K.; Buchanan, B. G.; Feigenbaum, E. A.; Lederberg, J., *Applications of artificial intelligence for organic chemistry: the DENDRAL project*. McGraw-Hill: New York, 1980.

[57] Doweyko, A. M. The hypothetical active site lattice. An approach to modelling active sites from data on inhibitor molecules. *J. Med. Chem.* 1988, *31,* 1396-1406.

[58] Cramer, R. D.; Patterson, D. E.; Bunce, J. D. Comparative molecular field analysis (CoMFA). 1. Effect of shape on binding of steroids to carrier proteins. *J. Am. Chem. Soc.* 1988, *110,* 5959-5967.

[59] Atkins, P.; De Paula, J., *Physical chemistry for the life sciences*. Oxford University Press: 2010.

[60] Basak, S. C.; Magnuson, V.; Niemi, G.; Regal, R. Determining structural similarity of chemicals using graph-theoretic indices. *Discrete Applied Mathematics* 1988, *19,* 17-44.

[61] Klebe, G.; Abraham, U.; Mietzner, T. Molecular similarity indices in a comparative analysis (CoMSIA) of drug molecules to correlate and predict their biological activity. *J. Med. Chem.* 1994, *37,* 4130-4146.

[62] Wold, S.; Sjöström, M.; Eriksson, L. PLS-regression: a basic tool of chemometrics. *Chemometr. Intell. Lab.* 2001, *58,* 109-130.

[63] Karelson, M.; Maran, U.; Wang, Y.; Katritzky, A. R. QSPR and QSAR models derived using large molecular descriptor spaces. A review of CODESSA applications. *Collect. Czech. Chem. Commun.* 1999, *64,* 1551-1571.

[64] Mauri, A.; Consonni, V.; Pavan, M.; Todeschini, R. Dragon software: An easy approach to molecular descriptor calculations. *MATCH Commun Math Comput Chem* 2006, *56,* 237-248.

[65] Schwab, C., ADRIANA. Code, 2.0. Molecular Networks GmbH, Erlangen, Germany: 2007.

[66] Todeschini, R.; Consonni, V., *Handbook of molecular descriptors.* John Wiley & Sons: 2008; Vol. 11.

[67] Yap, C. W. PaDEL-descriptor: An open source software to calculate molecular descriptors and fingerprints. *J. Comput. Chem.* 2011, *32,* 1466-1474.

[68] Glennon, R. A. Pharmacophore identification for Sigma-1 (σ1) receptor binding: Application of the deconstruction-reconstruction-elaboration approach. *Mini Rev. Med. Chem.* 2005, *5,* 927-940.

[69] Berman, H. M.; Westbrook, J.; Feng, Z.; Gilliland, G.; Bhat, T. N.; Weissig, H.; Shindyalov, I. N.; Bourne, P. E. The Protein Data Bank. *Nucleic Acids Res.* 2000, *28,* 235-242.

[70] Kuntz, I. D.; Blaney, J. M.; Oatley, S. J.; Langridge, R.; Ferrin, T. E. A geometric approach to macromolecule-ligand interactions. *J. Mol. Biol.* 1982, *161,* 269-288.

[71] Ewing, T. J.; Makino, S.; Skillman, A. G.; Kuntz, I. D. DOCK 4.0: search strategies for automated molecular docking of flexible molecule databases. *J. Comput. Aided Mol. Des.* 2001, *15,* 411-428.

[72] Moustakas, D. T.; Lang, P. T.; Pegg, S.; Pettersen, E.; Kuntz, I. D.; Brooijmans, N.; Rizzo, R. C. Development and validation of a modular, extensible docking program: DOCK 5. *J. Comput. Aided Mol. Des.* 2006, *20,* 601-619.

[73] Lang, P. T.; Brozell, S. R.; Mukherjee, S.; Pettersen, E. F.; Meng, E. C.; Thomas, V.; Rizzo, R. C.; Case, D. A.; James, T. L.; Kuntz, I. D. DOCK 6: combining techniques to model RNA-small molecule complexes. *RNA* 2009, *15,* 1219-1230.

[74] Goodford, P. J. A computational procedure for determining energetically favorable binding sites on biologically important macromolecules. *J. Med. Chem.* 1985, *28,* 849-857.

[75] Trott, O.; Olson, A. J. AutoDock Vina: improving the speed and accuracy of docking with a new scoring function, efficient optimization, and multithreading. *J. Comput. Chem.* 2010, *31,* 455-461.

[76] Friesner, R. A.; Banks, J. L.; Murphy, R. B.; Halgren, T. A.; Klicic, J. J.; Mainz, D. T.; Repasky, M. P.; Knoll, E. H.; Shelley, M.; Perry, J. K. Glide: a new approach for rapid, accurate docking and scoring. 1. Method and assessment of docking accuracy. *J. Med. Chem.* 2004, *47,* 1739-1749.

[77] Halgren, T. A.; Murphy, R. B.; Friesner, R. A.; Beard, H. S.; Frye, L. L.; Pollard, W. T.; Banks, J. L. Glide: a new approach for rapid, accurate docking and scoring. 2. Enrichment factors in database screening. *J. Med. Chem.* 2004, *47,* 1750-1759.
[78] Jorgensen, W. L. Rusting of the lock and key model for protein-ligand binding. *Science* 1991, *254,* 954-955.
[79] Hartshorn, M. J.; Verdonk, M. L.; Chessari, G.; Brewerton, S. C.; Mooij, W. T.; Mortenson, P. N.; Murray, C. W. Diverse, high-quality test set for the validation of protein-ligand docking performance. *J. Med. Chem.* 2007, *50,* 726-741.
[80] Kramer, B.; Rarey, M.; Lengauer, T. CASP2 experiences with docking flexible ligands using FlexX. *Proteins* 1997, *29,* 221-225.
[81] Spyrakis, F.; Cozzini, P.; Kellogg, G. E., Docking and scoring in drug discovery. In *Burger's Medicinal Chemistry and Drug Discovery*, 2009.
[82] Erlanson, D. A.; McDowell, R. S.; O'Brien, T. Fragment-based drug discovery. *J. Med. Chem.* 2004, *47,* 3463-3482.
[83] Hajduk, P. J.; Greer, J. A decade of fragment-based drug design: strategic advances and lessons learned. *Nat. Rev. Drug Disc.* 2007, *6,* 211-219.
[84] Congreve, M.; Chessari, G.; Tisi, D.; Woodhead, A. J. Recent developments in fragment-based drug discovery. *J. Med. Chem.* 2008, *51,* 3661-3680.
[85] Murray, C. W.; Rees, D. C. The rise of fragment-based drug discovery. *Nat. Chem.* 2009, *1,* 187-192.
[86] Kitchen, D. B.; Decornez, H.; Furr, J. R.; Bajorath, J. Docking and scoring in virtual screening for drug discovery: methods and applications. *Nat. Rev. Drug Disc.* 2004, *3,* 935-949.
[87] Shoichet, B. K. Virtual screening of chemical libraries. *Nature* 2004, *432,* 862-865.
[88] Alvarez, J.; Shoichet, B., *Virtual screening in drug discovery*. CRC press: 2005.
[89] Walters, W. P.; Stahl, M. T.; Murcko, M. A. Virtual screening, An overview. *Drug Discov. Today* 1998, *3,* 160-178.
[90] Baroni, M.; Cruciani, G.; Sciabola, S.; Perruccio, F.; Mason, J. S. A Common Reference Framework for Analyzing/Comparing Proteins and Ligands. Fingerprints for Ligands And Proteins (FLAP): Theory and Application. *J. Chem. Inf. Model.* 2007, *47,* 279-294.
[91] Varnek, A.; Tropsha, A., *Chemoinformatics approaches to virtual screening*. Royal Society of Chemistry: 2008.
[92] Adamson, G. W.; Bush, J. A. A method for the automatic classification of chemical structures. *Information Storage and Retrieval* 1973, *9,* 561-568.
[93] Adamson, G. W.; Bush, J. A. Method for relating the structure and properties of chemical compounds. *Nature* 1974, *248,* 406-407.
[94] Adamson, G. W.; Bush, J. A. A comparison of the performance of some similarity and dissimilarity measures in the automatic classification of chemical structures. *J. Chem. Inf. Comput. Sci.* 1975, *15,* 55-58.
[95] Carhart, R. E.; Smith, D. H.; Venkataraghavan, R. Atom pairs as molecular features in structure-activity studies: definition and applications. *J. Chem. Inf. Comput. Sci.* 1985, *25,* 64-73.
[96] Willett, P.; Winterman, V.; Bawden, D. Implementation of nearest-neighbor searching in an online chemical structure search system. *J. Chem. Inf. Comput. Sci.* 1986, *26,* 36-41.

[97] Clare, D.; Swaisgood, H. Bioactive milk peptides: a prospectus. *J. Dairy Sci.* 2000, *83,* 1187-1195.

[98] Pihlanto-Leppälä, A. Bioactive peptides derived from bovine whey proteins: opioid and ace-inhibitory peptides. *Trends Food Sci. Technol.* 2000, *11,* 347-356.

[99] Pripp, A. H.; Isaksson, T.; Stepaniak, L.; Sørhaug, T.; Ardö, Y. Quantitative structure activity relationship modelling of peptides and proteins as a tool in food science. *Trends Food Sci. Tech.* 2005, *16,* 484-494.

[100] Korhonen, H.; Pihlanto, A. Bioactive peptides: production and functionality. *Int. Dairy J.* 2006, *16,* 945-960.

[101] Hartmann, R.; Meisel, H. Food-derived peptides with biological activity: from research to food applications. *Curr. Opion. Biotechnol.* 2007, *18,* 163-169.

[102] Sarmadi, B. H.; Ismail, A. Antioxidative peptides from food proteins: a review. *Peptides* 2010, *31,* 1949-1956.

[103] Sharma, S.; SIngh, R.; Rana, S. Bioactive peptides: a review. *Int. J. Bioautomation* 2011, *15,* 223-250.

[104] Li, Y.; Yu, J. Research progress in structure-activity relationship of bioactive peptides. *J. Med. Food* 2015, *18,* 147-156.

[105] Nosho, Y.; Otagiri, K.; Shinoda, I.; Okai, H. Studies on a model of bitter peptides including arginine, proline and phenylalanine residues. II. Bitterness behavior of a tetrapeptide (Arg-Pro-Phe-Phe) and its derivatives. *Agric. Biol. Chem.* 1985, *49,* 1829-1837.

[106] Otagiri, K.; Nosho, Y.; Shinoda, I.; Fukui, H.; Okai, H. Studies on a model of bitter peptides including arginine, proline and phenylalanine residues. I. Bitter taste of di-and tripeptides, and bitterness increase of the model peptides by extension of the peptide chain. *Agric. Biol. Chem.* 1985, *49,* 1019-1026.

[107] Ishibashi, N.; Arita, Y.; Kanehisa, H.; Kouge, K.; Okai, H.; Fukui, S. Studies on flavored peptides. Part I. Bitterness of leucine-containing peptides. *Agric. Biol. Chem.* 1987, *51,* 2389-2394.

[108] Ishibashi, N.; Sadamori, K.; Yamamoto, O.; Kanehisa, H.; Kouge, K.; Kikuchi, E.; Okai, H.; Fukui, S. Studies on flavored peptides. Part II. Bitterness of phenylalanine- and tyrosine-containing peptides. *Agric. Biol. Chem.* 1987, *51,* 3309-3313.

[109] Ishibashi, N.; Ono, I.; Kato, K.; Shigenaga, T.; Shinoda, I.; Okai, H.; Fukui, S. Studies on flavored peptides. Part III. Role of the hydrophobic amino acid residue in the bitterness of peptides. *Agric. Biol. Chem.* 1988, *52,* 91-94.

[110] Ishibashi, N.; Kubo, T.; Chino, M.; Fukui, H.; Shinoda, I.; Kikuchi, E.; Okai, H.; Fukui, S. Studies on flavored peptides. Part IV. Taste of proline-containing peptides. *Agric. Biol. Chem.* 1988, *52,* 95-98.

[111] Kirimura, J.; Shimizu, A.; Kimizuka, A.; Ninomiya, T.; Katsuya, N. Contribution of peptides and amino acids to the taste of foods. *J. Agric. Food Chem.* 1969, *17,* 689-695.

[112] Takeuchi, T.; Yoshida, M.; Yoshii, H. Studies on peptides in miso and soy-sauce. VI-VII. *J. Ferment. Technol.* 1969, *47,* 496-501.

[113] Oka, S.; Nagata, K. Isolation and characterization of acidic peptides in soy sauce. *Agric. Biol. Chem.* 1974, *38,* 1195-1202.

[114] Habibi-Najafi, M. B.; Lee, B. H. Bitterness in cheese: a review. *Crit. Rev. Food Sci. Nutr.* 1996, *36,* 397-411.

[115] Aristoy, M. C.; Toldrá, F. Isolation of flavor peptides from raw pork meat and dry-cured ham. *Dev. Food Sci.* 1995, *37,* 1323-1344.

[116] Noguchi, M.; Arai, S.; Yamashita, M.; Kato, H.; Fujimaki, M. Isolation and identification of acidic oligopeptides occurring in a flavor potentiating fraction from a fish protein hydrolysate. *J. Agric. Food Chem.* 1975, *23,* 49-53.

[117] Adler-Nissen, J., *Enzymic hydrolysis of food proteins.* Elsevier Applied Science Publishers: 1986.

[118] Ney, K. H. Prediction of bitterness of peptides from their amino acid composition. *Eur. Food Res. Technol.* 1971, *147,* 64-68.

[119] Ney, K. H. In *Bitterness of peptides: amino acid composition and chain length*, ACS Symposium Series, 1979; pp 149-173.

[120] Ney, K. H., Recent Advances in Cheese Flavor Research. In *The quality of foods and beverages, Chemistry and Technology*, Eds. Charalambous G; Inglett G, Academic Press Inc: New York, 1981; Vol. 1, pp 389-435.

[121] Lemieux, L.; Simard, R., E. Bitter flavour in dairy products. II. A review of bitter peptides from caseins: their formation, isolation and identification, structure masking and inhibition. *Lait* 1992, *72,* 335-385.

[122] Bull, H. B.; Breese, K. Surface tension of amino acid solutions: a hydrophobicity scale of the amino acid residues. *Arch. Biochem. Biophys.* 1974, *161,* 665-670.

[123] Clegg, K. M.; Lim, C.; Manson, W. The structure of a bitter peptide derived from casein by digestion with papain. *J. Dairy Res.* 1974, *41,* 283-287.

[124] Clegg, K. M.; McMillan, A. D. Dietary enzymic hydrolysates of protein with reduced bitterness. *Int. J Food Sci. Technol.* 1974, *9,* 21-29.

[125] Schalinatus, E.; Behnke, U. The bitter peptides from casein and cheese. 1. Current state of knowledge. *Nahrung* 1974, *18,* 697-704.

[126] Schalinatus, E.; Behnke, U. Studies on the bitter peptides from casein and cheese. Part II. Formation and isolation of bitter peptides. *Nahrung* 1975, *19,* 433-445.

[127] Schalinatus, E.; Behnke, U. Studies on the bitter peptides from casein and cheese. Part III. Characterization of bitter peptides and studies on the elimination of bitter taste. *Nahrung* 1975, *19,* 447-459.

[128] Adler-Nissen, J., Relationship of structure to taste of peptides and peptide mixtures. In *Protein Tailoring for Food and Medical Uses*, Eds. Feeney , R. E.; Whitaker, J. R., Marcel Dekker Inc.: New York, 1986.

[129] Shinoda, I.; Nosho, Y.; Otagiri, K.; Okai, H.; Fukui, S. Bitterness of diastereomers of a hexapeptide (Arg-Arg-Pro-Pro-Phe-Phe) containing D-phenylalanine in place of L-phenylalanine. *Agric. Biol. Chem.* 1986, *50,* 1785-1890.

[130] Kanehisa, H.; Miyake, I.; Okai, H.; Aoyagi, H.; Izumiya, N. Studies of Bitter Peptides from Casein Hydrolyzate. X. Synthesis and Bitter Taste of H–Arg–Gly–Pro–Phe–Pro–Ile–Ile–Val–OH Corresponding to C-Terminal Portion of β-Casein. *Bull. Chem. Soc. Jpn.* 1984, *57,* 819-822.

[131] Kanehisa, H.; Okai, H. Studies of bitter peptides from casein hydrolyzate. V. Bitterness of the synthetic N-terminal analogs of des-Gly2-BPIa (Arg-Pro-Pro-Phe-Ile-Val). *Bull. Chem. Soc. Jpn.* 1984, *57,* 301-302.

[132] Kanehisa, H. Studies of Bitter Peptides from Casein Hydrolyzate. VI. Syntheses and Bitter Taste of BPIc (Val–Tyr–Pro–Phe–Pro–Pro–Gly–Ile–Asn–His) and Its Analogs and Fragments. *Bull. Chem. Soc. Jpn.* 1984, *57,* 97-102.

[133] Shigenaga, T.; Otagiri, K.; Kanehisa, H.; Okai, H. Studies of Bitter Peptides from Casein Hydrolyzate. VII. Bitterness of the retro-BPIa (Val–Ile–Phe–Pro–Pro–Gly–Arg) and Its Fragments. *Bull. Chem. Soc. Jpn.* 1984, *57,* 103-107.

[134] Asao, M.; Iwamura, H.; Akamatsu, M.; Fujita, T. Quantitative structure-activity relationships of the bitter thresholds of amino acids, peptides, and their derivatives. *J. Med. Chem.* 1987, *30,* 1873-1879.

[135] Gomez, M. J.; Garde, S.; Gaya, P.; Medina, M.; Nuñez, M. Relationship between level of hydrophobic peptides and bitterness in cheese made from pasteurized and raw milk. *J. Dairy Res.* 1997, *64,* 289-297.

[136] Gulyaeva, N.; Zaslavsky, A.; Chait, A.; Zaslavsky, B. Relative hydrophobicity of di- to hexapeptides as measured by aqueous two-phase partitioning. *J. Pept. Res.* 2003, *61,* 129-139.

[137] Buchbauer, G.; Klein, C. T.; Wailzer, B.; Wolschann, P. Threshold-based structure-activity relationships of pyrazines with bell-pepper flavor. *J. Agric. Food Chem.* 2000, *48,* 4273-4278.

[138] Tan, Y.; Siebert, K. J. Quantitative structure-activity relationship modeling of alcohol, ester, aldehyde, and ketone flavor thresholds in beer from molecular features. *J. Agric. Food Chem.* 2004, *52,* 3057-3064.

[139] Jonsson, J.; Eriksson, L.; Hellberg, S.; Sjöström, M.; Wold, S. Multivariate Parametrization of 55 Coded and Non-Coded Amino Acids. *Quant. Struct.-Act. Relat.* 1989, *8,* 204-209.

[140] Kim, H.-O.; Li-Chan, E. C. Quantitative structure-activity relationship study of bitter peptides. *J. Agric. Food Chem.* 2006, *54,* 10102-10111.

[141] Wu, J.; Aluko, R. E. Quantitative structure-activity relationship study of bitter di-and tri-peptides including relationship with angiotensin I-converting enzyme inhibitory activity. *J. Peptide Sci.* 2007, *13,* 63-69.

[142] Spasov, B.; Hall, L. H. Modeling Dipeptides as ACE Inhibitors and Bitter-Tasting Compounds by Means of E-State Structure-Information Representation. *Chem. Biodiv.* 2007, *4,* 2528-2539.

[143] Soltani, S.; Haghaei, H.; Shayanfar, A.; Vallipour, J.; Asadpour Zeynali, K.; Jouyban, A. QSBR Study of Bitter Taste of Peptides: Application of GA-PLS in Combination with MLR, SVM, and ANN Approaches. *BioMed Res. Intl.* 2013, *2013,* 1-13.

[144] Wu, S.; Qi, W.; Su, R.; Li, T.; Lu, D.; He, Z. CoMFA and CoMSIA analysis of ACE-inhibitory, antimicrobial and bitter-tasting peptides. *Eur. J. Med. Chem.* 2014, *84,* 100-106.

[145] Floriano, W. B.; Hall, S.; Vaidehi, N.; Kim, U.; Drayna, D.; Goddard, W. A., 3rd Modeling the human PTC bitter-taste receptor interactions with bitter tastants. *J. Mol. Model* 2006, *12,* 931-941.

[146] Dai, W.; You, Z.; Zhou, H.; Zhang, J.; Hu, Y. Structure-function relationships of the human bitter taste receptor hTAS2R1: insights from molecular modeling studies. *J. Recept. Signal Transduct. Res.* 2011, *31,* 229-240.

[147] Morini, G.; Bassoli, A.; Borgonovo, G. Molecular modelling and models in the study of sweet and umami taste receptors. A review. *Flavour Frag. J.* 2011, *26,* 254-259.

[148] Yousif, R. H.; Khairudin, N. B. A. Homology Modeling of Human Sweet Taste Receptors: T1R2-T1R3. *J. Med. Bioeng.* 2014, *3,* 84-86.

[149] Li, G.-H.; Le, G.-W.; Shi, Y.-H.; Shrestha, S. Angiotensin I–converting enzyme inhibitory peptides derived from food proteins and their physiological and pharmacological effects. *Nutr. Res.* 2004, *24,* 469-486.

[150] Ferreira, S. H.; Bartelt, D. C.; Greene, L. J. Isolation of bradykinin-potentiating peptides from Bothrops jararaca venom. *Biochemistry* 1970, *9,* 2583-2593.

[151] Maruyama, S.; Suzuki, H. A peptide inhibitor of angiotensin I converting enzyme in the tryptic hydrolysate of casein. *Agric. Biol. Chem.* 1982, *46,* 1393-1394.

[152] Kohama, Y.; Oka, H.; Matsumoto, S.; Nakagawa, T.; Miyamoto, T.; Mimura, T.; Nagase, Y.; Satake, M.; Takane, T.; Fujita, T. Biological properties of angiotensin-converting enzyme inhibitor derived from tuna muscle. *J. Pharmacobiodyn.* 1989, *12,* 566-571.

[153] Miyoshi, S.; Ishikawa, H.; Kaneko, T.; Fukui, F.; Tanaka, H.; Maruyama, S. Structures and activity of angiotensin-converting enzyme inhibitors in an α-zein hydrolysate. *Agric. Biol. Chem.* 1991, *55,* 1313-1318.

[154] Yokoyama, K.; Chiba, H.; Yoshikawa, M. Peptide Inhibitors for Angiotensin I-Converting Enzyme from Thermolysin Digest of Dried Bonito. *Biosci. Biotech. Biochem.* 1992, *56,* 1541-1545.

[155] Fujita, H.; Yokoyama, K.; Yoshikawa, M. Classification and Antihypertensive Activity of Angiotensin I-Converting Enzyme Inhibitory Peptides Derived from Food Proteins. *J. Food Sci.* 2000, *65,* 564-569.

[156] Pihlanto-Leppälä, A.; Koskinen, P.; Piilola, K.; Tupasela, T.; Korhonen, H. Angiotensin I-converting enzyme inhibitory properties of whey protein digests: concentration and characterization of active peptides. *J. Dairy Res.* 2000, *67,* 53-64.

[157] Wu, J.; Ding, X. Hypotensive and physiological effect of angiotensin converting enzyme inhibitory peptides derived from soy protein on spontaneously hypertensive rats. *J. Agric. Food Chem.* 2001, *49,* 501-506.

[158] Yoshii, H.; Tachi, N.; Ohba, R.; Sakamura, O.; Takeyama, H.; Itani, T. Antihypertensive effect of ACE inhibitory oligopeptides from chicken egg yolks. *Comp. Biochem. Physiol.* 2001, *128,* 27-33.

[159] Jia, J. Q.; Ma, H. L.; Wang, Z. B. Structure-Activity Relationship of Antihypertensive Peptides. *J. Chin. Cereals Oils Assoc.* 2009, *5,* 110-114.

[160] Alemán, A.; Giménez, B.; Pérez-Santin, E.; Gómez-Guillén, M.; Montero, P. Contribution of Leu and Hyp residues to antioxidant and ACE-inhibitory activities of peptide sequences isolated from squid gelatin hydrolysate. *Food Chem.* 2011, *125,* 334-341.

[161] Pripp, A. H.; Sørensen, R.; Stepaniak, L.; Sørhaug, T. Relationship between proteolysis and angiotensin-I-converting enzyme inhibition in different cheeses. *LWT-Food Sci. Technol.* 2006, *39,* 677-683.

[162] Wu, J.; Aluko, R. E.; Nakai, S. Structural requirements of angiotensin I-converting enzyme inhibitory peptides: quantitative structure-activity relationship study of di-and tripeptides. *J. Agric. Food Chem.* 2006, *54,* 732-738.

[163] Wu, J.; Aluko, R. E.; Nakai, S. Structural Requirements of Angiotensin I-Converting Enzyme Inhibitory Peptides: Quantitative Structure-Activity Relationship Modeling of Peptides Containing 4-10 Amino Acid Residues. *QSAR Comb. Sci.* 2006, *25,* 873-880.

[164] Gómez-Ruiz, J. A.; Recio, I.; Belloque, J. ACE-Inhibitory activity and structural properties of peptide Asp-Lys-Ile-His-Pro [β-CN f (47-51)]. Study of the peptide forms synthesized by different methods. *J. Agric. Food Chem.* 2004, *52,* 6315-6319.

[165] Kim, S.; Sung, H. 3D-QSAR of angiotensin-converting enzyme inhibitors: Functional group interaction energy descriptors for quantitative structure-activity relationships study of ACE inhibitors. *J. Biochem. Mol. Biol.* 1998, *31,* 459-467.

[166] Li, S. Z.; Fu, B.; Wang, Y.; Liu, S. On structural parameterization and molecular modeling of peptide analogues by molecular electronegativity edge vector (VMEE): estimation and prediction for biological activity of dipeptides. *J. Chin. Chem. Soc.* 2001, *48,* 937-944.

[167] Tian, F.; Zhou, P.; Li, Z. T-scale as a novel vector of topological descriptors for amino acids and its application in QSARs of peptides. *J. Mol. Struc.* 2007, *830,* 106-115.

[168] Foltz, M.; Van Buren, L.; Klaffke, W.; Duchateau, G. S. Modeling of the relationship between dipeptide structure and dipeptide stability, permeability, and ACE inhibitory activity. *J. Food Sci.* 2009, *74,* H243-H251.

[169] Cheung, H.-S.; Wang, F.-l.; Ondetti, M. A.; Sabo, E. F.; Cushman, D. W. Binding of peptide substrates and inhibitors of angiotensin-converting enzyme. Importance of the COOH-terminal dipeptide sequence. *J. Biol. Chem.* 1980, *255,* 401-407.

[170] Collantes, E. R.; Dunn III, W. J. Amino acid side chain descriptors for quantitative structure-activity relationship studies of peptide analogs. *J. Med. Chem.* 1995, *38,* 2705-2713.

[171] Pripp, A. H.; Isaksson, T.; Stepaniak, L.; Sørhaug, T. Quantitative structure-activity relationship modelling of ACE-inhibitory peptides derived from milk proteins. *Eur. Food Res. Technol.* 2004, *219,* 579-583.

[172] Pripp, A. H. Docking and virtual screening of ACE inhibitory dipeptides. *Eur. Food Res. Technol.* 2007, *225,* 589-592.

[173] Majumder, K.; Wu, J. A new approach for identification of novel antihypertensive peptides from egg proteins by QSAR and bioinformatics. *Food Res. Intl.* 2010, *43,* 1371-1378.

[174] Smith, R.; Waddell, W.; Cohen, S.; Fukushima, S.; Gooderham, N.; Hecht, S.; Marnett, L.; Porthogese, P.; Rietjens, I.; Adams, T. GRAS Flavoring Substances 25. The 25th publication by the Expert Panel of the Flavor and Extract Manufacturers Association provides an update on reent progress in the consideration of flavoring ingredients generally recognized as safe under the Food Additives Amendment. *Food Technol.* 2011, *65,* 44-75.

[175] Munro, I.; Shubik, P.; Hall, R. Principles for the safety evaluation of flavouring substances. *Food Chem. Toxicol.* 1998, *36,* 529-540.

[176] Smith, R. L.; Cohen, S. M.; Doull, J.; Feron, V. J.; Goodman, J. I.; Marnett, L. J.; Munro, I. C.; Portoghese, P. S.; Waddell, W. J.; Wagner, B. M. Criteria for the safety evaluation of flavoring substances: The Expert Panel of the Flavor and Extract Manufacturers Association. *Food Chem. Toxicol.* 2005, *43,* 1141-1177.

[177] Schrankel, K. R. Safety evaluation of food flavorings. *Toxicology* 2004, *198,* 203-211.

[178] Joint FAO/WHO Expert Committee on Food Additives World Health Organization, *Evaluation of certain food additives and contaminants: sixty-first report of the joint FAO/WHO expert committee on food additives*. World Health Organization: 2004; Vol. 61.

[179] Auerbach, S. S.; Shah, R. R.; Mav, D.; Smith, C. S.; Walker, N. J.; Vallant, M. K.; Boorman, G. A.; Irwin, R. D. Predicting the hepatocarcinogenic potential of alkenylbenzene flavoring agents using toxicogenomics and machine learning. *Toxicol. Appl. Pharmacol.* 2010, *243,* 300-314.
[180] Cronin, M. T.; Jaworska, J. S.; Walker, J. D.; Comber, M. H.; Watts, C. D.; Worth, A. P. Use of QSARs in international decision-making frameworks to predict health effects of chemical substances. *Environ. Health Perspect.* 2003, *111,* 1391.

In: From Medicinal Chemistry to Food Science
Editor: Pietro Cozzini
ISBN: 978-1-63483-587-9

Chapter 9

CONSTRAINT-BASED COMPUTATIONAL METHODS FOR INVESTIGATION OF PROTEIN FLEXIBILITY

Alessandro Dal Palù [1,*], ***Agostino Dovier*** [2†] ***and Enrico Pontelli*** [3‡]
[1]Dept. of Mathematics and Computer Science, University of Parma, Parma, Italy
[2]Dept. of Mathematics and Computer Science, University of Udine, Udine, Italy
[3]Department of Computer Science, New Mexico State University Las Cruces, New Mexico US

Keywords: constraint programming, fragment assembly, protein flexibility

Abstract

This chapter presents how constraint programming offers the base for computational methods that model and solve molecular structures. The distinctive features that differentiate this approach from the classical ones are the possibility to model structures by means of constraints (specified by the user) and the presence of a constraint solver, which computes solutions with no need for user intervention. After a gentle introduction on constraint solving, we present a set of constraints that help in characterizing flexible proteins and a real-case application, where a long range movement is recovered through the analysis of the flexible range of motion of a protein. To cope with the need for faster computation, we present an implementation of constraint programming techniques with parallel Graphic Processing Units.

1. Introduction

Computational methods for molecular structures are severely limited by the complexity of the model described and/or by the size of the search space. Usually each method is structured to provide a sensible trade-off between the quality of the computed solution, the amount of time needed to generate it and the cost per hour of parallel time available for

*alessandro.dalpalu@unipr.it
†agostino.dovier@uniud.it
‡epontell@cs.nmsu.edu

the computation. Currently there is still a high interest in improving model quality, while decreasing the computational time.

This chapter presents a computational method that offers interesting opportunities when applied to protein structure modeling. In particular *constraint programming* differs from traditional simulation methods included in popular software for computational modeling (e.g., molecular dynamics, genetic algorithms, etc). The peculiar feature, that is presented here in a brief introduction in Sections 2 and 3, is that the description of the molecules is performed through *constraints*, which are the native mean to program the properties of the domain of interest. The concept of constraint is very general and it has been applied to a number of different applications, including several applications in the area of bioinformatics (see, e.g., the proceedings of the Workshop on Constraint-based Methods for Bioinformatics [35]). Constraint-based approaches are also very intuitive, in the sense that the properties described are normally characterized by simple relations and they can be easily combined in order to form complex descriptions of the model. For example, a constraint can specify the range of an interatomic distance between two specific atoms, while another may impose a maximal amount of free energy in the protein. The clear description of a model through constraints allows one to make use of constraint programming techniques to automatically *reason* about constraints. The reasoning phase is performed by the resolution engine and requires no intervention by the end-user. This phase consists of exploiting the information expressed by constraints to infer additional properties and facts, that are used by the engine to efficiently explore the solutions of the model. As an example, one could describe the geometry of a protein-ligand interaction and let the solver search the protein poses that fulfill the constraints.

The goal of this chapter is to highlight the potentialities of constraint programming techniques and to stimulate the reader with a new perspective that may ease the description of a particular model and simplify the computation of solutions.

In this chapter, we present an introduction to constraint programming (Section 2) and we detail one of the models used in the literature to describe a flexible protein (Section 3). As an example of application of constraint technology, we report on an exploration of the flexible motion of a protein to recover a long range motion (Section 4). The chapter is concluded with some consideration about the computational advantages offered by constraint programming in terms of transparent exploitation of parallelism (Section 5), by mapping the resolution process on a variety of parallel architectures (e.g., Graphic Processing Units). We also describe a graphical user interface that allows a semi-automated workflow for the exploration of protein properties (Section 6).

2. Constraint Programming

2.1. Introduction

Constraint Programming (CP) is a programming paradigm that has gained significance because of its strengths in supporting the design of declarative, yet efficient, solutions to complex problems—especially problems of combinatorial nature. Grounded in strong theoretical foundations, CP has been the focus of the development of both general as well as ad-hoc resolution techniques, leading to highly efficient implementations that have been used to tackle complex real-world applications [11,30].

Intuitively, CP relies on modeling a problem in terms of properties that each solution of the problem needs to satisfy (*constraints*). A constraint is simply a relation among the unknowns of the problem, used to limit the combinations of acceptable values that can be assigned to the unknowns involved in the relation. Constraints are purely declarative entities, as they focus on the description of a the problem without specifying any specific computational process to determine the values of the unknowns. Thus, the core idea is to solve problems by describing the constraints about the problem and, consequently, determining the solution(s) satisfying all the constraints.

The foundations of CP can be traced back to Waltz's scene labeling algorithm [39], which is often recognized as the first instance of a consistency-based constraint solving approach. The formalization of general and application-independent frameworks for problem solving based on CP was fueled by the research in the field of logic programming, which led to the development of the *Constraint Logic Programming* paradigm, along with a variety of implementations [24].

In recent years, we have witnessed the development of a number of stand-alone efficient constraint solving platforms, either as independent libraries (e.g., [1,23,27]) or embedded in general programming languages (e.g., [8,32]).

2.2. Constraint Satisfaction and Optimization

A *Constraint Satisfaction Problem (CSP)* is described by a collection of variables and the constraints among such variables. Formally, a CSP $\mathcal{P}$ is described by a tuple $\langle X, D, C\rangle$ where X is a finite collection of variables $\{X_1, \ldots, X_n\}$, D is a function mapping each variable X_i to its corresponding *domain* $D(X_i)$, and C is a collection of constraints $\{C_1, \ldots, C_m\}$. Each constraint C_j has a scope $scope(C_j) = \langle X_{j_1}, \ldots, X_{j_k}\rangle$, where $\{j_1, \ldots, j_k\} \subseteq \{1, \ldots, n\}$ and $C_j \subseteq D(X_{j_1}) \times \cdots \times D(X_{j_k})$.

Given a constraint C_j with scope $scope(C_j) = \langle X_{j_1}, \ldots, X_{j_k}\rangle$, the constraint is *unary* if $k = 1$; it is *binary* if $k = 2$. We will often use the term *global constraint* to refer to a constraint where $k > 2$.

An assignment $\sigma : X \mapsto \bigcup_{i=1}^{n} D(X_i)$ is a function that satisfies the condition $\sigma(X_i) \in D(X_i)$ for $i = 1, \ldots, n$—i.e., each value assigned to an unknown is "legal." A *solution* of $\mathcal{P}$ is an assignment that satisfies all the constraints, i.e., if $scope(C_j) = \langle X_{j_1}, \ldots, X_{j_k}\rangle$ then $\langle \sigma(X_{j_1}), \ldots, \sigma(X_{j_k})\rangle \in C_j$ for each $j = 1, \ldots, m$. A CSP $\mathcal{P}$ is *consistent* if it admits at least one solution. We will denote with $SOLN(\mathcal{P})$ the set of all solutions of the CSP $\mathcal{P}$.

For the sake of simplicity, in the rest of this work we will focus on *finite domains CSPs*, where $D(X_i)$ is a finite set for each $X_i \in X$.

A *Constraint Optimization Problem (COP)* $\mathcal{Q}$ is a CSP where we are seeking solutions that meet certain optimality criteria. Formally, $\mathcal{Q}$ is described by a tuple $\langle X, D, C, f\rangle$ such that $\langle X, D, C\rangle$ is a CSP and $f : D(X_1) \times \cdots \times D(X_n) \mapsto \mathcal{R}$ is the objective function of the optimization problem. An element $\sigma \in SOLN(\langle X, D, C\rangle)$ is a solution of $\mathcal{Q}$ if there is no other element $\sigma' \in SOLN(\langle X, D, C\rangle)$ such that $f(\sigma'(X_1), \ldots, \sigma'(X_n)) < f(\sigma(X_1), \ldots, \sigma(X_n))$.

2.3. Constraint Solvers

A typical CSP/COP solver makes use of a search procedure to explore the space of assignments in search of solutions. Two steps are typically part of CSP/COP solvers:

Labeling: a labeling step consists of selecting a variable $X_i \in X$ such that $|D(X_i)| > 1$ and assigning to it one of the values in $D(X_i)$.

Propagation: a propagation phase removes from the domains of the variables values that cannot participate in any solution of the CSP/COP.

The propagation step allows us to reduce the choices of the values in the next labeling step. If during this task it is discovered that the current assignment violates some constraint (typically, by observing that the domain of some variable is empty), the search continues by changing the previous assignments made, in reverse chronological order (backtracking). Algorithm 1 shows a basic backtracking solver for CSPs that makes use of the steps above.

Algorithm 1 search(X,D,C)

$D' \leftarrow propagation(X, D, C)$
if $(\exists X_i \in X. D'(X_i) = \emptyset)$ **then return** *Failed* ▷ Backtracking
else if $(\forall X_i \in X. |D'(X_i)| = 1)$ **then return** D' ▷ Found a solution
else if $(\exists X_i \in X. |D'(X_i)| \geq 2)$ **then** ▷ Non-deterministic choice
  for $a_i \in D'(X_i)$ **do**
    $r = search(X, D', C \cup \{X_i = a_i\})$
    if $(r \neq Failed)$ **then return** r
    end if
  end for
  return *Failed* ▷ All labelings have failed
end if

Different forms of propagation have been designed; they often build on the use of notions of *consistency*—a process that makes use of content of the domains of certain variables in the scope of a constraint to identify irrelevant values from the domains of other variables. For example, *arc-consistency* considers binary constraints: if $scope(C_i) = \langle X_j, X_k \rangle$ and $D(X_j) = \{a\}$, then arc-consistency removes from $D(X_k)$ all values b such that $\langle a, b \rangle \notin C_i$.

Global constraints often introduce dedicated forms of consistency, which rely on the specific semantics of the constraint (e.g., the widely studied *all-different* constraint [18,37]).

In presence of particularly large and complex search spaces, complete search (even with the aid of propagation techniques) is infeasible; this has led researchers to explore the use of heuristic search and stochastic search techniques (e.g., [22,28,36]—see also Section 5).

3. Fragment Assembly and Constraints

3.1. Modeling Proteins in a Constraint Setting

A variety of approximated representation of a protein structure and its spatial model have been proposed to enable the use constraint-based methods to address studies of structure conformations and dynamics.

The simplest protein representation considered in the CP literature is the HP model [3, 17]. In this model, amino acids are represented in terms of the class they belong to (hydrophobic or polar), and each amino acid is encoded as a single point; in terms of constraints, this is represented by a variable whose value encode the spatial coordinates of the C_α atom of the amino acid. This representation allows the use of spatial constraints to restrict the possible placements of amino acids in space—e.g., enforcing a distance of about 3.81Å among consecutive amino acids of a protein: $|dist(P_i, P_{i+1}) - 3.81| \leq \epsilon$ for some small ϵ, where $dist(X, Y)$ is the traditional Euclidean distance. Additional spatial constraints, i.e., based on distances between coordinates assigned to the amino acids, can be introduced to capture features like secondary structure patterns, torsional angles, and disulfide bonds.

A natural refinement of this representation allows the encoding of the 20 different types of amino acids as distinct entities (instead of using the two large classes of the HP model), while still encoding each amino acid in a structure as a single point in space (i.e., the coordinate of the C_α atom). While the spatial constraints are essentially the same as in the HP model, the C_α model enables the use of more precise energy function to score solutions (e.g., [13]).

The C_α representation ignores the atomic structure of the amino acids, which may lead to conformations that are physically infeasible (due to the more complex atomic structure). A first step in addressing this concern can be achieved by acknowledging the presence of a side-chain; the model discussed in [15] expands the C_α representation by adding a second component which models the center of mass of the side-chain of the amino acid (the side-chain's centroid). This representation allows us to capture the different distances and volumes of side-chains, enabling also the use of more refined energetic models.

More precise constraints capturing the dihedral angles and other atomic-level properties have been proposed, ranging from the 5@ model [5] , which represents each amino acid by the position of four atoms—the nitrogen (N), the carbon alpha (C_α), the carbon (C') and the oxygen (O)—and the centroid (see Fig. 1), to full atoms representations (see [9, 33] for a survey).

As mentioned in the section of constraint solving, each unknown used in the modeling process needs to be associated to a *domain*, that identifies the set of legal values that can be assigned to the unknown. Considering that the unknowns predominantly represent spatial positions, the domains capture parts of the 3D space where the (components of) amino acids can be placed—this is typically represented as a *cuboid* in the 3D space that encloses the legal coordinates that can be used to place the point in the space [15].

The literature has explored different representations of these domains. The main distinction between the different approaches proposed in the literature derives from the level of discretization used. On one hand, one can use continuous domains, where each 3D coordinate (within a give range) is allowable [26]. The use of continuous domains has not been used as extensively in the context of CP approaches. The other alternative is represented by the use of *discrete domains*, where each domain contains only a finite number of points. This allows the direct use of problem solving techniques derived from CSP; the use of discrete representations of the space does not necessarily compromise precision of the solutions, either by choosing a very fine degree of discretization (e.g., points at a distance of 0.1Å or less) or by relaxing the solution by means of an external tool in order to

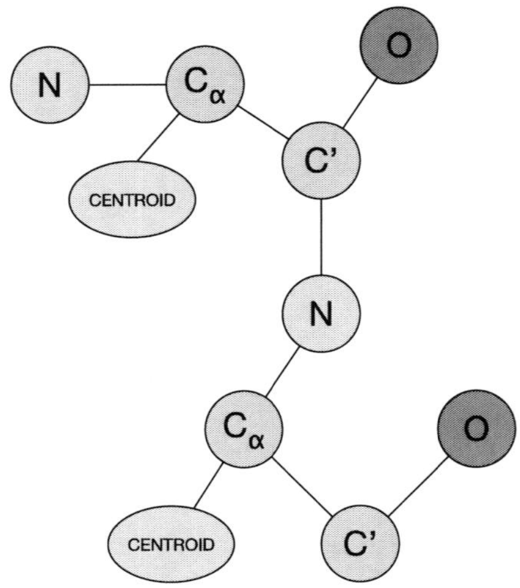

Figure 1. A simple view of the 5@ representation.

compensate approximation errors. In order to facilitate reasoning about spatial properties in the 3D space, several works have adopted a structured discrete representation of the space, where only points satisfying a regular structure are taken into consideration. The regularity simplifies the representation of domains and the implementation of procedures to solve spatial constraints. In particular, crystal lattice structure organizations have been extensively used in the context of structural studies of proteins, with emphasis on cubic lattices [15,21], triangular lattices [2] and face-centered cubic lattices [2, 3, 15, 25].

3.2. Fragment Assembly as Constraint Solving

3.2.1. Fragment Assembly

A promising approach to address the problem of protein structure prediction is the *fragment assembly* method; this is at the basis of popular prediction systems like Rosetta [29]. The fundamental idea is to consider each protein as an assembly of components (*fragments*), drawn from a finite dictionary of possible fragments. Each fragment describes a possible spatial arrangement of a given collection of amino acids, and the repository provides all statistically significants arrangements for each sequence of amino acids of a given length. Thus, the fragments offer a model for a local segment of a protein. Using this perspective, it is possible to study the structure of a protein as a problem of composing fragments for the different segments of the protein. This problem can be naturally viewed as a COP, since

- Multiple candidate fragments for every part of the protein requires a combinatorial search among the possible choices; spatial constraints have to be respected during the assembling process, e.g., to ensure that no distinct atoms are place too close to each others;
- Criteria need to be established to score different assemblies, typically based on energetical considerations—thus, the assembling process can be viewed as an optimization problem.

In the specific constraint-based approach explored in [5], fragments of k amino acids[1] represent spatial atomic arrangements according to the 5@ representation of amino acids—thus, it represents the spatial arrangements of sequences of $4k+3$ atoms of the type

$$C'O\underbrace{(NC_\alpha C'O)}_{k\ times}N$$

Two fragments can be assembled if the second fragment can be rotated in space in such a way that the first n points of the fragment perfectly overlap the last n points of the first fragment, for a fixed value of n. A typical selection of n is 3, which means that we are expecting the last $C'ON$ atoms of the first fragment to overlap the first $C'ON$ atoms of the second fragment. Given two fragments P and Q, we use the notation $P \frown Q$ to denote the fact that there exists a rotation of Q that overlaps the first three atoms of Q with the last three atoms of P. With a slight abuse of notation, we will also denote with $P \frown Q$ the structure resulting from the assembling of the two fragments.

Figure 2 shows a simple example in the two-dimensional space (for simplicity, we use numbers instead of atom names). The first case (top row) shows a situation where, regardless of how we rotate the second fragment, we cannot overlap its first three atoms with the last three atoms of the first fragment. The other two cases show a situation where the rotation is possible. The third case, though, illustrates another problem: even though a rotation of the second fragment is possible, the assembling is infeasible as it would cause the fourth atom of the second fragment to clash with the position of the second atom of the first fragment. This implies the need of introducing spatial constraints to guide the assembling process.

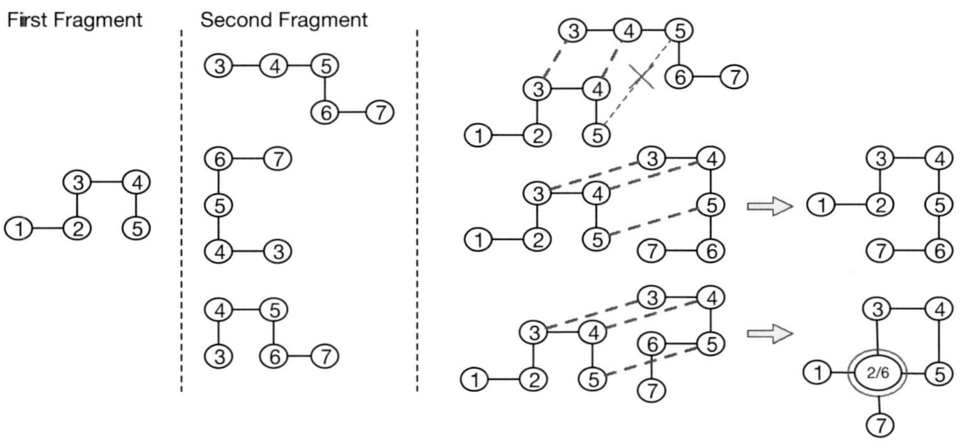

Figure 2. Overlapping fragments.

The assembly problem is then reduced to the problem of partitioning the protein sequence in blocks $S_1, \ldots, S_n$ of k amino acids each, determining from the chosen fragments repository the set of fragments $\mathcal{B}_i$ for each amino acid subsequence S_i, and then identifying

[1]The method described in that work allows the parametric derivation of fragment repositories for any length k. Moreover, although developed for the 5@ model, it can be easily extended to more detailed models.

one fragment $B_i \in \mathcal{B}_i$ for each subsequence such that: **(1)** $B_i \frown B_{i+1}$, i.e., each fragment can be rotated and overlapped with the successive one, and **(2)** $B_1 \frown \cdots \frown B_n$ is a consistent structure—i.e., it satisfies spatial constraints relevant for a protein structure.

3.2.2. Mapping Fragment Assembly to CSP and COP

The fragment assembly process can be described as a COP, where the CSP component captures the ability of overlapping fragments and the protein-related spatial constraints (e.g., lack of spatial clashes), while the optimization component is used to score a consistent assembling using a statistical energy function. We will not elaborate on the energy function used—the reader is referred to [4, 14, 34] for a discussion of possible energy functions.

The CP modeling of the fragment assembly problem makes use of two sets of variables:

- Variables associated to each subsequence S_i of the primary structure; the variable is used to represent which fragment is selected for the amino acids in S_i; the domain of the variable is the set of possible fragments $\mathcal{B}_i$ for the sequence of amino acids S_i;
- Variables representing the specific position in space assigned to each atom.

The intuition is that the two sets of variables are linked to each other—i.e., determining the fragment to be used for S_i will fix the spatial positions of the atoms in S_i, and vice versa, determining the position in space of the atoms will fix the fragment used for S_i.

The following are the type of constraints that can be applied to guide the assignment of values to the problem variables:

1. Given a subsequence S_i, the constraint *fragment* $(F_i, \vec{P_i})$ ensures that the fragment assigned to S_i (denoted by the variable F_i) is consistent with the spatial positions assigned to the atoms involved in S_i (described by the sequence of position variables $\vec{P_i}$);

2. Point constraints:

   (a) Distance Constraints: these are used to create spatial boundaries between the placements of atoms in the space. In particular, constraints are used to both limit how close atoms can be placed—$dist(P_i, P_j) \geq d_i + d_j$ where the distances d_i, d_j capture the size of an ideal sphere around each atom/centroid (see, e.g., [19, 40])—and how far they can be (see also Fig. 3)—e.g., C_α atoms have a distance of approximately 1.52Å from the following C' atoms, etc.

   (b) Centroid Constraints: these constraints take advantage of the fact that the position of the centroid is uniquely determined once the positions of the N, C_α and C' atoms have been fixed, and such position is dependent on the type of amino acid considered (i.e., structure of the corresponding side chain). This spatial relation is captured by constraints of the form *centroid*$_a(P_N, P_{C_\alpha}, P_{C'}, P_{Centroid})$, where a is the specific type of amino acid considered. Observe that the distance constraints will interact with the centroid constraints, ensuring that adequate space is reserved for the centroid in the placement of the atoms.

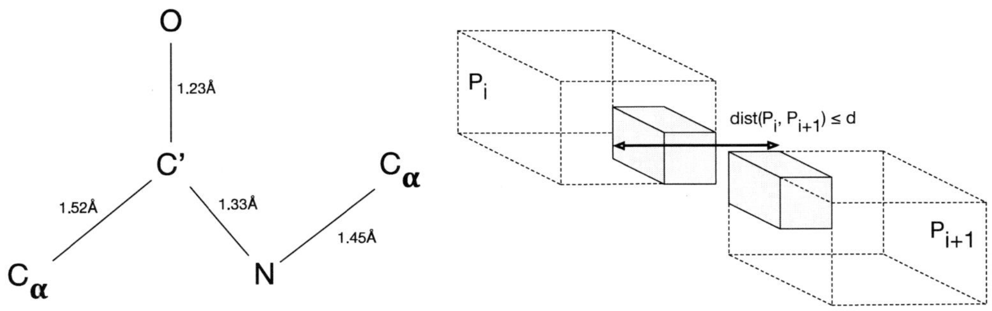

Figure 3. Atomic distances for distance constraints.

3. Fragment Constraints:

   (a) Joint Constraints: the most natural constraint for the fragments is the requirement of fragments associated to consecutive blocks of the protein to be compatible on their first/last three atoms (i.e., $B_i \frown B_{i+1}$). This constraint (expressed as *joint* (F_i, F_{i+1}), where F_i and F_{i+1} are variables associated to two consecutive subsequences of the protein. The meaning of this constraint is to verify the existence of a transformation matrix [20] describing the roto-translation of the second fragment to overlap the first fragment in the three atoms in common. In simple terms, we are seeking three rotation angles ϕ, θ, ψ such that $P^i_A = X \cdot Y \cdot Z \cdot (P^{i+1}_A)^T$, where

   - $A \in \{C', O, N\}$, P^i_A denotes the last occurrence of the atom A in the subsequence S_i, and P^{i+1}_A denotes the first occurrence of the atom A in the subsequences S_{i+1}.
   - the rotation matrices are

$$X = \begin{bmatrix} 1 & 0 & 0 \\ 0 & cos\phi & sin\phi \\ 0 & -sin\phi & cos\phi \end{bmatrix} \quad Y = \begin{bmatrix} cos\theta & 0 & sin\theta \\ 0 & 1 & 0 \\ -sin\theta & 0 & cos\theta \end{bmatrix} \quad Z = \begin{bmatrix} cos\psi & sin\psi & 0 \\ -sin\psi & cos\psi & 0 \\ 0 & 0 & 1 \end{bmatrix}$$

   (b) Rigid Blocks: The idea of fragment can be extended to capture more complex structures (e.g., secondary structure components), where the fragment is actually a longer and more complex pattern, that still has to be rotated and translated to properly connect with the previous fragment. This has been described in [16] as the *block* $(P_i, \ldots, P_n)$ constraint, and is a natural generalization of the joint constraint described above.

In order to make these constraints effective, they are often combined in order to better exploit their interactions during the propagation process. For example, in [5], the constraint *JM* is described, which combines the *joint* and *distance* constraints mentioned earlier.

The constraints provide two directions of propagation:

1. Direct Propagation: constraints among points (e.g., distance constraints) allow us to reduce the number of feasible locations where atoms can be placed; for example, if the domain of a point variable is represented as a cube (which includes all the points that can be used), then distance constraints enable "shrinking" such cube to ensure that distance restrictions are satisfied (see, e.g., Fig. 3).

2. Indirect Propagation: the link between fragments and points allows us to propagate across types of variables; for example, fixing the position of several atoms in space may rule out certain fragments from being used (as no rotations are possible to match the position of such points), and choosing a fragment (even before linking it to the preceding ones) will establish precise restrictions concerning the relative positions in space of the points covered by the fragment.

4. An Example of Application: Protein Flexibility

In this section we present an example where constraint modeling and search can help the study of a complex scenario such as the protein flexibility. For more details, we refer the interested reader to [12]. When dealing with flexibility, proteins can show different kind of movements, ranging from side-chain adjustments to large scale motions of the entire domain. The study of protein-ligand interactions benefits from flexible models that account for local modifications during the binding of the compounds. Even if some experimental techniques (e.g., NMR) are able to partially give some hints about the folding during the binding process, they are limited by several practical restraints. On the other side X-ray crystallography can deliver clear snapshots of some specific states of the compounds, but it does not shed any light about what is in between. The most popular computational technique that is used to fill the gap is the Molecular Dynamics (briefly, MD), which is able to faithfully model the flexibility of the system, at the cost of increasing computational complexity. In fact, MD can be successfully applied for rather small protein adjustments. Other approximated version of MD can be applied to model large-scale protein motions [31].

The framework presented in [12] shows a viable alternative, based on Constraint Logic Programming, that offers the possibility to geometrically investigate the flexible range of motions of a protein and to investigate interactions with multiple ligands with a single protein computation analysis (the demanding part of computation). As depicted in Figure 4, the methodology is based on a generation of potential protein conformations that fulfill the *geometric constraints* of the model and a post processing phase where the most likely candidates are combined to form a plausible trace. In this approach, the protein flexibility is rendered as a choice between multiple admissible fragments that sample the mobility of the backbone for each specific amino acid. Combining different choices of fragments for each amino acid, results in different spatial structures.

As discussed in Section 3, according to the specific protein, different spatial constraints and fragments describe the range of motion and the forbidden volumes for each amino acid. In particular, flexible coils are modeled by a sequence of joint constraints that cover the coil length. Fragments are three amino acid long, in order to limit the number of possible combinations. Fragment templates are sampled from the Protein Data Bank according to matching amino acid sequence. For each position, multiple choices of fragments are available and the flexibility of the coil is reproduced by selecting and assembling different fragments. Moreover, in the presence of α-helices surrounded by flexible coils, a rigid block constraint imposes the ideal helical structure, which will be simply rotated and translated in each generated solution. In case slight helix bending has to be taken into account, the rigid block constraint can be replaced by some joint constraints that cover the block and by corresponding fragments that include minimal variations respect to the perfect helical fragment. Distance constraints for avoiding overlaps are enforced for pairs of atoms, except for those belonging to the same rigid block. For faster search, only centroids of side chains are accounted for and corresponding distance constraint from other atoms and/or centroids are added. During the search of solution to the constraint problem, specific algorithms for constraint propagation avoid in advance the unnecessary exploration of areas of the conformational space that correspond to some constraint violation (e.g., steric clashes).

One peculiarity of the search engine that explores the conformational space is the possibility to use the information carried by the explicit presence of constraints to deduce how to direct the search. Compared to a random choice of fragments to be assembled, where self overlapping of the protein is rather likely, constraint resolution is able to exclude fragment choices that bring towards such failures *during* the search.

Another advantage when using constraints such as the JM constraint, not yet used in the approach presented in [12], is that it is possible to control the distribution of the searched conformations. Let us compare with a naive sampling, based on the assembly of random fragments. In this scenario, a random fragment is equivalent to the definition of the backbone pseudo-torsional angles. The assembly procedure terminates with a structure that has a very complex relation (which is difficult to invert) between such angles and the actual atoms coordinates. The presence of a JM constraint on top of a random selection of fragments, can select those representative conformations that describe a group of equivalent structures. The most important aspect is that no extra time is spent in searching equivalent proteins. This has the immediate impact of search efficiency and it allows to spend more time for the generation of interesting samples.

The generation of the pool of conformations can be used for various analyses. In [12] a geometrical and least expensive path has been identified, in order to reconstruct the putative trace between two poses determined by x-ray crystallography. In particular, the shortest path algorithm has been run over the graph induced by conformations and their pairwise RMSD. A shortest path between two poses corresponds to identify in the pool, a sequence of conformations that are pairwise similar in shape and form a short and smooth transition between the two ends. The pool has to be rich enough to contain a sufficient number of conformations, in order to avoid large changes between two conformations close-by in the path. However the number of conformation has to be kept low for computational issues: in [12] we used 1.2M conformations and the need for non redundant samples drove the search towards the use of constraints such as the JM.

This analysis can be performed over the same pool for different ligands, with no need for the recomputation of the pool. Instead, a simple conformation screening is performed before running the shortest path algorithm, in order to identify possible protein-ligand overlapping which is to be kept minimal. Another advantage of this framework is the possibility to include energetic scoring in the shortest path evaluation.

A possible different use of the constraint model and solution generation, is a more interactive one. The goal is to avoid the complete generation of the pool of the solution and then the shortest path optimization. The idea is to exploit the model to generate solutions that lie in the spatial neighborhood of the ones being analyzed by e.g., the shortest path algorithm. This reminds of a genetic algorithm which combines different elements of the same offsprings. In our case, however the challenge is more difficult, since the request is to generate conformations that share a similar shape in the space within some distance with respect to the current one. This can be easily expressed by constraints and the propagation inside the solver is able to produce an answer in a short time.

It is reasonable to embed this kind of iterative search to explore neighborhood with the guidance of the shortest path algorithm, where specific constraints are added each time to the original model. The main advantage of this procedure is that there is no need to generate the whole sampling of the conformations pool, but conformations are generated when necessary.

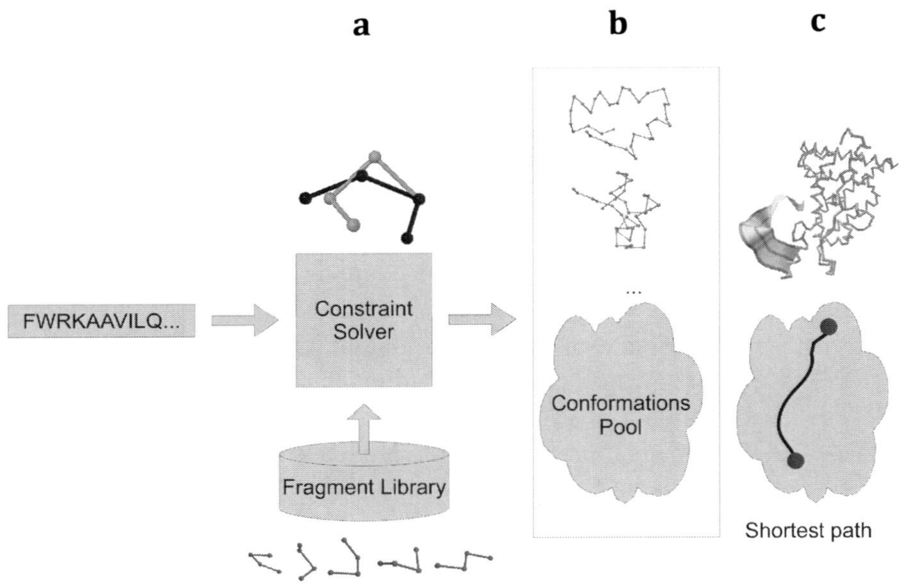

Figure 4. The framework based on constraint fragment assembly used in [12].

5. Parallelism

The availability of several processors in a single computer suggests the user the possibility of splitting a task among them in order to speed-up the computation. Several kinds of parallelism have been analyzed in literature for searching the (best) solution to a constrained problem.

The first one is called *search parallelism.* Search parallelism can be implemented by assigning the values of some of the variables in all possible ways at the beginning, and executing in different processors the reduced problems; the solutions computed by different branches are compared and the best one is retrieved. This can be implemented in every computer where multi-core CPUs are available. However, the number of cores is typically 2 or 4, therefore this is also the maximum theoretical speed-up one can obtain.

Let us focus on a second, less known, method to exploit parallelism, usually known as *consistency parallelism.* In Section 3 we have briefly reviewed the tool FIASCO for modeling protein reconstruction using fragment assembly and constraints. During the searching stage there are a lot of activities that are repeated several times, such as the computation of the free energy of the protein that requires (in particular) the computation of the distance between some atoms of each pair of protein's amino acids. Just to fix the ideas, assume that (a piece of) the energy contribution is defined by the following formula:

$$E = \sum_{i=1}^{n-1} \sum_{j=i+1}^{n} f(a_i, a_j, \langle x_i, y_i, z_i \rangle, \langle x_j, y_j, z_j \rangle)$$

where a_k, for $k = i$ or $k = j$, is the 'kind' of the corresponding amino acid in the protein and $\langle x_i, y_i, z_i \rangle$ its spatial coordinates. This would require $\frac{n(n-1)}{2}$ computations of the function f and $n-2$ sums of the partial results. If $n-1$ parallel computing cores were available, one might imagine to parallelize the computation of E as follows: for $i = 1, \ldots, n-1$ in parallel, core i computes $\sum_{j=i+1}^{n} f(a_i, a_j, \langle x_i, y_i, z_i \rangle, \langle x_j, y_j, z_j \rangle)$. Core i needs to compute the function f at most $n-1$ times. Then the sum of all the results is done. This can be done in time proportional to $\log_2 n$ (see Figure 5).

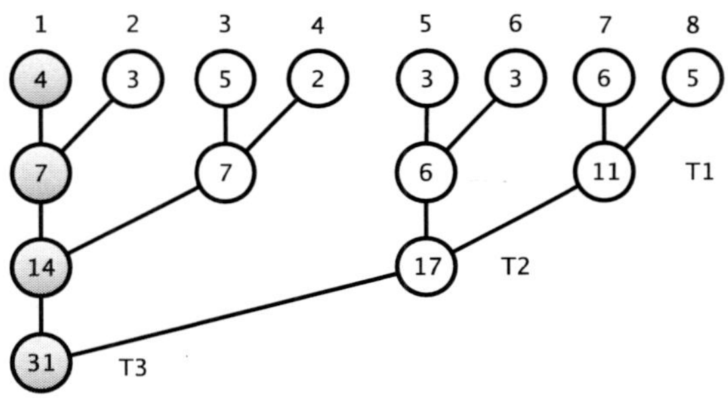

Figure 5. Logarithmic sum of 8 elements resulting from computations in 8 cores in $3 = \log_2 8$ instants of time (T1, T2, and T3).

Assume $n = 256$, and each single operation takes time 1 (e.g., 1 ns), then with the first approach we need time $32640 + 254 = 32894$, with the second approach, instead, we only need time $255+8 = 263$. Of course this can be done only if 254 communicating computing cores are available. This is a typical feature of a modern General Purpose Graphics Processing Unit (briefly GPU). GPUs are hardware cards normally available in personal computers (either laptop or desktop). Their common use is to speed-up computations for graphical applications and they offer a large number of cores (394 is a common number for a new laptop/desktop); this architecture is however not suited for search parallelism (computing cores have little and slow "private" memory: they can help each other in global tasks, communicating with a shared memory). Moreover one can buy a graphic card dedicated only to computing (not involved in the video rendering), with thousands of cores at a reasonable price.

In [7] it is shown that GPUs can be very effective for fragment-assembly constraint-based approach to protein structure prediction. GPUs are used to parallelize the energy function as told above; this kind of parallelism is defined as *consistency parallelism*. However, they can be used in parallel to predict secondary structure elements. An external tool (JNET [10]) is called to select the portions of the protein where secondary structures might be located. Then these piece are joined together and a local search procedure is invoked to build the a shape that minimize the energy. In the same tool developed in [7], also known as COCOS, a *local search* stage is also parallelized with GPUs. In local search, given a current solution, the value of a randomly selected (small) set of variables is assigned with all possible values for them. The set of these assignments is called *neighborhood*. The assignment satisfying the constraints that returns the best value for the cost function is returned and the procedure continues from this new assignment. As far as local search is concerned, two forms of parallelism are usually employed: *single walk* vs *multi-walk* [38]—see Figure 6. Single walk methods exploits parallelism for testing concurrently several of these values. Multi-walk methods, instead, parallelize the random selection of the variables to be re-assigned, thus executing several single walk in parallel. The latter method requires shared-memory and synchronization primitives, but GPUs behave very well for these requirements.

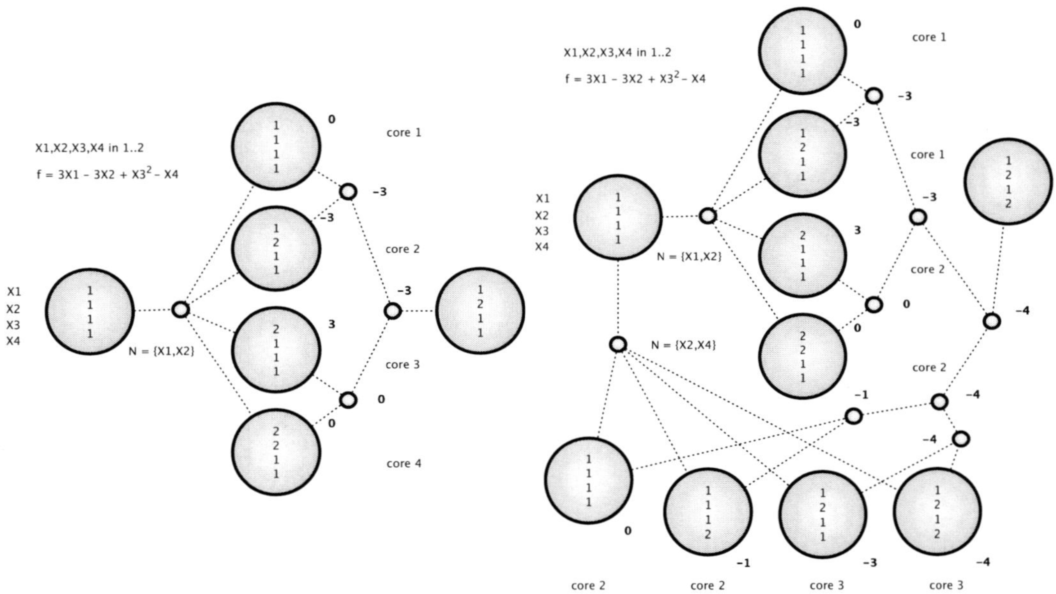

Figure 6. Simple example of local search: single walk parallelism vs multi-walk parallelism. In the second case two neighborhoods are analyzed in parallel. The best result obtained will be the starting point of the next stage.

Using a GPU NVIDIA GeForce GTX TITAN, a speed-up of 20–40 is obtained w.r.t. a CPU AMD Opteron 6376, 2.3GHz. Details can be found in [6]. This tool will be refered to as COCOS in the remaining part of the paper.

6. A Graphical Tool

The current version of COCOS (Section 5) works with a general fragment database that allows a fast prediction of the secondary structures and an approximated prediction of the overall tertiary structure in relatively short time. FIASCO (Section 3) is instead optimized to compute (a set of) more precise closure of the "loops" between known parts of the protein. The idea is to use the two tools together. COCOS for an initial prediction, then FIASCO to refine it. However, the tool should be able to exploit expert human knowledge for refining the predictions. For instance one might want to "move" one or more secondary structures w.r.t. the predicted shape. For doing an interface should be available; we are developing a graphical interface (with a slight abuse of notation, it will be referred to as to the FIASCO interface) that we briefly described below (a screenshot is reported in Figure 7).

The tool helps the user in installing the desired software (basically, COCOS, either in a CPU or in a GPU version, if a NVIDIA GPU is available in the PC, and FIASCO). Compilation is made by clicking the "Tools" options. A protein can be selected either as pre-computed (PDB format) or as a list of aminoacids (FASTA format) — this is made by clicking the "File" option. In the last case, COCOS can be launched (with some parameters, either on GPU or on CPU). A possible 3D shape is computed (see the window on the left). Then, either using the mouse or clicking "Select", some parts of the proteins are selected. With the "Assemble" option, these parts are copied in the central window. Here, again using the mouse or the options one might change the relative position of these parts. Then, with "Solve" the FIASCO solver is called (several options are available, but it runs with the default one, as well) and after a running time bounded by a timeout a new shape is obtained (actually other solutions are computed and can be viewed clicking on "Extract"). The process can be iterated moving the solution on the left window (e.g., saving and reloading it).

The fact that we are using constraint-programming is crucial here. Constraints are added graphically, but there is no need for the user to know about pre-processing, normal forms, ad-hoc heuristics, while he/she can just simply add the constraints and let the solver work.

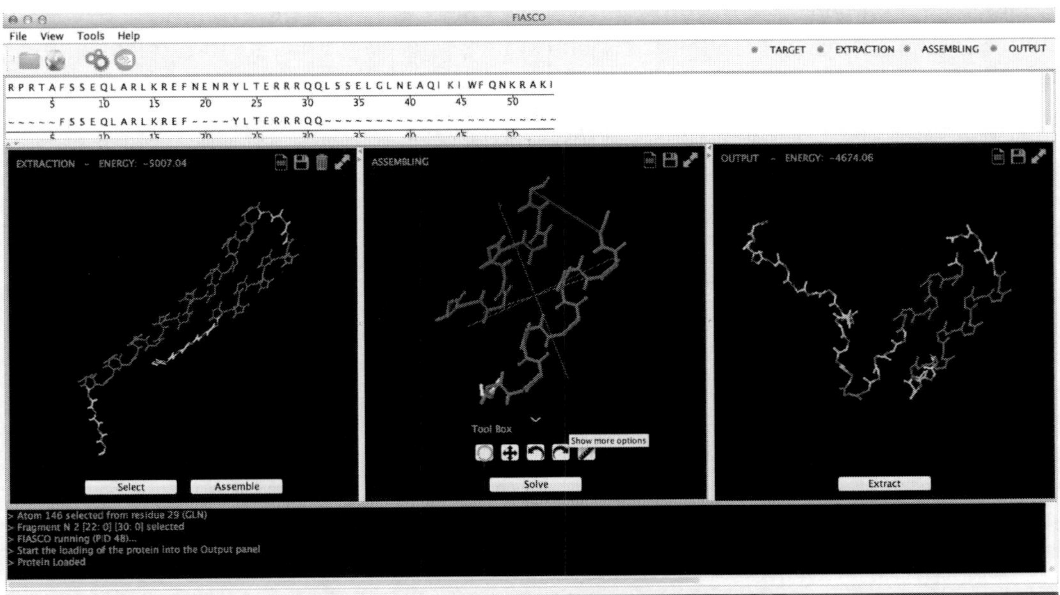

Figure 7. A screenshot of the FIASCO interface.

A fully functional version of the tool is available on the web http://cs.nmsu.edu/fiasco.

We are expanding the tool with other capabilities and therefore we refer the reader to the website for the latest version.

7. Conclusion

In this chapter, we provided an overview of the use of constraint-based methodologies to guide studies of protein flexibility. Constraints offer a high level modeling of proteins and their components, along with efficient methodologies to derive protein structures. The combination of dedicated procedures for pruning infeasible structures and the use of parallelism to distribute part of the resolution process to different processors allow constraint programming to achieve excellent performance.

We showed details about specific constraints that can be combined together to describe a general model for protein flexibility and discussed an example of application to a real-case investigation. The framework can be adapted to several case analyses, where only the declaration of constraint is modified, while the actual constraint solver requires no modifications. The constraint definition allows a clean, simple yet powerful abstraction of the model and it allows to focus on critical aspects of the investigation (e.g., a specific geometric relationship between some amino-acids). Moreover the solution search does not necessarily depend on energetic scoring, as opposed to MD. This can be helpful in contexts where multiple energetic model are to be compared and where the search should not be biased by a specific energy function.

The addition of parallel computing capability on GPUs shows that this recent technology is suited for speeding up the search at a affordable price, compared to traditional cluster of processors.

The methodologies described in this chapter have been implemented in the context of different systems, available to the community, and enhanced with effective graphical user interfaces.

Acknowledgments

The authors wish to acknowledge the help, suggestions and contributions of Federico Campeotto, Ferdinando Fioretto, Federico Fogolari, and Matteo Dusefante. The research has been partially supported by INdAM GNCS 2014 and 2015 grants and NSF grants CBET-1401639, HRD-1345232, CNS-1337884, and DGE-0947465. Hardware partially supported by NVIDIA.

References

[1] GECODE: Generic Constraint Development Environment. `http://www.gecode.org`, 2014.

[2] R. Agarwala, S. Batzoglou, V. Dančík, S. Decatur, S. Hannenhalli, M. Farach, S. Muthukrishnan, and S. Skiena. Local rules for protein folding on a triangular lattice and generalized hydrophobicity in the hp model. *Journal of Computational Biology*, 4:275–296, 1997.

[3] R. Backofen and S. Will. A Constraint-Based Approach to Fast and Exact Structure Prediction in 3-Dimensional Protein Models. *Constraints*, 11(1):5–30, 2006.

[4] M. Berrera, H. Molinari, and F. Fogolari. Amino acid empirical contact energy definitions for fold recognition in the space of contact maps. *BMC Bioinformatics*, 4(8), 2003.

[5] F. Campeotto, A. Dal Palù, A. Dovier, F. Fioretto, and E. Pontelli. A Constraint Solver for Flexible Protein Model. *Journal of Artificial Intelligence Research*, 48:953–1000, 2013.

[6] F. Campeotto, A. Dovier, and E. Pontelli. Declarative Concurrent System for Protein Structure Prediction on GPU. *Journal of Experimental & Theoretical Artificial Intelligence*, 1:292–300, 1995.

[7] F. Campeotto, A. Dovier, and E. Pontelli. Protein Structure Prediction on GPU: a Declarative Approach in a Multi-agent Framework. In *International Conference on Parallel Processing (ICPP)*, pages 474–479. IEEE Computer Society Press, 2013.

[8] M. Carlsson and P. Mildner. SICStus Prolog: the First 25 Years. *Theory and Practice of Logic Programming*, 12(1-2):35–66, 2011.

[9] C. Clementi. Coarse-grained Models of Protein Folding: Toy Models or Predictive Tools? *Curr Opin Struct Biol*, 18:10–15, 2008.

[10] Christian Cole, Jonathan D. Barber, and Geoffrey J. Barton. The jpred 3 secondary structure prediction server. *Nucleic Acids Research*, 36(Web-Server-Issue):197–201, 2008.

[11] A. Dal Palù, A. Dovier, F. Fogolari, and E. Pontelli. Protein Structure Analysis with Constraint Programming. In Pietro Cozzini and Glen E Kellogg, editors, *Computational Approaches to Nuclear Receptors*, chapter 3, pages 40–59. The Royal Society of Chemistry, 2012.

[12] A. Dal Palù, F. Spyrakis, and P. Cozzini. A New Approach for Investigating Protein Flexibility Based on Constraint Logic Programming: The First Application in the Case of the Estrogen Receptor. *European Journal of Medicinal Chemistry*, 49:127–140, 2012.

[13] Alessandro Dal Palù, Agostino Dovier, and Federico Fogolari. Constraint Logic Programming Approach to Protein Structure Prediction. *BMC Bioinformatics*, 5:186, 2004.

[14] Alessandro Dal Palù, Agostino Dovier, Federico Fogolari, and Enrico Pontelli. Exploring Protein Fragment Assembly Using CLP. In Toby Walsh, editor, *Proceedings of the International Joint Conference on Artificial Intelligence, IJCAI*, pages 2590–2595. IJCAI/AAAI, 2011.

[15] Alessandro Dal Palù, Agostino Dovier, and Enrico Pontelli. A Constraint Solver for Discrete Lattices, its Parallelization, and Application to Protein Structure Prediction. *Software Practice and Experience*, 37(13):1405–1449, 2007.

[16] Alessandro Dal Palù, Agostino Dovier, and Enrico Pontelli. Computing Approximate Solutions of the Protein Structure Determination Problem using Global Constraints on Discrete Crystal Lattices. *International Journal of Data Mining and Bioinformatics*, 4(1):1–20, 2010.

[17] K.A. Dill. Dominant Forces in Protein Folding. *Biochemistry*, 29:7133–7155, 1990.

[18] I.P. Gent, I. Miguel, and P. Nightingale. Generalised arc consistency for the AllDifferent constraint: An empirical survey. *Artificial Intelligence*, 172(18):1973–2000, 2008.

[19] M. Häckel, H-J. Hinz, and G.R. Hedwig. Partial molar volumes of proteins: amino acid side-chain contributions derived from the partial molar volumes of some tripeptides over the temperature range 10–90oC. *Biophysical Chemistry*, 82(1):35–50, 1999.

[20] R.S. Hartenberg and J. Denavit. A Kinematic Notation for Lower Pair Mechanisms Based on Matrices. *Journal of Applied Mechanics*, 77:215–221, 1995.

[21] V. Heun. Approximate protein folding in the HP side chain model on extended cubic lattices. *Discrete Applied Mathematics*, 127(1):163–177, 2003.

[22] H.H. Hoos and E. Tsang. Local Search Methods. In *Handbook of Constraint Programming*. Elsevier, 2006.

[23] IBM. IBM ILOG. `http://www-01.ibm.com/software/info/ilog/`, 2014.

[24] J. Jaffar and M. Maher. Constraint Logic Programming: A Survey. *Journal of Logic Programming*, 19/20:503–581, 1996.

[25] A. Kolinski, P. Rotkiewicz, B. Ilkowski, and J. Skolnick. Protein Folding: Flexible Lattice Models. *Progress of Theoretical Physics*, 138:292–300, 2000.

[26] Ludwig Krippahl and Pedro Barahona. Psico: Solving protein structures with constraint programming and optimization. *Constraints*, 7(3/4):317–331, July 2002.

[27] K. Kuchcinski and R. Szymanek. JaCoP: Java Constraint Solver. `http://jacop.osolpro.com`, 2014.

[28] D. Poole and A. Mackworth. *Artificial Intelligence: Foundations of Computational Agents*. Cambridge University Press, 2010.

[29] Srivatsan Raman, Robert Vernon, James Thompson, Michael Tyka, Ruslan Sadreyev, Jimin Pei, David Kim, Elizabeth Kellogg, Frank DiMaio, Oliver Lange, Lisa Kinch, Will Sheffler, Bong-Hyun Kim, Rhiju Das, Nick V. Grishin, and David Baker. Structure Prediction for CASP8 with All-atom Refinement using Rosetta. *Proteins*, 77(Suppl. 9):89–99, 2009.

[30] Francesca Rossi, Peter van Beek, and Toby Walsh. *Handbook of Constraint Programming*. Elsevier Science Inc., 2006.

[31] J.A. McCammon S.A. Adcock. Molecular dynamics: survey of methods for simulating the activity of proteins. *Chem. Rev.*, 106:1589–1615, 2006.

[32] J. Schimpf and K. Shen. ECLiPSe - from LP to CLP. *Theory and Practice of Logic Programming*, 12(1-2):127–156, 2011.

[33] A. Shehu. *Conformational Search for the Protein Native State*, pages 431–452. John Wiley & Sons. Inc., 2010.

[34] M. Shen and A. Sali. Statistical Potential for Assessment and Prediction of Protein Structures. *Protein Sci*, 15:2507–2524, 2006.

[35] Univ. of Udine, CLP LAB. Workshop on Constraint-based Methods for Bioinformatics. `http://clp.dimi.uniud.it/wcb/`, 2015.

[36] P. Van Hentenryck and L. Michel. *Constraint-Based Local Search*. The MIT Press, 2005.

[37] W-J. van Hoeve. The alldifferent constraint: A survey. In *Proceedings Sixth Annual Workshop of the ERCIM Working Group on Constraints*, 2001.

[38] M.G.A. Verhoeven and E.H.L. Aarts. Parallel local search. *Journal of Heuristics*, 1:43–65, 1995.

[39] D. Waltz. *Understanding Line Drawings of Scenes with Shadows*. McGraw-Hill, 1975.

[40] A.A. Zamyatin. Protein Volume in Solution. *Prog. Biophys. Mol. Biol.*, 24:107–123, 1972.

In: From Medicinal Chemistry to Food Science
Editor: Pietro Cozzini ISBN: 978-1-63483-587-9

Chapter 10

A Quick Guide to Running Classical Molecular Dynamics on High Performance Computers

*Andrew Emerson**
Cineca Supercomputing Centre, Casalecchio di Reno, Italy

Abstract

Molecular dynamics (MD) simulations have been have been employed on computer systems for at least 50 years but it took at least a couple of decades before computers were powerful enough to make significant contributions to the modelling of molecules relevant to biological systems. Nowadays, researchers using MD programs, and in particular those performing simulations of biomolecular systems, represent one of the most important communities of High Performance Computing (HPC) facilities. However, the rapid development of HPC, which has seen an evolution from single processor machines to complex, multi-core hybrid clusters, means that modern facilities provide challenging environments to researchers and code developers alike. In this Chapter we aim to help researchers to get started with classical molecular dynamics simulations for biomolecular systems, with a particular emphasis on how they are implemented on current parallel supercomputers. We also describe some of the difficulties which might be encountered, in particular how the requirement for very high computational parallelism is sometimes incompatible with the projects which researchers wish to pursue.

PACS: 05.20.-y, 02.70.Ns

Keywords: Statistical Mechanics, Molecular dynamics, HPC

Introduction

Just about any organisation or project involved in supercomputing, or more generally, high performance computing (usually referred to as HPC), needs to know what applications its users or clients are using on their systems. Data giving information on which programs

*E-mail address: a.emerson@cineca.it

users are running, or at least wish to run, are important input in hardware procurement and for tailoring the services provided by the organisation. There are various mechanisms for obtaining this information, e.g. with user surveys, from successful grant applications or direct monitoring on the computer systems, and often computer centres publish such data on their web sites. From these results it is clear that one of the most important classes of programs used on HPC systems is that of molecular dynamics (MD), and in particular classical MD, i.e. where the underlying physics is based on Newtonian mechanics rather than on quantum mechanics. In fact, popular programs such as GROMACS and NAMD have for years been two of the most commonly used HPC applications as indicated by many sources. This is not to undervalue the importance of quantum chemistry or other fields in computational science (astrophysics, bioinformatics, engineering, etc) but just to emphasise how closely linked classical molecular dynamics and HPC have been and will continue to be in the foreseeable future. Indeed it is common to include MD programs in so-called benchmark suites during hardware procurement to ensure a certain level of performance will be attained (for an example, see the PRACE benchmark suite [1]). Such suites are useful because the usual metrics quoted by manufacturers, such as the peak performance based on hardware data (e.g. CPU clock speed), or standard performance analysis packages, such as LINPACK which performs linear algebra calculations, do not accurately reflect the performance a typical user may expect from their own projects. Knowing the expected performance of an application code before embarking on a project is important for many reasons not least because resource funding agencies require this information before granting computing time.

Given that modern MD programs are very versatile, capable of modelling a wide variety of chemical and physical systems (e.g. biological, inorganic, polymeric, etc) do we know what users are doing with their MD applications? This information is harder to obtain because we do not have any automatic procedures which tell us so we must rely on project applications and user feedback alone, but we can say that on supercomputers classical MD is important and in particular those involving projects of biological or biochemical relevance (e.g. medicinal chemistry). In the remainder of this chapter we will try and understand why these topics, i.e. molecular dynamics of biomolecular systems and HPC, are so inextricably linked and along the way give pointers to the beginners on how they can get started with running classical molecular dynamics simulations on HPC systems. But we will also stress the limitations which the field faces and the challenges arising from developments in HPC systems.

Why Use Molecular Dynamics?

There are many excellent articles and books which can give an introduction to classical molecular dynamics [2, 3] as well as on-line resources but it is recommended that for those users wishing to embark seriously on an simulation project a good background of statistical mechanics is essential, a point that may not be welcome to those researchers more used to working in the laboratory. In fact, a quite surprising number of support requests sent to application mailing lists are actually basic questions about statistical mechanics. On the other hand classical MD is one of those fields where the essential principles can be understood without mastering too many formulae or difficult concepts since we are just trying to model an atomic system as it evolves in time. This could be contrasted with the Monte Carlo simulation method, for example, which is quite difficult to follow without a basic comprehension of statistical mechanical concepts such as ensembles, phase space, ergodicity, etc.

Although a wealth of thermodynamics data are available from the numerical output of even simple MD applications, most researchers in medicinal chemistry focus on what happens to the arrangements of atoms or the molecular structures which are being modelled, information which may be difficult to obtain from experimental techniques. In fact, probably the most important result from a molecular dynamics simulation is the *trajectory*, essentially a file, or a sequence of files, containing a series of snapshots of the atomic coordinates taken at regular intervals in time. From the trajectory it is thus possible to discern what happened during the simulation run to our molecular system (e.g. perhaps conformational or structural changes) in the corresponding time interval. Given that we know that biological processes may require long time scales compared to many chemical reactions (e.g. milliseconds, seconds or longer for protein folding, for example), an important question to ask is whether we can simulate on a time scale long enough to study the processes we are interested in. This is important because even with supercomputers, trajectory lengths rarely extend beyond the hundred nanosecond regime (we recall that $1\text{ns}=10^{-}9\text{s}$). Indeed, this chapter is centred around to what extent developments in algorithms and computational science can help us to generate a trajectory as long as possible.

Why HPC?

Before looking at the role of HPC in MD it is worth considering the history of the technique and some of the milestones along the way. MD is a discipline with a history which stretches back almost to the introduction of the first supercomputers with the hard sphere simulations of Alder and Wainwright in 1959 [4]. A few years later the first simulation of a continuous potential (i.e. Lennard-Jones particles) was performed [5] but because of the computational power required to model more complex molecules, the first reported simulation of a biomolecule had to wait until 1977 with an MD study of the Bovine Pancreatic Trypsine inhibitor by McCammon, Gelin and Karplus [6]. Other milestones include the first microsecond simulation in 1998 [7], the MD of the complete satellite tobacco mosaic virus (STMV) in 2006 [8] and in the same year the 500 microsecond run of the Villin Headpiece protein by the Folding Home project [9]. In recent years, a supercomputer dedicated to running molecular dynamics built by D.E. Shaw Research has performed a number of millisecond trajectories [10].

Computers are clearly needed for MD but given the current capabilities of modern laptops, tablets or even multi-core smartphones, and assuming that we are not aiming at some sort of record or milestone, do we still need supercomputers? If we just want to see if the MD application with our molecular system runs then a standard personal computing device is normally sufficient. The reason is that unlike other computational science calculations (e.g. quantum mechanics), memory and I/O requirements are very modest and with the exception of very large input data sets, the simulation usually can be run easily within the hardware resources of a laptop, for example. This is a big advantage of MD - you can experiment first on a simple personal computer before starting on the HPC system. The advantage provided by HPC is therefore primarily of massive number-crunching power, allowing the researcher to run the simulation for long times and, therefore, be able to model more physical or biological phenomena. A secondary advantage is that of being to able to model very large systems, perhaps with millions of atoms, but from our experience few researchers in medicinal chemistry want to do this even though hardware developments actually favour larger atom counts over simulating smaller systems for longer periods, as we shall see. Obviously, having powerful computers does not compensate for poor science but it is undeniable that researchers with better resources will have a competitive edge over those with fewer.

Getting Started with Molecular Dynamics

In this section we give a very brief outline of the steps needed to pursue a simulation project. More complete information, often in the form of tutorials, is readily available, particularly from user guides and forums of the major software packages.

Overall Workflow

A schematic workflow describing the overall procedure of a simulation project is given in figure 1. There are essentially three phases in this procedure:

1. System preparation.
2. Running the simulations to generate trajectories.
3. Analysis of the trajectories and other results.

The system preparation stage does not need to be done on a HPC system, in fact it is probably easier not to, while for the analysis it may well be more convenient to perform it on the cluster where the trajectories were produced, to avoid large file transfers. In the following we describe in more detail some of the more important aspects of these stages.

Choosing the Software and Force Field

One of the first decisions to make at the beginning of a molecular dynamics project involves the software to be used. One reason for making the decision early on lies in the fact that not only do the various software programs vary greatly in capabilities and features but also because there is usually limited compatibility between the different MD build and run systems. The incompatibilities arise from the lack of standardisation in file formats which in practice means it is difficult, if not impossible, to use files generated for one program with a different one. It should be mentioned that there have been numerous attempts to present common formats for representing chemical information, the XML-based Chemistry Markup Language (CML) format [11, 12] being one such example, but at least in the field of computational chemistry there has never been widespread adoption of any particular format. The closest we have to a standard is the Protein Database (PDB) file originally designed for storing the coordinates of atoms in a protein measured by X-ray diffraction experiments,

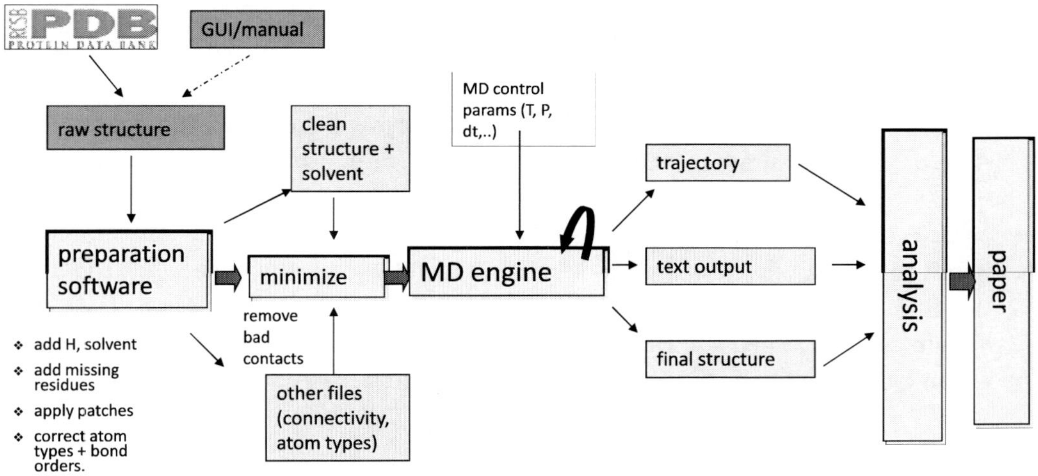

Figure 1. Schematic workflow of the procedure required to prepare and run a molecular dynamics simulation.

but the format is rigid and limiting and, although many programs may be able to read PDB files at the beginning of a run, they are rarely used later on in the simulation. For other types of information needed for or generated by MD simulations such as molecular topologies, coordinate files, trajectories, etc, each software has its own format for representing the data.

So knowing that we must choose early on what sort of factors determine the software we should use for our project? Four general considerations include:

- compatibility with the system we want to use (e.g biological, metallic, polymers, etc);
- includes the protocol or level of theory (e.g. ensemble method, implicit solvent, etc);
- HPC suitability (e.g. single processor performance, parallel scalability, GPU compatibility)
- cost, availability of source code, ease-of-use and support.

An important consideration as regards system compatibility is whether the software supports the *force field* necessary to model the interactions in the system. Force fields are carefully curated sets of formulae and parameters designed to represent accurately as possible the interactions within a molecular system. Being based on molecular mechanics, rather than quantum mechanics, the energy of the system is based on the nuclear positions only, ignoring the motions of the electrons. Thus, properties depending on the electronic distribution cannot be calculated, but force fields require considerably less computer time than quantum mechanical methods which are still too expensive for most, practical biomolecular simulations (except perhaps for small regions of the biomolecule, such as in hybrid QM/MM simulations). Force fields tend to come in "families", designed for particular classes of molecules (e.g. AMBER for proteins or nucleic acids) or situations (e.g. MMF for small molecules in organic or protein environments[13]). They are also derived in different ways, often employing both experimental and computational (quantum mechanical) data; being a sensitive component of the simulation process a user should not attempt modifications of the force fields that come with the software. in biomolecular simulations the most common force field families are Charmm, Amber, GROMOS and OPLS.

It is worth commenting also on the fourth point above since most MD packages are either free or available at a relatively low cost, at least for non-profit research. In an HPC environment, source code availability is essential in order to obtain the best performance and compatibility for a particular hardware and software set-up. Software vendors, on the other hand, rarely provide the source code with the result that the applications are restricted to a very limited set of computer environments (e.g. personal workstations). The downside of course of free, open source applications, is that installation may be non-trivial and support can only usually be sought from forums or mailing lists.

System Preparation

This is probably the phase where expert help is most needed and can be quite complex but the procedure is broadly as follows: all classical molecular dynamics simulations need to start from a set of atomic coordinates and usually also data structures which describe the "connectivity", i.e. how they are bonded together to form molecules. The coordinates of atoms used in biomolecular simulation are normally derived from X-ray diffraction databases, such as the RCSB Protein Data Bank (or simply PDB). However, structures from the PDB may contain errors, such as in stereochemistry or peptide bond configuration, for example, and so some manipulation of the data may be necessary. Once the correct molecular structures have been generated, the structures are then "force field typed", i.e. processed with the chosen force field to ensure that all the possible interactions can actually be modelled during the simulation. It is then usual to solvate the system, by adding water molecules, and if simulating physiological conditions, add Na+ and Cl- ions to the correct concentration (e.g. 0.1M). If the Particle Mesh Ewald method is used (see later in this Chapter) then the numbers of Na+ and Cl- ions need to be chosen to ensure overall electrical neutrality - a necessary requirement for this method. It is then essential before starting the dynamics to run a *minimisation* protocol in order to remove bad overlaps, i.e prevent atoms from being too close to each other, which in an MD run can generate large forces and not only generate errors in the structure, but also cause the program to crash. Long minimisation runs should be avoided, however, as they can bring structures too far away from the structures likely to be expected under normal MD simulation conditions.

Typical Simulation Parameters

Once the system has been prepared, the user will have a number of data files containing information on the atomic coordinates, connectivity and possibly also force field parameters. The next step is then to construct the command file used by the program to run the simulation. The commands or parameters will of course be program-dependent and there may be many of them, but they are generally grouped in sections and control simulation conditions and some of these are described below.

- **Simulation box and periodic boundary conditions.** In one section of the command file parameters concerning the spatial dimensions of the simulation will be specified. In brief, atoms and molecules are simulated in a simulation "box" and usually with so-called periodic boundary conditions (PBC), which are used to approximate an infinitely large system by surrounding the main box with replicas of the system (or *images*). Thus, if an atom leaves the box through one side, it will reappear on the opposite side of the box. With PBC it is normal to apply the *minimum image convention* where each atom in the simulation interacts with the closest image of the remaining particles in the system. It is for this reason that for simulations of a single, large molecule such as a protein in water it is necessary to surround the molecule with a layer of water which is thick enough to prevent the protein from interacting with itself. Note that is not obligatory to use Periodic Boundary Conditions, it is possible to simulate a molecule in a sphere of water, for example, but they are needed if PME is used for the electrostatic interactions.

- **Temperature and pressure.** For biomolecular simulations, the temperature and pressure are normally set to ambient conditions, i.e 298K and 1 atmosphere pressure, and various algorithms ("thermostats" or "barostats") for achieving these will be available. Note however that it is generally inadvisable to use temperatures far from 298K since most force fields are validated near this temperature and so strictly speaking will not be valid.

  For most packages it will also be possible choose the statistical mechanical ensemble of the simulation, i.e. the canonical or NVT ensemble with constant number of particles (N), volume (V) and temperature (T) or the isothermal-isobaric or NPT ensemble, where instead of the volume the pressure P is kept constant. The utility of the NPT ensemble is that it allows the box size and hence the density to change. Thus, since immediately after the system preparation phase the system density may not be correct for the simulation conditions, it is usual to equilibrate the density with NPT before running with NVT (which may be faster). Constant pressure simulations are used also where the box shape may change appreciably during the simulation, such as in a protein unfolding experiment.

- **Time step and related parameters.** MD simulations generate atomic trajectories by solving Newton's laws in discrete cycles or steps, where the time interval between each cycle is given by the *time step*. From physical considerations, the time step has to be shorter than the time related to the highest frequency motion in a molecule, which is normally a carbon-hydrogen bond vibration, so it is of the order of a femtoseconds ($1\text{fs} = 10^{-13}\text{s}$). Since it is desirable to have a time step as long as possible, often a value of about 1-2fs is used.

- **Electrostatics.** Most applications allow the user to decide on the protocol to use for calculating the electrostatic interactions in the simulation. The most commonly used option is the PME method and is described in more detail later in this chapter. If water is involved, as is commonly the case in biomolecular simulations, then a user might wish to apply an *implicit solvent* method, such as the Generalized Born solvent, which replaces the explicit water interactions with a model which approximates the solvent environment [14]. The advantage of implicit solvent models is that without explicit waters, the simulations run much faster. However, they are considered less accurate than explicit models and in MD their use is generally discouraged, particularly where the water is expected to play a structural role [15].

- **I/O management.**

  Here the user will decide various options regarding the reading of input and writing of output from the simulation. Most options will be fairly self-explanatory but for HPC users it is worth remarking that although most MD simulations are not I/O intensive, writing frequently to trajectory files is not recommended. Not only is it not necessary, since atomic configurations taken from successive time steps are anyway very highly correlated, but the post-processing of large trajectories containing many gigabytes of data is difficult and time consuming. In addition, many trajectory files use legacy,binary formats which do not allow the storage of *metadata*, i.e. important information describing how the simulation was done, rather than just the atomic coordinates and velocities. Thus, a user will probably have to use with great care descriptive file and directory names to record data such as the simulation temperature, pressure, molecule name and so on.

Analysis Strategy

As in many computational science fields, analysis of the simulations can continue for a long time after the last simulation has finished. For molecular dynamics it is usual to analyse directly the trajectory file or files produced by the simulation. The most obvious analysis method is simply to visualize the trajectory with an appropriate tool (e.g. VMD, Pymol, etc) to ensure that the molecules have at least behaved in a physically plausible manner. This is one reason why, it is generally advisable not to generate extremely large trajectory files if they have to be downloaded or transferred in order to be visualized. Many computer centres, on the other hand, provide remote visualization facilities to avoid such large data transfers[16]. Other analyses can usually be performed rapidly directly on the HPC system.

Moving up to HPC

As mentioned above, with the exception of very large systems, the first few tests with the newly prepared system can be run on a personal computer but serious work needs to be transferred to the HPC system. The transition from personal workstation to a HPC cluster can be traumatic. The hurdles a researcher is likely to meet include:

- The need to learn an unfamiliar user interface such as a UNIX (usually Linux) command line interface.
- Obtaining familiarity with a potentially complicated computing environment and architecture. For example, it is essential to know details such as how many nodes are available, the number of cores and memory per node, etc. Many modern computer systems even have nodes of different types or memory capacities, e.g. so-called "thin" or "fat" nodes or "islands" with accelerators.
- The need to use a batch system to perform even small tests.

- Other subtleties of HPC systems and local administration policies on system access (perhaps with personal certificates), use of file systems and archiving data.
- Non-availabiity of the computer systems due to hardware or software problems (HPC hardware is generally less reliable than the services used in commercial environments).

It is worth mentioning of course that these difficulties are common to all those wishing to use HPC, not just researchers in molecular dynamics, but since there are few problems in MD which cannot be adequately handled with non-HPC resources all those wishing to work in the field will be forced to confront them. MD has the advantage, as mentioned earlier, that small test runs can be run on a laptop but for meaningful results production needs to be switched to a supercomputer.

The many obstacles to the exploitation of HPC by scientific researchers, particularly by those more used to working in a laboratory environment, has long been recognised and seems particularly acute now that the first exposure to computing for many young researchers is through smart devices such as tablets and smart-phones, as the sales of traditional PCs decline. One of the great hopes for removing the barriers to HPC has been Grid Computing with its aim to provide seamless access to heterogeneous computing resources, hiding complexities of operating systems, batch systems, etc. via more user-friendly interfaces such as web-portals. Despite considerable investment, Grid computing as a direct replacement for accessing HPC never really became successful, instead mutating into Cloud services mainly aimed at businesses and later for backing up personal data from smart devices. Another example is the UNICORE system [17] which has been adopted in EU flagship projects such as the Human Brain Project [18] and found particular relevance in projects such as the MMM@HPC initiative where it was used to construct complex workflows useful for multiscale materials modeling [19].

But despite the availability of say UNICORE at Cineca and other PRACE centres, for example, virtually all usage to the academic resources is via the command line SSH protocol to access the Unix environment and this is mirrored at other sites within the PRACE project. Thus, the standard procedure for running an MD simulation on a supercomputer is normally as follows:

1. Log on to the supercomputer which is usually via an SSH client from the user's workstation and with a username and password provided by the computer centre or resource provider. However, note that some centres, for security reasons, may require a different protocol (e.g. gsissh + X.509 certificate) and/or registration of the IP address of the user's terminal.
2. If not already present, copy or transfer the files necessary for the MD run (e.g. for the atomic coordinates, topology, force field, etc) to the HPC resource.
3. Prepare and submit the batch job.
4. When the batch job has completed, check carefully the output to make sure the simulation ran as expected and start the analysis or re-submit further jobs as necessary

In this procedure the third step can be the most difficult and frustrating for many users: the difficulty comes from having to know what resource requirements to ask for in the batch job (e.g. total elapsed time or "walltime", number of processor cores, memory, etc), while the frustration can come from having to have your job wait days in batch "queue" only to discover some error in the input or the job preparation which causes the MD run to terminate prematurely.

The need for a good estimate of the walltime is particularly crucial since asking for too much time will probably give the job a lower priority as most batch systems are configured to favour jobs which require shorter walltimes (and fewer hardware resources such as processor cores). Requesting too small a walltime, however, can result in the batch system simply "killing" the run if elapsed time is insufficient for the time steps specified, with the possible loss of results and data. Even though most MD simulation programs allow restarts by periodically writing the current configuration to file (e.g. every 20 time steps), it is advisable to choose the batch walltime and number of time steps such as the run finishes correctly before the end of the job.

The other main difficulty is the choice of the number of the processors to use for the run. We know that to exploit HPC we need parallel programs and all common MD programs used on supercomputers are parallel, but the optimum number of processor cores to use may not be obvious. In principle, for a program which for a given input is "embarrassingly parallel", i.e. where for example the performance P_N doubles on doubling the number of processors for any value of N, the minimum time-to-solution would be obtained by asking the maximum N available on the hardware. However, when there is a batch system this may not be the optimal choice because the system will certainly be configured to prioritize jobs requiring fewer processors. Indeed many users make the mistake of submitting single jobs requesting high numbers of processor cores and long walltimes when in most cases it would be more efficient to submit many jobs requiring fewer cores with short walltimes. Since in MD we usually want to simulate a trajectory for a particular molecular system, these jobs are not independent and must be run in sequence, but most queuing systems permit so-called "job chains" where one job is run only after the successful completion of the previous one. The optimum number of processors to use also depends on the parallel scalability of the simulation since for a given input no MD program is embarrassingly parallel. Consider, for example, the parallel scaling curve given in figure 2. This variation of performance with the number of cores N, i.e. a nearly linear scaling regime for small N followed by an attenuation to a maximum, is typical for most MD programs.

We shall discuss the reasons for this later but for the moment we notice that $N = 64$ would not be a good choice since this would give a performance lower than the maximum possible which occurs for 32 cores. But even $N = 32$ may not be the best choice since the relative increase in performance is fairly small compared to the number of processors needed to produce the increase from $N = 16$. This is is important because all computational projects have an associated CPU time budget (usually in terms of core hours) and a researcher needs to consider not only the fastest way of achieving the project's goal but also whether it can be achieved within the budget available.

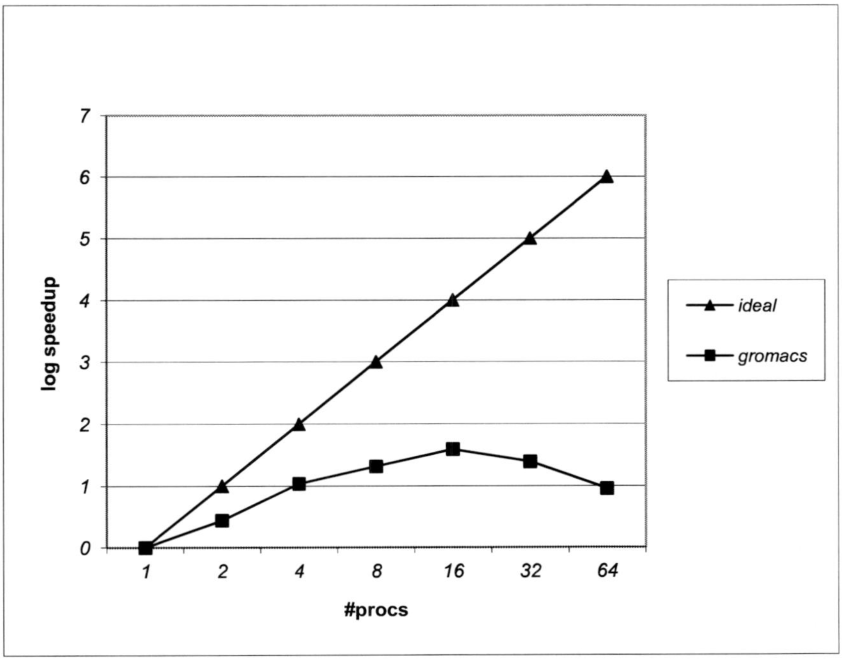

Figure 2. Parallel Scaling results for GROMACS v.3.3. The triangles indicate perfect ("ideal") scaling behaviour.

In fact computational scientists would describe the point P_{64} as having a low *parallel efficiency* E_N which is defined as

$$E_N = \frac{P_N}{N \times P_1}$$

where P_1 is the performance at one core. In this example the parallel efficiency drops to only about 19% at 16 cores. on going from N=64 to N=32 cores. [1] This metric is worth calculating since although in molecular dynamics a researcher is usually more interested in the absolute performance calculate in terms of ns/day, in order to gauge how much time will be needed to generate a trajectory, the parallel efficiency is domain-independent and is often required by funding agencies in project proposals. This example demonstrates how important it is to benchmark a simulation *before* production runs are started and also why many funding agencies require this information in the project proposal.

Efficient Use of HPC - How MD Programs Are Parallelised

Nowadays, with the availability of high quality, parallel MD programs, which are often free and open source, researchers are rarely called upon to write software from scratch or even

[1]We must stress that current versions of GROMACS perform much better than the old version used in this example.

modify existing packages. A danger though of using these packages is that while the researcher may investigate the features available, the algorithms utilised are not studied - this is equivalent to a lab researcher not understanding the how a particular piece of equipment works. Thus the incorrect use of the program can lead to inefficient use of time (both "real" and CPU time) and poor science deriving from incorrect results.

Given the HPC focus of this chapter, here we shall illustrate some of the algorithms relevant to implementing MD programs on supercomputers with a particular emphasis on parallelisation. Most researchers will eventually seek a more detailed knowledge of current parallelisation paradigms such as message passing (e.g. MPI) and/or shared memory (e.g. OpenMP), but to start it is sufficient to understand the concepts rather than be able to write programs using them. Here we merely give an outline -for a full discussion the reader is referred to the many resources (e.g. [20]).

Non-bonded Interactions

In a serial, molecular dynamics program, often 70-90% or more of the total CPU time required is consumed in the calculation of the energies and forces between the atoms, in particular the non-bonded interactions. These come in two flavours: dispersion-type interactions, often represented by a form of the Lennard-Jones potential, and electrostatic forces, which will be discussed in the next section. For a system of N interacting atoms without electrostatic forces (e.g. liquid argon), and assuming pairwise interactions, there would be in principle $N(N-1)/2$ interactions to calculate in a double loop; this N^2 dependence explains the difficulty in simulating large numbers of atoms. For biomolecular systems this dependence is especially important because of the need to simulate in physiological conditions which means including a large number of water molecules. However, in practice the total time can be reduced and the N^2 dependence brought closer to an N dependence by the use of potential cutoffs, methods such as united atoms or holonomic constraints such as Shake [3] and specialised water models, as well as "housekeeping" tools such as neighbour lists and linked cells, which make use of the cutoffs to reduce the number of interactions to calculate. The water molecules can be removed altogether by implicit solvent models, but as mentioned earlier this approximation is not widely accepted. An additional approach to reducing the number of N interacting sites, is to use a so-called coarse-grain approach, such as the Martini force field [21]. Here groups of atoms are combined into "beads", thereby reducing considerably the number of atoms to be modelled. Force field Models such as MARTINI have been particularly useful in large systems such as lipid bilayers but clearly they cannot be used where atomic-level detail is necessary (e.g. protein folding).

Atom and Force Decomposition

Regardless of optimisations, the "core" of a MD program will contain an energy-force loop for calculating up to N^2 pair interactions. One of the earliest implementations of parallelism in MD was straightforward and acted on the force loop. It was called atom or particle decomposition because the N iterations of the loop are divided equally the among the P processors. Thus each processor calculates the interactions between its own set of N/P atoms and all the others. It was thus also called replicated data because each processor requires a copy of the entire system, i.e. N sets of atomic coordinates, which need to be updated after every step. Early parallel MD programs which used atom decomposition were easy to program and worked quite well but once higher number of cores became available, the failings of the algorithm became apparent: parallel scaling would stop at relatively low core counts, perhaps 8 or 16. One deficiency of the method is the memory necessary to replicate the data on each processor, but the other main problem would be familiar to parallel programmers, namely the communication step needed to update all the processors. The communication required is called "all-to-all" since each processor must contact every other processor. In parallel programs, communications between processors eventually stops parallel scaling in most cases and the all-to-all type communications needed for the updating in atom decomposition are particularly expensive. Because of this atom decomposition is rarely used nowadays, except for very small systems.

An evolution of atom decomposition was force decomposition which recognised that computing the double force loop was equivalent to solving an $N \times N$ interaction matrix. In analogy to the parallelisation of matrix problems, force decomposition divides the square matrix into P sub-matrices and assigns a processor to each one. With respect to atom decomposition, each processor calculates fewer total interactions ($=N^2/P$) and no longer needs to replicate the entire system. In this way the communication costs are reduced and the method fares better than atom decomposition but still the parallel scaling is relatively poor and like atom decomposition, is rarely used.

Domain Decomposition

The atom and force decomposition both suffer from the fact that they do not exploit the *locality* of interatomic interactions, many of which are short-ranged. Thus, for the Lennard-Jones, variants of which are often used to simulate non-bonded, dispersion interactions, the energy of interaction can be close to zero after only few atomic diameters. This explains why a cutoff of a few angstroms can be safely used in MD simulations for such forces. This also means that for a simulation box with sides of say, 100 Å the vast majority of the $N(N-1)/2$ pairwise interactions can be safely neglected. Spatial or domain decomposition exploits this fact by dividing up the three dimensional space into sub-spaces such that each region has sides greater than the cutoff r_c (see figure 3).

If a processor is assigned to each region then each processor will only need to compute the interactions of the atoms within its own space and with those in the nearest-neighbour spaces. In three dimensional space, each processor therefore only needs to communicate with at most 26 other processors of the nearest-neighbour domains. Performance and parallel scalability can be enhanced significantly as demonstrated by the graph in figure 4 which compares two versions of the GROMACS program: the version 3.3.3 which used force-decomposition and version 4.0, where domain decomposition has been implemented. Even allowing for improvements in other areas of the program, the increase in performance and parallel scalability are notable and can mainly be attributed to the superior domain decomposition algorithm.

It should be mentioned that from a programming point of view, domain decomposition is significantly more complex, but since the algorithm is well known in computational science and used in many other fields (e.g. astrophysics, finite element codes, etc) parallel libraries such as MPI have many facilities for writing these codes and are well optimised.

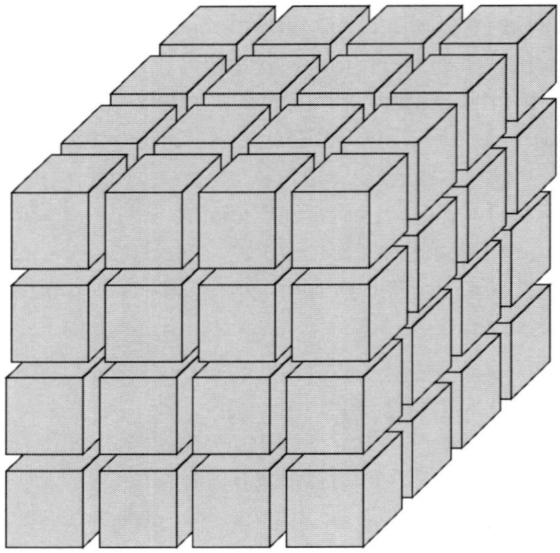

Figure 3. Schematic representation of domain decomposition. The simulation space is divided into sub-domains, where the side of each sub-domain r_d is less than the cutoff, i.e. $r_d < r_c$. Each sub-domain is then assigned to a processor core.

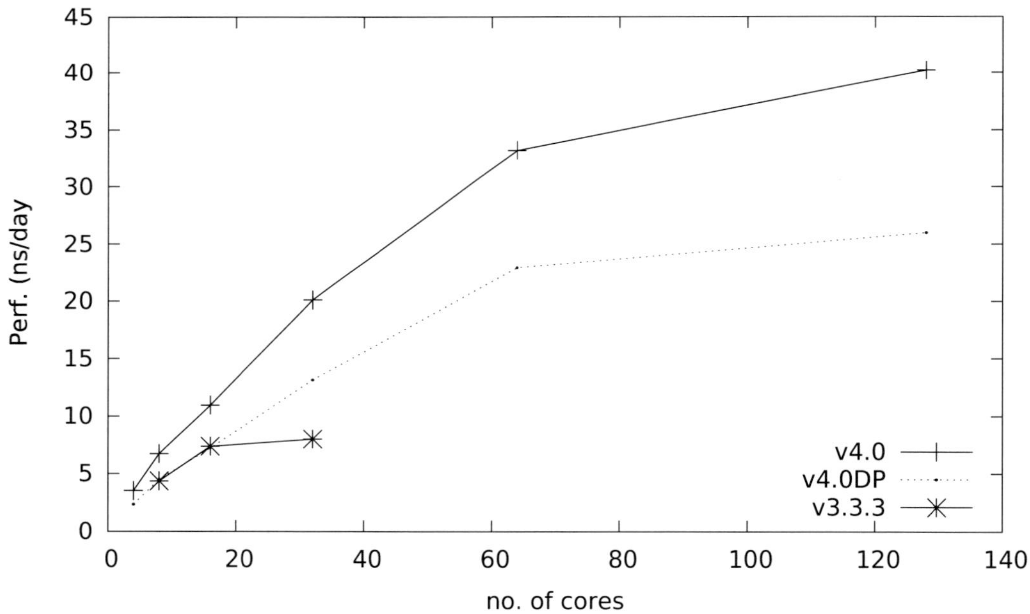

Figure 4. Comparison of GROMACS versions 3.3.3 and 4.0. The dotted line shows the results for the double precision version of GROMACS.

Nowadays, all MD programs offer domain decomposition as a parallelisation scheme, usually as the default option.

One weakness of the method is apparent when the atomic system is inhomogeneous, perhaps during a simulation of a liquid and its vapour, such that the sub-domains have very unequal numbers of atoms. If this happens, then corresponding processors have unequal workloads and the system is not "load-balanced", leading to loss of performance. Modern programs such as Desmond, GROMACS and NAMD, offer a variety of more sophisticated DD algorithms together with *dynamic load balancing* procedures which monitor and if necessary re-allocate atoms to processors to ensure a good load balancing. Another point of course is that the system must have sufficient atoms to distribute to the processors but we shall discuss this further later in this chapter.

Electrostatics and Particle Mesh Ewald

With the exception perhaps of the modeling of simple, atomic liquids (e.g. liquid argon) it is hard to imagine a simulation which does not take account of electrostatic forces. However, while non-bonded interactions can be modelled fairly well with a Lennard-Jones type potential and parallelised efficiently in a domain decomposition scheme, forces arising from electrostatic interactions present a different set of problems. This is quite a complex topic so we shall only illustrate the most common approach and additionally, only quite formulae when necessary to understand the concepts.

In many simulation procedures the electrostatic field of a molecule is represented by a set of partial charges which depend on the atom and on the environment. These partial charges form part of the force field and are usually assigned during the preparation of the simulation input. In principle, it should be quite straightforward to compute the interactions with these partial charges since the interaction energy between two isolated point charges is available from the coulomb potential,

$$V_{ij} = \frac{q_i q_j}{4\pi\epsilon_0 r_{ij}}$$

where q_i and q_j are the charges on atoms i and j, r_{ij} the separation between the two atoms and ϵ_0 is the electric constant The force is easily obtained from

$$F_{ij} = \partial V_{ij} / \partial r_{ij}$$

The problem though is that these interactions are long-ranged, varying as $1/r$ and therefore decay to zero quite slowly - this behaviour should be compared to the non-bonded dispersion forces which vary as $1/r^6$. In fact, for all usual box sizes, the interaction energy is significantly different to zero at the box edges and the electrostatic potential cannot be truncated without resulting in non-physical behaviour such as cutoff "artifacts". A solution to this problem for periodic systems was discovered in 1920 by the German physicist Paul Ewald who was studying ionic crystals. He started with the expression for the interaction of all other particles, including their images:

$$V = \frac{1}{2} \sum_{n} \sum_{i=1}^{N} \sum_{j=1}^{N} \frac{q_i q_j}{|\mathbf{r_{ij}} + \mathbf{n}|}$$

where

$$\mathbf{n} = (n_x L, n_y L, n_z L)$$

and $\mathbf{n}$ is a lattice vector for a box of side L. This pairwise summation converges slowly, but by assuming gaussian charge distributions around each charge it can be converted into faster converging real space (short range) and reciprocal space (long range) sums:

$$V = \text{real space sum} + \text{reciprocal space sum} + \text{constant corrections}$$

The real space term can be calculated quite readily with standard libraries while the reciprocal space term consists of a double loop over N:

$$V = \frac{1}{2} \sum_{k \neq 0} \sum_{i=1}^{N} \sum_{j=1}^{N} \frac{4\pi^2 q_i q_j}{L^3 k^2} \exp\left(\frac{-k^2}{4\alpha^2}\right) \cos(\mathbf{k}.\mathbf{r_{ij}})$$

This series converges quite quickly but is nevertheless computationally expensive because of the N^2 dependence on the number of charges. The calculation of the formula can however be accelerated by invoking an efficient computational procedure known as the Fast Fourier Transform (or FFT) which converts a quadratic dependence to a logarithmic one, namely $N \log N$. A complication is that the FFT routines available are *discrete*, i.e. they need the data values to be sampled from a regularly spaced grid or lattice, but this can be achieved without too much difficulty by interpolating the charge distribution onto a grid before passing the data to the FFT library. The resulting algorithm is called "Particle Mesh Ewald" or PME, and is a standard option in most MD packages. From a performance point of view, PME consumes about 50% of the time during a typical simulation, mainly in the FFT library.

Since the FFT is crucial to performance, the parallel implementation of PME relies heavily on the parallelisation of this library. This is possible and is usually performed by a domain decomposition-type algorithm. Unfortunately, all known FFT parallelisation strategies involve an all-to-all type communication step which means, as we mentioned above, that performance is lost, particularly at high processor counts. This will have consequences to the overall parallel scaling behaviour of the simulation as we shall see in the next section.

The Scaling Problem and Performance Limits

As researchers in the field soon discover, even with domain decomposition and dynamic load balancing there will come a point when adding more processor cores to the simulation will not result in a significant increase in performance. i.e. the program stops scaling. In fact, not only may the performance level off with increasing core count but could also even drop with the result that increasing the parallelisation could actually make the program go slower (as in figure 2). This is one reason why it is essential to benchmark the MD program with the molecular system right at the beginning of the project to ensure that computer and "real" time is not being wasted by using too many processor cores. Why does the domain decomposition MD code stop scaling? From the discussion in the previous section about the need to minimise communication between the processor cores one might expect that the scaling limit is reached when the benefit obtained by adding another core is outweighed by the resulting increase in inter-core communications. This is true for classical MD, and indeed for many parallel programs, but it should be emphasised that it may not be the only factor affecting parallel scaling: the program may also be limited by I/O, i.e. the reading and writing of files, particularly if performed by each processor participating in the parallel program. However, for the vast majority of MD projects, normally only one processor or task is responsible for reading and writing files, and as we have mentioned since they are usually small (at most a few Gb during a whole run) the time taken for I/O is negligible.

In order to understand the reasons for the breakdown in parallel scaling at high core counts a computational scientist can usually call upon a wide variety of parallel performance utilities and tools to discern the communication pattern between the processors. However, if it is known that domain decomposition is being used then the source of one performance limit is immediately obvious - in order to have at least one atom per core, the number of cores cannot exceed the number of atoms. One might expect also there should be perhaps tens of atoms per core, in order to give each core sufficient work to do, but the one atom-per-core ratio certainly represents a hard limit for parallel scaling. The parallel scaling is thus system-size dependent with the result that larger systems, i.e. atomic or molecular systems with more atoms, should scale better and be able to exploit more processor cores than smaller systems.

This limit might be considered to be not too serious given that the number of atoms in most MD projects is likely to be much higher than the cores available to users' batch jobs. Thus, for a common simulation where a globular protein is simulated in an aqueous solvent, we might expect between 20-150 thousand atoms to be in our simulation box while for membranes or more complicated set-ups, there would be many more interacting particles. An exception to this rule might be for IBM BlueGene supercomputer architecture, where potentially tens or even hundreds of thousands of cores are be available, but even here average job sizes tend to be of a few thousand cores. Unfortunately, in practice it is found that for most MD simulations scaling stops at a few hundred cores and, as a rule of thumb, it is inefficient to have fewer than 150-200 atoms/core (depending on the precise simulation conditions).

This much lower limit derives not from the domain decomposition of the short range interactions but instead from the calculation of the long-range electrostatic forces via the Particle Mesh Ewald (PME) method, and in particular the parallelisation of the Fast-Fourier Transform algorithm FFT on which the PME relies. As mentioned above, parallel FFT unavoidably uses all-to-all communication steps between participating processor cores, leading to poor parallel scaling behaviour. To demonstrate this we show in figure 5 a comparison of a simulation performed with an implicit solvent, i.e. without PME, and the same molecule with explicit waters and PME.

It can be seen from the figure that for the explicit water simulation a maximum performance of about 35 ns/day is reached at just over 100 cores; adding further cores does not bring about an increase in performance. For the implicit model, however, the simulation is still scaling and has yet to reach a maximum performance.

Attempts have been made to reduce the influence of the all-to-all communication by, for example, allocating only a subset of the available cores to the PME (as in GROMACS) or by introducing novel decomposition schemes in the FFT parallelisation[23], but as far as we know, no one has yet come up with an alternative to the current algorithm which eliminates the costly all-to-all communication step. Alternatives to PME for the calculation of long range forces, such as fast multipole methods[24], are attracting interest but are yet

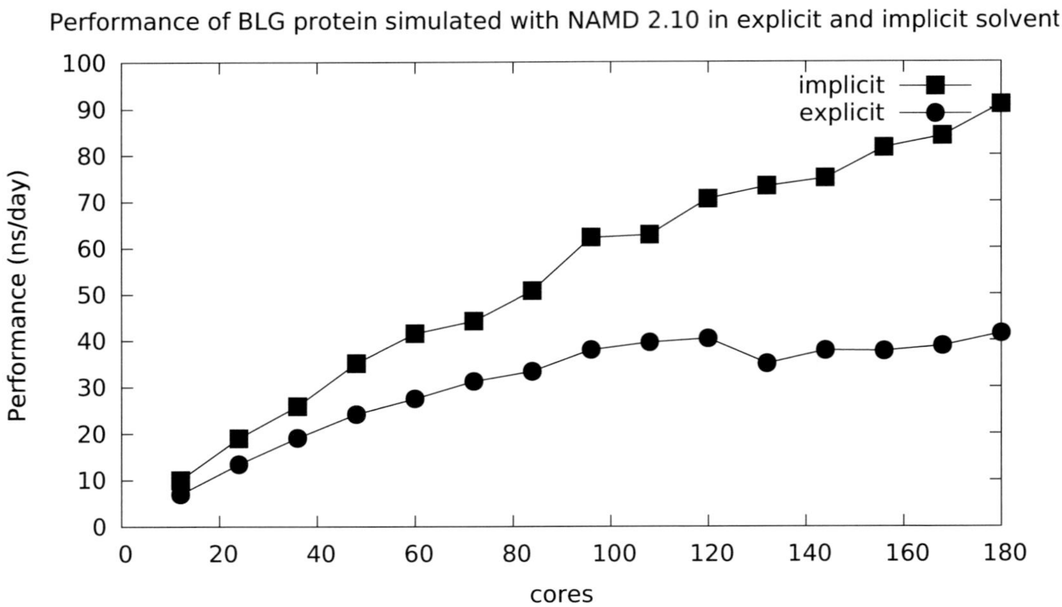

Figure 5. Comparison of implicit solvent and explicit water with PME for a small protein, β-lactoglobulin (BLG) in aqueous solvent (ca. 30,000 atoms), simulated with NAMD 2.10. For more details of the BLG system see ref.[22].

to be implemented by the main software developers and it is not clear if they will improve upon PME.

Beating the Scaling Limit with Accelerators

The system-size dependence of the parallel scaling limit is a problem for those researchers wishing to run long trajectories of systems with relatively few atoms since the maximum number of processors that can be efficiently used will be no more than a few hundred - using more processor cores may even slow down the simulations.

But what if it were possible to increase significantly the performance by "accelerating" how a single task does its calculations, rather than by just adding more processor cores? Moore's law and hardware vendors tell us that we can expect no significant increase in processor clock speed due to engineering limits such as transistor density, power consumption and heat dispersion. Instead, hardware vendors are concentrating on multi-core strategies or perhaps faster memories.

An alternative strategy is to couple to the standard CPU a separate, specialized device able to take some of the computational load from the main processor and hopefully perform the calculations more efficiently. This is a reasonable idea because most processors in a cluster are general purpose, designed not just to run user programs but also interact with the operating system, filesystem, for example. In fact this is not a new idea and in computational science there has been long history of using co-processors and accelerators attached in some way to the main CPU. In HPC, one proposed idea for an accelerator was to link a conventional cluster with re-programmable circuits called FPGAs (Field Programmable Gate Arrays) but because of the difficulty in programming them, they never achieved widespread adoption as computing units in supercomputer architectures.

Instead the most popular solution has been GPUs but also others such as Xeon PHI are also attracting attention and both of these are described below.

GPU

Graphics Processing Units, or simply GPUs, are devices which are able to manipulate computer graphics and perform image processing very efficiently. They work differently compared to conventional CPUs, being designed to process large blocks of data in parallel. Because of the difficulty in programming them for non-graphics applications there was little interest in HPC until NVIDIA released the CUDA programming language in 2007, allowing them to be used for general purpose processing (hence the term General Purpose GPU, or GPGPU). They have aroused considerable interest in HPC circles because they potentially provide considerable computing power for a relatively low cost and power consumption. In addition, in some application fields coupling a GPU to a standard processor can result in speed-ups of many tens or even hundreds of times compared to a single CPU alone (although in MD we rarely see such performance differences). It should be emphasised though that these improvements usually require a significant investment in application porting, not only because of the need to program in the probably unfamiliar CUDA language to achieve the best efficiency, but also because the program logic will probably have to change in order to adapt to the different CPU-GPU architecture. Another point worth noting is that the data connection between the GPU and the rest of the HPC hardware is via the PCI Express (or PCIe) Bus, a standard interface for personal computers, but quite slow in HPC terms. This means frequent data transfers between the GPU and the main memory have to be avoided.

MD Implementation on GPUs

Many of the major MD packages, including AMBER, GROMACS and NAMD, now have GPU-enabled versions of their applications. Although there are differences, the implementations in all these cases basically rely on offloading the expensive, non-bonded interactions (e.g. the PME calculation) onto the GPU, while leaving other operations on the CPU. For AMBER in particular there has been considerable investment in GPU-porting, and has one of the most efficient GPU implementations currently available. As an example of the performance increases available with GPUs consider the scaling results shown in figure 6 for AMBER (version 12).

We can see from the figure that for low core counts there is a good increase in performance, with a speed-up of about a factor of 4 for the accelerated code when compared to the non-accelerated version. At high core counts the speed-up disappears in this case but the possibility of getting close to the maximum performance with fewer resources (i.e. fewer computing nodes) is clearly evident. For GROMACS and NAMD we also observe similar good speed-ups at most core counts with the former application proving particularly sucessful at high parallelization. figure 7 and figure 8.

In these examples we notice that we still get speed-ups at many core counts, so it is undeniable that GPUs are worth having, but they are less dramatic than some of the in-

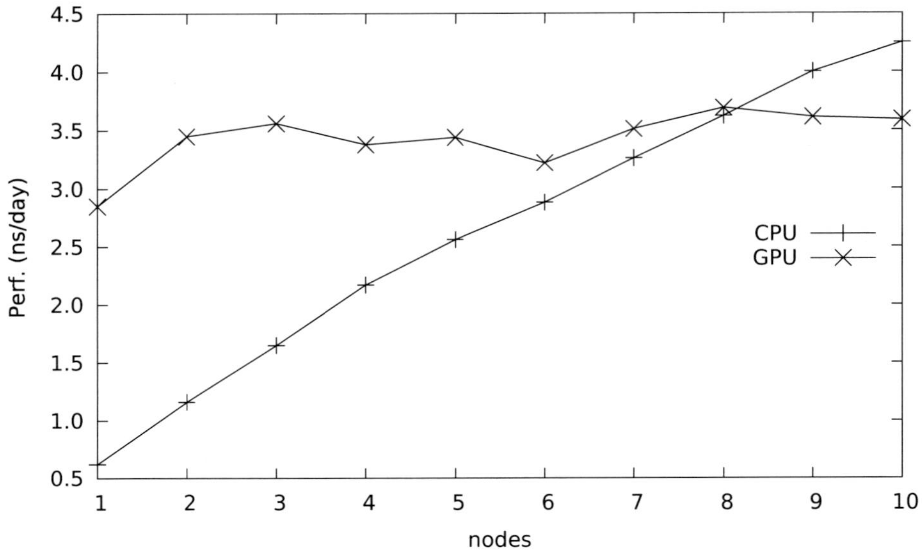

Figure 6. Performance of Amber12 using GPUs compared to CPUs only for the AMBER benchmark system Cellulose NVE (408,609 atoms) on Cineca's PLX computer.

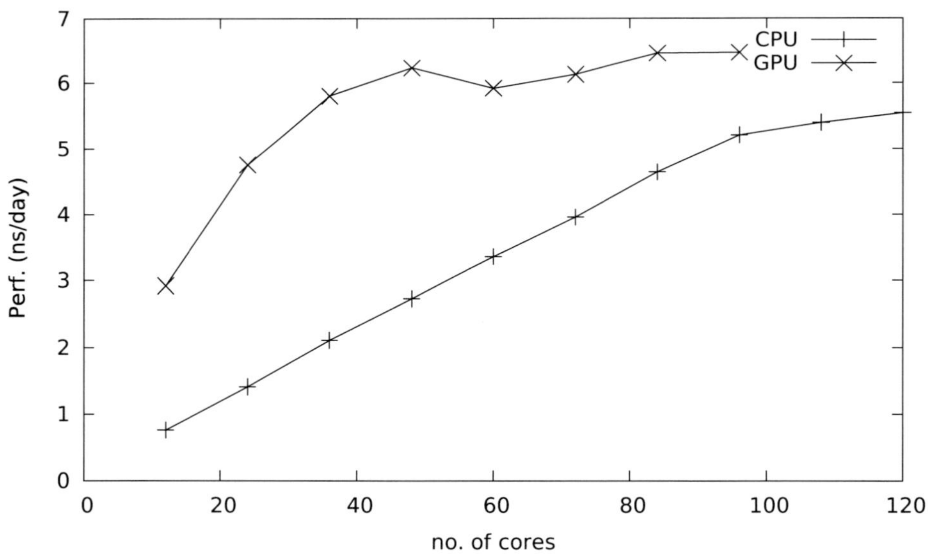

Figure 7. Performance of NAMD2.10 using GPUs compared to the CPU-only version for the NAMD benchmark system APOA1 (ca. 96,000 atoms). The simulations were performed on Cineca's PLX computer.

creases reported for other applications (e.g. a factor of 10, 20 or even 100-200). A possible explanation for the speedup differences is that the highest speed-ups occur when the original code is not so well highly optimised, whereas for already high performing codes, a GPU porting results in a less spectacular performance gain. This is plausible since many

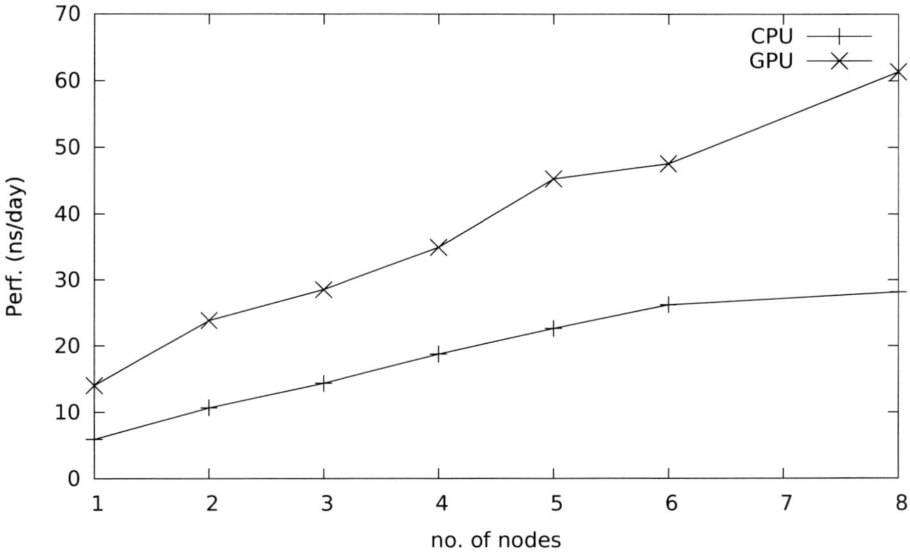

Figure 8. Performance of GROMACS 4.6.2 using GPUs compared to the CPU-only version for the GROMACS benchmark system DPPC (ca. 120,000 atoms). The simulations were performed on Cineca's PLX computer with 2 OpenMP threads and 6 MPI tasks per node for both benchmarks.

molecular dynamics code are highly optimised, not just for parallelism but also for single processor performance (particularly in the case of GROMACS).

In addition, we also note that for AMBER and NAMD the overall maximum performances are similar, or even lower for GPUs. The reason is evident if we recall that also GPUs accelerate the calculation by using parallelism, admittedly with different hardware compared to the multi-core scenario, but still relying on concurrency to perform the calculations. Since we have a fundamental performance limit due to the parallel algorithms we use in MD, we cannot hope to massively advance our simulations towards the millisecond scale with accelerators alone. But what we can do, and in the end this represents probably the best advantage of GPUs, is to reach a certain performance with fewer resources, hopefully more cheaply and with less energy. For GROMACS we see an improvement also in the maximum performance obtained, but even here it eventually tails off.

We should emphasise though that GPU application technology is advancing rapidly: the latest version of AMBER (AMBER 14), for example, is now able to exploit communications between GPUs which do not pass through the CPU, i.e. they communicate directly through the PCIe bus. The PCI Express interface itself is also going to be upgraded: NVIDIA has recently released details of a new communications interconnect, NVLINK, which will provide much high data transfer rates. The new technology is expected to be introduced in the next generation of hybrid CPU-GPU supercomputers.

Xeon PHI (MIC) Accelerator

In the last few years GPUs as accelerators have been joined by devices based on a technology invented by Intel called Many Integrated Core or MIC architectures and commercialised as Xeon Phi [25]. They are similar to GPUs in so far as they can be used as accelerators by linking them to Xeon processors via the PCIe express connection, although future devices, such as the so-called Knights Landing version, will be able to be used in "standalone mode", i.e. without needing a host CPU to boot them up [26]. In addition, like GPUs, they contain many processing elements for performing parallel computations. However, unlike GPUs the hardware architecture within a Xeon PHI is not too dissimilar from a standard Intel CPU so in principle programs (whether written in C, C++ or FORTRAN) do not need to be modified to run on the accelerator, just re-compiled with the standard Intel compilers and perhaps additional compiler options. In practice, some modifications will need to be made to the main program in order to use the accelerators efficiently but the porting process should be much easier than the equivalent GPU version. For this reason there is currently considerable interest in the HPC community in the use of MIC cards for scientific applications, even if, at the current time, there are few MD programs with stable MIC versions. Some results available from Intel for NAMD and LAMMPS suggest speed-ups which are sometimes lower than those obtained for GPUs [27], but it should be recalled that, unlike the more mature GPU community, few HPC programmers are familiar with the technology and porting efforts are generally incomplete. Given that new HPC systems are increasingly being installed with Xeon PHi-equipped nodes, this situation will change in the very near future.

Conclusion

In this chapter we have attempted to provide the tools and background information for those users wishing to embark on a molecular dynamics project on an HPC resource such as a supercomputer. In some respects this has not been too difficult because the actual procedure of running an MD program on a powerful computer has changed very little for perhaps over twenty years. Thus, despite the massive increase in computer power, an HPC user performs more or less the same activities as a researcher did two decades ago. One difference is that high quality graphics are readily available on common devices used by researchers whereas previously dedicated workstations were needed but otherwise the workflow as described above is the same. This is not to say that grids, portals, workflow systems, and similar enabling technologies have been complete failures, but their influence hasn't been substantial – the most common access method at Cineca and other PRACE centres is still via SSH and username/password and users generally need some programming experience, even if only in a scripting language (e.g. UNIX shell scripts or Python).

Another constant feature of HPC is the need for the application field to adapt to the latest advances in technology. In some respects, classical molecular dynamics is currently faring worse than some other disciplines because of the emphasis on parallelism. The drive towards parallelism is exemplified by IBM's BlueGene architecture where hundreds of thousands of computing cores are potentially available. The problem though is computer centres typically insist on a minimal parallel scalability of 2048 or even more cores and, for the reasons we have seen above, MD simulations rarely reach this limit. This has forced researchers to adopt "ensemble" methods such as REMD (Replica Exchange Molecular Dynamics), metadynamics or other free-energy procedures where replicas or copies of the system can be used to increase the parallelism needed in resource applications, but the result is that in recent years molecular dynamics researchers have been under-represented on such machines.

Accelerators such as GPUs or Xeon PHIs offer an alternative to BlueGene-style parallelism but it should be recalled that performance still derives from many processes or threads – it is just that now the parallelism is within the processor or device rather than between processors. The use of accelerators in HPC hardware though is becoming increasingly relevant. Since IBM have announced that the BlueGene line will not be continued further, it is reasonably certain that all supercomputers will soon be "hybrid", i.e. conventional cores together with accelerators such as GPUs or Xeon PHIs. MD researchers will therefore have to learn how to use these devices, if they have not started already. However, unless someone manages to come up with more efficient, alternative parallel algorithms to domain decomposition and FFT parallelisation, great breakthroughs will not be made soon in advancing maximum trajectories regularly in the microsecond regime or higher. It is possible instead that classical force fields will first be replaced by more accurate quantum mechanical calculations.

Acknowledgments

The author would like to thank the developers of free or low-cost, open-source molecular dynamics packages such as GROMACS, NAMD, AMBER, LAMMPS,DL_POLY and others, for making the use of MD simulation on HPC possible. Thanks are also due to numerous co-workers at Cineca and I. Eberini of the University of Milan for his help in preparing the BLG simulation data set and assistance in general in the practice of molecular dynamics.

Supplementary Information

Unless otherwise indicated in the text, all the simulation data reported in this Chapter were performed by the author on Cineca's PLX GPU supercomputer which has 512 nodes, each with 2 Intel Westmere 2.40GHz processors (12 cores) and 2 NVIDIA Tesla M2070 GPUs.

References

[1] PRACE Unified European Applications Benchmark Suite. http://www.prace-ri.eu/ueabs, 2013.

[2] A. Leach. *Molecular Modeling, Principles and Applications*. Pearson, 2001.

[3] M.P. Allen and D.J. Tildesley. *Computer Simulations of Liquids*. Oxford Science Publications, 1989.

[4] B.J. Alder and T.E Wainwright. Studies in molecular dynamics. i. general method. *J. chem Phys.*, 31:459–466, 1959.

[5] A. Rahman. Correlations in the motion of atoms in liquid argon. *Phys. Rev*, 136A:405–411, 1964.

[6] J.A. McCammon, B.R. Gelin, and M. Karplus. Dynamics of folded proteins. *Nature*, 267(5612):585–90, 1977.

[7] Y. Duan and P.A. Kollman. Pathways to a protein folding intermediate observed in a 1-microsecond simulation in aqueous solution. *Science 23*, 282(5389):740–744, 1998.

[8] Peter L. Freddolino, Anton S. Arkhipov, Steven B. Larson, Alexander McPherson, and Klaus Schulten. Molecular dynamics simulations of the complete satellite tobacco mosaic virus. *Structure*, 14:437–449, 2006.

[9] V. Vishal2 Guha Jayachandran1 and Vijay S. Pande. Using massively parallel simulation and markovian models to study protein folding: Examining the dynamics of the villin headpiece. *The Journal of Chemical Physics*, 124:164902, 2006.

[10] Anton supercomputer for biomolecular simulation . https://www.psc.edu/index.php/computing-resources/anton.

[11] P. Murray-Rust and H.S. Rzepa. Chemical markup, xml, and the worldwide web. 1. basic principles. *Journal of Chemical Information and Modeling*, 39(6):928–942, 1999.

[12] Chemistry Markup Language. https://www.xml-cml.org.

[13] Thomas A. Halgren. Merck molecular force field. i. basis, form, scope, parameterization, and performance of mmff94. *Journal of Computational Chemistry*, 17:490–519, 1996.

[14] D. Qui, P. Shenkin, F. Hollinger, and W. Still. The gb/sa continuum model for solvation. a fast analytical method for the calculation of approximate born radii. *J. Phys. Chem. A.*, 101(16):3005–3014, 1997.

[15] Y.M. Rhee, E.J. Sorin, G. Jayachandran, E. Lindahl, and V.S. Pande. Simulations of the role of water in the protein-folding mechanism. *Proc. Natl. Acad. Sci. U.S.A.*, 101:6456–6461, 2004.

[16] Marzia Rivia, Luigi Calori, Giuseppa Muscianisi, and Vladimir Slavnic. In-situ visualization: State-of-the-art and some use cases. Technical report, PRACE, 2012.

[17] UNICORE. http://www.unicore.eu/unicore.

[18] Human Brain Project. https://www.humanbrainproject.eu.

[19] Multiscale material modelling on high performance computer architectures. https://https://www.multiscale-modelling.eu/.

[20] S.J. Plimpton. Fast parallel algorithms for short-range molecular dynamics. *J Comp Phys*, 117:1–19, 1995.

[21] Siewert J. Marrink, H. Jelger Risselada, Serge Yefimov, D. Peter Tieleman, and Alex H. de Vries. The MARTINI Force Field: Coarse Grained Model for Biomolecular Simulations. *J. Phys. Chem. B*, 111:7812–7824, 2007.

[22] Eberini I., Emerson A., Sensi C., Ragona L., Ricchiuto P., Pedretti A., Gianazza E., and Tramontano A. Simulation of urea-induced protein unfolding: a lesson from bovine β-lactoglobulin. *J. Mol. Graph Model*, 30:24–30, 2011.

[23] Massimiliano Guarrasi, Ning Li, Sandro Frigio, Andrew Emerson, and Giovanni Erbacci. Testing and implementing some new algorithms using the fftw library on massively parallel supercomputers. In *Volume 25: Parallel Computing: Accelerating Computational Science and Engineering (CSE)*, Advances in Parallel Computing, pages 375–386, 2014.

[24] J. Kurzak and B. M. Pettitt. Fast multipole methods for particle dynamics. *Mol. Simul*, 32(10-11):775–790, 2009.

[25] Intel Xeon Phi Product Family. http://www.intel.com/content/www/us/en/processors/xeon/xeon-phi-detail.html.

[26] Dynamical Exascale Entry Platform. http://www.deep-project.eu/.

[27] NAMD for Intel Xeon Phi Coprocessor. https://software.intel.com/en-us/articles/namd-for-intel-xeon-phi-coprocessor.

EDITOR'S CONTACT INFORMATION

Dr. Pietro Cozzini,
Senior Researcher
Molecular Modelling Lab,
Department of Food Science,
University of Parma,
Parco Area delle Scienze,
17/A 43124 PARMA, Italy
Email: pietro.cozzini@unipr.it

INDEX

#

A

D

E

J

K

L

M

N

O

P

Q

R

S

T

U

V

W

X

Y

Z